PHYSICALISM

The Philosophical Foundations

JEFFREY POLAND

CLARENDON PRESS • OXFORD
1994

Oxford University Press, Walton Street, Oxford OX2 6DP
Oxford New York Toronto
Delhi Bombay Calcutta Madras Karachi
Kuala Lumpur Singapore Hong Kong Tokyo
Nairobi Dar es Salaam Cape Town
Melbourne Auckland Madrid
and associated companies in
Berlin Ibadan

Oxford is a trade mark of Oxford University Press

Published in the United States
by Oxford University Press Inc., New York

British Library Cataloguing in Publication Data
Data available

Library of Congress Cataloging in Publication Data
Poland, Jeffrey Stephen.
Physicalism, the philosophical foundations / Jeffrey Poland.
Includes bibliographical references and index.
1. Materialism. 2. Knowledge, Theory of. 3. Science—Philosophy.
I. Title.
B825.P63 1994 146'.3—dc20 93–5500
ISBN 0–19–824980–2

1 3 5 7 9 10 8 6 4 2

Typeset by Graphicraft Typesetters Ltd., Hong Kong
Printed in Great Britain
on acid-free paper by
Bookcraft (Bath) Ltd.
Midsomer Norton, Avon

ACKNOWLEDGEMENTS

Special thanks to John Post, who has been a constant source of support over the past five years while the manuscript was in preparation. He has provided faithful and tough-minded criticisms on numerous drafts and other ancestors of the book, and he has provided much encouragement for which I am deeply grateful.

Barbara Von Eckardt has read the entire manuscript and offered invaluable comments and criticisms. Her love and friendship have made it possible to see this project through to completion.

Edward Becker has recently read the penultimate draft and made useful and challenging comments. He helped me to avoid a number of errors, although no doubt he did not catch them all.

The groundwork for this book was my doctoral dissertation, which was supervised by Jerry Fodor. His wit, wisdom, and inspiration have had a lasting influence upon this project.

I wish also to thank Ned Block, Richard Boyd, Dan Little, Pat Manfredi, and Vic Mansfield for their helpful and encouraging comments on early versions of various chapters and for numerous conversations on the issues discussed in this book.

Thanks also to Jerome Balmuth, Phillip Bricker, Sylvain Bromberger, Al Casulo, Hartry Field, Michael Friedman, R. I. G. Hughs, Jon Jarrett, David Lewis, Sidney Morgenbesser, and Scott Soames for helpful conversations at various stages along the way.

CONTENTS

Introduction

This is a book concerning physicalism—a programme for meta-physical and epistemological system-building guided by the view that everything is a manifestation of the physical aspects of existence. The physicalist programme, promising as it does to deepen our understanding of ourselves and the world in which we live, has its roots in antiquity, roots that have held fast through the millennia (changing with the times, but not in fundamental ways) until today, when it appears to many to be the only metaphysical programme worthy of pursuit. Two primary objectives of the book are to clarify what it means to be a physicalist and to come to an understanding of the issues bearing upon whether physicalism is indeed a worthy programme. There are many critics who doubt that it is, and they have their reasons. But there are also many physicalists who both dispute the objections of their opponents and feel confident in holding on to the doctrine. Thus there is tension in the air and it needs to be more deeply understood. More importantly, some understanding is called for regarding why such matters are of significance. But before pursuing these lofty objectives, some background is in order.

The project on which this book is based arose out of my interest in issues concerning mental representation, specifically issues concerning the realization of mental representations by some physical systems. I was initially seeking an answer to the question, 'What are the differences in virtue of which some physical systems do, while other physical systems do not, realize mental representations?' I made two background assumptions in formulating this question: first, that indeed some physical systems do realize mental representations and, second, that certain kinds of physical differences are ultimately pertinent, hence my emphasis upon *physical* systems. This is to say that I viewed this question from a physicalist perspective. As it turned out, the full thrust of what it means to be a physicalist became the topic on which I focused. However,

my initial stance was to be a physicalist, somewhat vaguely conceived, and to believe that mental representations are real aspects
of some physical systems and that psychology is a legitimate branch
of science that is in the business of studying them.

This stance, however, is not without its detractors, of whom not
the least adamant are some full-blooded physicalists. Over the
years, Quine's arguments for the indeterminacy of translation,
reference, and 'mentalistic psychology' have become more and
more clearly dependent upon physicalist assumptions.[1] As is well
known, the bottom line of those arguments is that, among other
things, mentalistic psychology and hence representational theories
of the mind do not count as legitimate natural science concerned
with objective matters of fact. Although I was not entirely clear
about what Quine meant by the claim that mentalistic psychology
does not concern objective matters of fact, I realized that Quine
had to be dealt with if my belief that I could be both a physicalist
and a supporter of representational theories of the mind was to be
vindicated.

As I understood Quine, his contention was that physicalism
provides standards for what is to count as an objective matter of
fact and for what is to count as a legitimate branch of knowledge
concerned with such facts. And he contended that mentalistic
psychological theories and their alleged subject matter failed to
meet these standards. It was less than clear to me why this should
be so; I found none of Quine's supporting arguments compelling,
and his formulation and defence of physicalist standards of objectivity and scientific legitimacy were, at the least, in need of further
clarification. On the other hand, I had no positive physicalist
account in hand to show that there were indeed facts concerning
mental representation and that representational psychology was
indeed legitimate science. None the less, despite my doubts concerning the significance of physicalist-based objections to representational psychology and despite my lack of a positive response to
them, I was, and continue to be, convinced of the importance of

---

[1] See Quine (1979). More precisely, one should distinguish between the indeterminacy of translation relative to behavioural facts and the 'infactuality' of translation. It is the latter thesis (i.e. that there is no fact of the matter in cases where
translation is not determined by behavioural facts) that depends upon physicalist
assumptions in Quine's discussions. Thanks to Ed Becker for emphasizing this
point.

physicalism, representational theories of the mind, and the project of locating such theories and their subject matter within a physicalist framework.

At this juncture my attention turned away from trying to vindicate representational theories directly and towards a consideration of the main source of the predicament. That is, I began to look at what the physicalist standards of objectivity and scientific legitimacy were. My goal was to clarify and defend physicalism in a way that would prepare the ground for a thorough defence of mental representations and representational theories. However, what I found as I surveyed the situation of physicalism was considerable difference and disagreement in virtually every aspect of the subject.

There was, for example, disagreement over whether physicalism matters and for what. Despite its long history, there was no clear and detailed statement of why physicalism is important; and there was no shortage of critics who claimed, for various reasons, that physicalism is not of much consequence. Further, there was little agreement about how physicalism ought to be formulated, and there was no focused discussion of what counts as a better or worse formulation of physicalist principles. Nor was there any systematic articulation and defence of the nature of physicalist principles: for example, their scope, modal status, epistemological status, methodological role, and relations to other philosophical positions such as realism, naturalism, and empiricism. Finally, there was considerable disagreement over the current standing of physicalism with regard to supporting or disconfirming evidence and argumentation. The philosophical horizon featured, and continues to do so, both confident proponents and confident opponents, a sure sign of a lack of clarity and of misunderstanding. The evidential status of physicalism had remained virtually unstudied since the landmark paper of Oppenheim and Putnam (1958). And there were numerous unanswered objections, some of which are irrelevant and do not require a detailed response, while others hit at the core of the physicalist programme and call for refutation by serious physicalists.[2] At the bottom of this disagreement about the standing of physicalism appeared to be a glaring lack of exploration

---

[2] E.g. the problem of identifying 'the physical' and the problem of the self-applicability of physicalist principles.

and understanding of just how physicalism ought to be evaluated. There was no clear articulation of standards of assessment and no consensus regarding what constitutes a legitimate counter-example to the doctrine. Given the disagreement and the lack of study concerning motives, formulation, and deeper philosophical under-standing of the principles, this is not surprising.

Such a dark picture is not meant to suggest that there was no important work being done by physicalists around this time (*circa* 1976): the work of Boyd, Field, Fodor, Friedman, and Hellman and Thompson should be noted.[3] Clarification of many of the points just mentioned was emerging, although no systematic and thorough-going analysis yet existed. Such an atmosphere gave rise to my own research into the foundations of physicalism, which culmi-nated in my Ph.D. thesis in 1983. Further, there have been sig-nificant developments in many of the areas initially found deficient, developments that I shall discuss extensively below. These include systematic study of the relations of supervenience and determination (for example, by Hellman, Horgan, Kim, Post, Teller), David Lewis's focusing on the idea of a minimal formulation of physicalism, and John Post's most important book, *The Faces of Existence*. Despite these developments, the situation is far from satisfactory; many issues remain either untouched or inadequately resolved (for example, issues regarding minimal formulations, the physical bases, the nature of the doctrine, and its broader signifi-cance). The aim of the present book is to rectify many of these deficiencies by systematically exploring and developing the foun-dations of physicalism.

A number of questions must be addressed if this aim is to be achieved. At a minimum such questions include the following:

- What are the central concerns, motivations, values, and ideas of the physicalist?
- What counts as a proper formulation of physicalist princi-ples?
- What problems must be solved if a proper formulation of physicalism is to be given?
- What counts as a formulation worthy of acceptance?
- What is the current status of the evidence and other argu-mentation bearing upon acceptance?

[3] See the text below for specific references to the work of these philosophers.

- What are the principal obstacles to the acceptability and success of physicalism?
- What is the scope of physicalist principles?
- What sort of principles are they? (i.e. what is their epistemological and modal status?)
- What methodological roles do physicalist principles play in the pursuit of knowledge?
- How does physicalism relate to other philosophical programmes and positions (for example, realism, empiricism)?
- What is the broader significance of physicalism for philosophy, science, society, and the individual?

Since I shall not address all these questions in full detail, this book should be viewed as part of the larger undertaking of systematically studying and working out the details of a fully developed physicalist philosophy. The focus will be on those questions concerning the formulation, nature, and assessment of physicalist doctrine, while the questions concerning the broader significance of physicalism will only be clarified to a limited extent. This does not reflect a lesser estimation of the importance of these latter questions; rather, it reflects limitations of time, space, and my current understanding.

As I have indicated above, my approach to the present task is that of one who is sympathetic to the physicalist programme. However, there is real value for anyone in seriously examining just how far such a programme can be taken and defended. To achieve this requires comprehensive and systematic development in order to clarify what physicalism is and what it is not, thereby eliminating any opposition premised upon misunderstanding and error. Only with such clarification can the true limitations, if any, of the programme be discerned. The spirit of this project, then, is to some extent exploratory.

In Chapter 1, my focus will be upon identifying the basic motivations, areas of concern, values, and ideas that constitute the physicalist's outlook. I shall distinguish these from those motives, concerns, values, and ideas that are best labelled 'gratuitous associations'. Such identifications and discriminations bear crucially upon the proper formulation, evaluation, and estimation of significance of physicalist principles. And they clarify just what it is that makes one a physicalist. As a consequence of this discussion, I shall frame criteria for assessing formulations of physicalism with regard to their *adequacy* and *acceptability* and for assessing the *success* of the programme based on them. A key issue will emerge involving the trade-offs that exist between how much is expressed by physicalist principles (i.e. their degree of adequacy) and how well they stand up in the face of evidence and argumentation (i.e. how well they meet the tests of acceptability). The basic formulation problem for the physicalist is that of developing principles that are sufficient for expressing the core ideas and values of the programme *and* that do not fail the criteria of acceptability. A major theme of this book will be that recent physicalists, in their retreats from the apparently unacceptable doctrine of physicalist reductionism, have retreated too far: in their search for acceptable formulations of physicalism they have compromised the core content of the programme.

In Chapter 2, I shall review the most prominent recent formulations of physicalist doctrine to determine how well they fare relative to the criteria of adequacy set out in Chapter 1. I shall argue that every such formulation fails in one or another crucial respect. In particular, most recent formulations fail to properly address physicalist concerns about emergence and explanation. The purpose of this chapter, however, is ultimately constructive. My aim is to clarify further the various constraints and obstacles to which any adequate and acceptable formulation of physicalism must be responsive.

In Chapters 3 to 6, I turn to the task of developing and assessing a version of physicalist doctrine that will fare better than those previously advanced. Chapter 3 is devoted to tackling the vexed problem of identifying the physical bases required for formulating physicalist theses. After reviewing several defective strategies and identifying important features that a successful strategy must have, I clarify and pursue a strategy that is more promising than those reviewed. Three presuppositions of this strategy are identified and a series of objections directed against them are considered. Although some of the most serious objections to physicalism concern the characterization of 'the physical', I shall argue that these objections amount to no more than identification of important problems that any viable form of physicalism must solve, problems which there is no good reason to believe cannot be dealt with successfully. However, another important theme of the book begins to emerge at this point: there are numerous unanswered objections (for example, that the bases are indeterminate, too easily modified, and not privileged in the proper way) that require serious attention by physicalists. Recent attempts to fend off the objections concerning the physicalist bases (for instance by Hellman (1985) and Post (1987)) are not effective; and the general complacency of most other physicalists in this area is not defensible. Physicalism can only be put on firm foundations if the many problems concerning identification of the physical are solved.

The main task of Chapter 4 is to formulate a set of theses that ultimately will satisfy the criteria of assessment set up in Chapter 1. Thus they must be theses that are sufficient for expressing the core ideas of the physicalist programme while avoiding the pitfalls encountered by other formulations. They must be capable of withstanding the numerous objections launched against the programme by its opponents. And they must fare well in the face of relevant evidence. Consistent with the negative assessment of the formulations considered in Chapter 2, I shall formulate theses that are responsive to physicalist concerns about emergence and explanation. Later on, in Chapter 6, I shall discuss a number of objections to the theses (for example, that they are utopian, trivial, monopolistic, and not self-satisfying).

In Chapter 5, the focus turns to consideration of four metatheses concerning the physicalist theses developed in Chapter 4. The significance of these considerations should not be underestimated as they bear upon proper understanding and assessment of physicalist doctrine as well as upon how physicalism fits into a larger philosophical framework. The first metathesis concerns the scope of the doctrine, and I will endorse a mildly restricted scope: specifically, physicalism applies to everything that is real, true, or possible *in nature*. Although some more severe restrictions of scope (for example, to natural science) may meet with success even if certain broader construals of the programme fail, I will argue that a much wider scope can and should be entertained. Two further metatheses concern the modal and epistemological status of the theses: just what sort of claims are they? Are they logically necessary, metaphysically necessary, contingent? Are they overarching empirical hypotheses, regulative ideals, both, or neither? Clarification here will bear directly upon how one ought to conceive of the assessment of the theses. Finally, and regardless of how the issues concerning scope and status are resolved, physicalist principles can and do exert certain kinds of regulative pressure in science and philosophy. In discussing the nature of this regulative role, I shall attempt to clarify some of the issues concerning when and how physicalist principles may be legitimately employed in the conduct of inquiry.

As will become evident, gaining a defensible understanding of the scope of the doctrine, the kinds of principles it involves, how they are to be assessed, and what methodological roles they play in the pursuit of knowledge raises questions regarding a number of broader philosophical commitments and, indeed, regarding the nature of philosophy itself. As the discussion proceeds, the issue of how physicalism fits into a larger fabric of philosophical understanding will become acute. Necessarily, I will be more concerned at such points with identifying the issues and outlining the options than with resolving them definitively in one way or another. Thus, for example, how physicalism is related to realism, actualism, and empiricism must await an ultimate decision. However, I will strongly suggest that physicalism is not committed to many versions of these other 'isms'.

In Chapter 6, the criteria of assessment developed in Chapter 1 will be applied to the physicalist doctrine as developed in Chapters 3 to 5. I will argue that the theses constitute adequate expressions of physicalist thought, and I will defend them against a wide range of objections, many of which are premised upon dated or improper formulations of physicalism. Of special importance will be the identification of numerous unnecessary monopolistic overtones characteristic of many past formulations of the doctrine. Further, I will discuss, in a preliminary way, the evidence bearing upon acceptance or rejection of the theses. And I will identify the principal obstacles to the success of the physicalist programme. What will emerge will be a version of physicalism that is stronger than most extant versions, that is arguably defensible in the light of objections, and that does not fly in the face of obviously insurmountable obstacles. My conclusion will be that the physicalist programme is alive and well, though, like any programme concerned with building, rather than decreeing, significant systems of knowledge, it faces many difficult obstacles to its successful completion.

In the final chapter, I will survey aspects of the significance of the physicalist programme for philosophy, science, culture, and the individual. This will be a less systematic and less comprehensive discussion than that which precedes it, but it is, none the less, important for gaining some insight into why physicalism matters and for setting the stage for future work. The overall perspective developed in Chapter 7 is that physicalism provides a significant framework for effectively integrating facts and values in the pursuit of our deepest and most important cognitive and non-cognitive objectives. It underwrites much activity in philosophy, science, and society. And it provides a framework for integrating the many ways in which we understand ourselves and our world, a framework that has significant potential for promoting human well-being. This book begins with the objective of understanding more fully what the nature of this framework is, how viable it is, and how it relates to those things of importance in the various domains in which we function. If successful, the coming chapters will carry us a significant distance towards this objective, and the problems posed at the end will help to define some of the additional work required to carry us further.

The Problem of Formulation

The physicalist and other constructionalists are trying to serve some purpose not served by physics and the other sciences; but they cannot formulate the difference very clearly, and often seem unaware of it.

Nelson Goodman (1972: 13)

Many a metaphysical dispute is really a value dispute in disguise.

John Post (1987: 22)

1.1. PRELIMINARIES

Physicalism is a programme of unification which accords special privilege to physics. It is directed towards development of a system of knowledge within which all aspects of reality have a place and are related to physics in certain specifiable ways. This rough statement is, of course, vague and it admits of a wide variety of more specific re-formulations. The goal of this chapter is to make explicit the relevant considerations for developing a detailed statement of what the physicalist programme involves.

To begin with, a number of distinctions must be made in order to identify what type of programme physicalism is and what it is not. First, there is an important difference between issues concerning the growth of knowledge and issues concerning the structure of knowledge. The former have figured centrally in, for example, discussions of physicalist reduction as a form of scientific development. They involve such matters as how successor theories are, or ought to be, related to predecessor theories and what the significance of such relations is. Such issues are orthogonal to the second class of issues that pertain to the structure of knowledge at any given stage of development. These latter issues are of primary concern within the physicalist programme. For example, physicalists are interested in the question of what logical and epistemological relations ought, ideally, to exist between theories in different areas of inquiry. Such an ideal can constitute a goal state towards which progress ought to be aimed; and, within a programme accepting it, this goal state will play a major methodological role in directing inquiry and evaluating developments. In this way, growth issues and structural issues, though distinct, bear upon each other. In what follows, I shall be principally concerned with physicalism taken as a programme focusing on the structure of knowledge and reality; in Chapter 5, however, I shall examine the bearing of such structure on growth issues.

Second, I distinguish between 'unitary science' and 'unified science'.[1] To view science as *unitary* is to view it as ideally consisting of a single, total theory in one language and as embodying a set of basic principles that are sufficient for all theoretical purposes. For a physicalist, unitary science is usually conceived of as physics alone constituting the total explanatory system of science. The physical vocabulary and the explanatory principles couched in that vocabulary are supposed to suffice for all scientific description and explanation. What is essential to this view is that science does not consist of any branches other than physics; the special sciences serve only heuristic purposes that in principle could be served by an ideally completed physics. Thus, unitary science is the expression of a very strong, *eliminative* physicalist position. As it bears upon the structure of science, the ideal of unitary science expresses the view that science is monolithic and embodied completely in physics. As this ideal pertains to growth issues, scientific activity ought to be directed towards the incorporation of all explanation into physics and the gradual elimination of the special sciences.

In contrast to unitary science is *unified science*, the view that there are genuine divisions among branches of science and that there are principles which specify relations that all branches must bear to some basic branch. Thus the unification consists, not in the eliminability of all branches in favour of one, but in there being one branch to which all others are related in a particular way. Such a view does not preclude the possibility that a given branch might be eliminable; but what the view affirms is that such

---

[1] See Feigl (1963) for a discussion of this distinction; my development of it differs somewhat from his.

elimination is neither the general case nor the goal of scientific activity. Physicalism, as I view it, embodies that spirit of unified science, although it concerns more than just the sciences. Thus in what follows I shall be conceiving of the physicalist programme as a unification programme which aims at building a system of knowledge that is grounded in physics and is structured in accordance with a set of principles which characterize significant unifying relations.

Two points here deserve special emphasis. On the one hand, the brand of physicalism under scrutiny in this project is non-eliminative in character. It contrasts sharply with radical eliminative physicalism, which purports to show that the world is entirely physical and that physics says all there is to say about that world. On such a view, all other languages and theories play the role of useful placeholder or convenient shorthand for the real (i.e. physical) truth. However, as I see it, physicalism fully acknowledges the necessity of incorporating non-physical aspects of existence and non-physical truths into the system. On the other hand, it should be kept clearly in mind that the spirit of physicalism does not preclude other unification programmes; many unifying strategies can be attempted and many can succeed without putting physicalism in a bad light. In general, unification programmes are distinguished in terms of the basic class of unifying entities[2] they postulate, the unifying relations that all elements of the system must bear to the basic class, and the goals towards which the unification is directed. There is no reason to suppose that only one such programme is of importance or that only one can succeed.

A third important preliminary distinction to make is between *general* and *specific* programmes of physicalist unification. The general programme involves identifying and clarifying relations that every branch of knowledge must bear to physics. A specific programme, on the other hand, will focus on one branch (for example, psychology), and identify the ways in which it satisfies the demands of the general programme. For the most part, it is the general programme with which I shall be concerned. I assume no

---

[2] I use 'entities' here in a very broad sense that includes individuals and attributes as well as such items as sentences, terms, and propositions. Unification programs can be grounded upon base classes that include many different sorts of entity. Which entities are actually in the base of a system depends upon the goals of the unification programme aimed at building the system in question.

particular views about the problems of consciousness, intentionality, values, and other phenomena traditionally thought of as presenting problems for physicalism. My main objective in the first several chapters of this book is to give an adequate rendering of what physicalism, conceived of as a general programme of unification, involves. Only later, in Chapter 6, will I give some attention to specific problematic phenomena and branches of knowledge.

The following distinctions will also facilitate discussion. The *doctrine* of physicalism consists of a set of three types of claim, as follows: *theses*, which are explicit statements concerned with basic entities and unifying relations (for example, all facts are determined by the physical facts); *presuppositions*, which are claims presupposed by the theses (for example, there are principled divisions between physical and non-physical domains); and *metatheses*, which are claims concerning the theses (for example, the theses are *a posteriori*). The *programme* of physicalism is the enterprise consisting of: formulating the doctrine and defending it against objections; verifying the doctrine by assessing relevant evidence; and constructing a system of knowledge which exhibits the properties described by the theses. This latter task will be referred to as 'the working out of the programme'.

Given the above distinctions, I can now state somewhat more specifically what kind of programme physicalism is: it is a general programme of non-eliminative, structural unification that accords a certain sort of privilege to physics. Such a statement is still vague, omitting as it does all details regarding the nature of the unifying entities and relations and the particular sort of privilege physics enjoys. Performing the task of formulating specific physicalist theses requires a response to these shortcomings. Once these tasks are accomplished and one has in hand such theses, and given an adequate philosophical defence of physicalist doctrine as a whole, the programme proceeds via the construction of a system of knowledge that exhibits the structure specified by the theses.

But the shortcomings of the above characterization of physicalism are severe. After all, many sorts of unifying entities and relations are possible, and different sorts of unification can have very different significance. To carry out properly the task of formulating physicalist doctrine, it is therefore essential to have some idea of what fundamentally motivates the physicalist. That is, before one can effectively decide which principles are appropriate for the physicalist programme, one must have a good idea of the unifying function they are supposed to perform. For this, one must have a clear conception of what ideas and values are central to physicalist thought.

In the remainder of this chapter I shall identify the fundamental ideas and motivations that fuel the physicalist's thinking and that will guide my subsequent efforts at properly formulating physicalist theses. To this end, I shall begin by stating what, in my view, are the core ideas and values of physicalism. Then I shall explicitly identify a number of 'gratuitous associations' to the physicalist programme: that is, ideas and values which are not essential aspects of the programme even if they might have been, at some time or another, associated with it. Finally, I shall codify much of the discussion in the form of criteria of assessment which I shall employ in later chapters for assessing my own and other formulations of physicalist doctrine.

1.2. CORE IDEAS AND VALUES

The fundamental idea of contemporary physicalism is that physics occupies a certain privileged status in relation to all other branches of knowledge. Physicalists who have been explicit on this point write as follows:

Mathematical physics, as the most basic and comprehensive of the sciences, occupies a special position with respect to the overall scientific framework. In its loosest sense, physicalism is a recognition of this special position. (Hellman and Thompson 1975: 551.)

It [the claim that there is no mental difference without a physical difference] is a way of saying that the fundamental objects are the physical objects. It accords physics its rightful place as the basic natural science without venturing any dubious hopes of reduction of other disciplines . . . (Quine 1979: 163.)

A review of physicalist writings reveals that there is considerable variation in just how this position of privilege is characterized. There is, however, widespread agreement on three areas that the privilege concerns: ontology, objectivity, and explanation.

First, with respect to ontology, the physicalist holds that everything that exists or occurs in nature is *ontologically dependent upon* the physical domain: that is, nothing can exist or occur in the absence of physical objects, attributes, and events. As Haugeland forcefully puts it, 'if you took away all the atoms, nothing would be left behind' (Haugeland 1982: 96). The relation of dependence here is supposed to be asymmetric in that, although nothing can exist in the absence of physical things, physical things can exist in the absence of non-physical (for example, mental) things. Taking away all the non-physical things and their attributes does not guarantee that nothing will be left.

This idea regarding ontological dependence, although essential to the physicalist position, is not sufficient to express the full sense in which the ontology of physics is thought to enjoy privilege. In addition, the primacy of the physical ontology consists in its providing a basis for the 'supervenience' of all attributes[3] upon physical attributes: that is, roughly, once all the physical attributes of things are fixed, then so are all the attributes of things.[4] Such a determinative relation goes beyond the mere ontological dependence of all phenomena upon physical phenomena, since ontological dependence is compatible with a failure of supervenience. However, since this relation of supervenience or determination is only a relation of systematic covariation between classes of attributes, even combining it with the relation of ontological dependence does not fully capture the deepest sense in which the physical ontology is primary.

From an ontological perspective, physicalists have always been concerned to rule out 'ontologically emergent' phenomena of any sort (for example, spirits). Thus the physicalist holds the view that the ontology of physical theory is exhaustive of all that exists in the sense that there are no objects that are neither basic physical objects nor objects 'built up' out of, or 'realized by', basic physical objects. Of course, the notion of being built up out of basic physical objects requires clarification, as does the notion of realization. And it is clear that the ways in which objects are built up out of, or realized by, the elementary physical objects are not restricted to those which are typically proposed (for example, mereological sums, structured wholes). But such residual indistinctness does not obscure the main point: all objects that exist do so in virtue of physical objects. There are no ghosts!

---

[3] By attributes I mean to include both properties and relations.
[4] The relation of supervenience I have in mind here is what has been referred to as 'global supervenience' in the literature. See Ch. 2 for further discussion.

Further, the physicalist is particularly concerned to rule out emergent attributes. It is important to recognize that the requirement that all attributes be ontologically dependent upon[5] and supervenient upon the physical base does not accomplish this aim. In a world populated only by physical objects and objects that are built up from them, possessed of attributes that both ontologically depend upon and supervene upon physical attributes, there could be attributes which were simply instantiated in a mysterious way: such attributes would be, *in a sense to be clarified below*, independent of the attributes in the physical base.[6] As such, they have no rightful place in a physicalist picture of things. Attributes that ontologically depend upon and supervene upon the physical base must do so in virtue of the physical attributes in that base. And the forms of covariation embodied by ontological dependence and supervenience are not enough, even in conjunction, to capture this physicalist idea. I shall say that all attributes, in addition to being ontologically dependent upon and supervenient upon physical attributes, must be 'realized by' physical attributes.[7] Clarification of this idea of realization is essential if the physicalist programme is not to founder upon obscurity. Here, I shall outline the key features of the idea.

To say that an attribute, N, is realized by a class, R, of physical attributes is to say that N is instantiated *in virtue of* the instantiation of the members of R. This means, in part, that there is a distinction between those physical attributes that are *relevant to* the instantiation of N and those that are not: all and only the relevant physical attributes are members of R, where the relevant attributes are those that do some work in making it the case that

---

[5] Attributes are ontologically dependent upon the physical with respect to their instantiation in nature: i.e. no attribute can be instantiated in nature unless some physical objects exist and/or some physical attributes are instantiated. As abstract objects, however, attributes may of course exist even if they are not instantiated in this, the actual, world; and as such, they are not ontologically dependent upon the physical ontology.

[6] Of course, such attributes can rightly be said to depend upon physical attributes in the sense that they ontologically depend upon and supervene upon them. But my contention is that there is also an important sense in which attributes could fail to be dependent upon physical attributes even if they exhibited these two sorts of dependence. The point of saying that such attributes would be instantiated in a mysterious way is to flag this other sort of dependence.

[7] Thus, in at least one sense, an attribute is 'emergent' if it is instantiated but it fails to be realized by some class of physical attributes.

N is instantiated. Although N might be realized by total states of the world or by total states of a particular system, neither of these possibilities undermines the distinction between relevant and irrelevant physical attributes regarding N's realization (i.e. the attributes relevant to the instantiation of N need not be highly localized). In situations where there is some borderline fuzziness of this class of relevant attributes, such borderline fuzziness should be managed on a case-by-case basis: there is no need for a *general* principle to handle all such cases.

In addition, the notion of realization central to physicalist thought requires that the co-instantiation of the members of R is, at least, nomologically sufficient for the instantiation of N. This much modal force is required if the physical attributes are to be more than just accidentally associated with the instantiation of N: so-called 'trivial physicalization' of N. Realization requires that the members of R jointly make it the case that N is instantiated, and nomological sufficiency captures this to some extent.

It is important to note, however, that the nomological sufficiency of a class of attributes for the instantiation of N is not enough to capture the notion of realization.[8] This can be seen when one considers that there could be a class, R', of nomic equivalents of the members of R, such that the co-instantiation of the members of R' is nomologically sufficient for N. But the members of R' do not constitute a realization of N, even though they are *ex hypothesi* equivalent to a class of attributes that do. How can one express what more is required?

I think one way this can be understood is as follows: the instantiation of the non-physical attribute, N, is *constituted by* the instantiation of the physical attributes in the class R. What does 'constituted by' mean? The idea is, roughly, that the attribute, N, has a certain nature or essence which can be instantiated by the specific configuration of physical objects and attributes that results when the members of R are instantiated. The relevance of the members of R consists in their contributing to the constitution of N in this sense.

For example, the property of *transparency* of an object is es-

---

[8] My contention below shall be that supervenience and ontological dependence relations between N and the members of R, in addition to nomological sufficiency, are consistent with the physical attributes in R not doing any work with regard to N's being instantiated: such relations are also too weak.

sentially that of *being capable of being seen through*, and there are many different physical configurations that constitute this essence. That is, there are many different physical configurations that can be seen through and hence constitute an instantiation of *transparency*. Such configurations are what make it the case that transparency is instantiated in physical domains. The generalization of this idea to all attributes is what is required by any version of physicalism that aims at precluding ontologically emergent attributes. Thus the physical ontology enjoys further privilege in that all attributes must be realized by physical attributes in the sense that configurations of physical objects and attributes constitute the instantiations of all attributes that are, or can be, instantiated in nature.

To summarize the central ontological idea of physicalism, it is that everything has a place in an ontological structure grounded in the ontology of physics: the physical ontology is most basic and comprehensive relative to all that there is in the sense that all objects and attributes are dependent upon, supervenient upon, and realized by physical objects and attributes. Thus everything that exists is either an element of the physical basis or is constituted by elements in that basis. I shall say that everything that exists is, in this sense, 'ontologically grounded' in the physical domain.

It should be clearly understood that the primacy of physics in ontological matters does not mean that everything is an element of a strictly physical ontology: the version of physicalism I am developing here allows for non-physical objects, properties, and relations. The primacy of the physical ontology is that it grounds a structure that contains everything, not that it includes everything within itself. In my opinion, with regard to ontological matters, physicalism should not be equated with the identity theory in any of its forms. In particular, I prefer the idea of a hierarchically structured system of objects grounded in a physical basis by a relation of *realization* to the idea that all objects are token identical to physical objects. I am not, however, letting ghosts back in when I say this: ghosts are not physically realized objects. The important point is that such evidently non-physical objects as works of art and social institutions are real and have a place within a physicalist system: they *are* ontologically grounded in the physical domain.

Further, there is no reason to suppose that there is just one sort of realizing relation that all entities bear to the physical entities that realize them (for example, composition of causal powers). The world is a rich, hierarchically organized, and multifaceted place containing a wide range of different objects and attributes, varying in complexity and constituted in diverse ways. The realization of this world by a purely physical basis may well be the result of many different sorts of modes of constitution. The diversity of what exists and the narrowness of the physical basis is what makes the physicalist ontological view both interesting and challenging. And since it is a paramount goal of the physicalist not to leave anything real out of account, physicalists are best advised not to handcuff themselves at the outset by unnecessarily and inappropriately restricting the sorts of structures which can be erected upon the physical foundation. Thus the sorts of realizing relations consistent with a physicalist outlook ought to be the subject of ongoing inquiry. The ontological structures found in the world, though grounded in the physical domain, need not be conceived in overly simplistic terms.

The second respect in which physics is accorded privilege within a physicalist system is as follows: physical truths and facts provide the conditions for all *objective* truths and facts. This has been expressed in recent years by such claims as that the physical facts determine all the facts, that there is no difference between individuals without there being a physical difference between them, and that the physical truth determines all the truths. The idea is that for there to be objective matters of fact and truth in some domain and for there to be objective sameness and difference in some respect between individuals, there must be certain appropriately related *physical* facts, truths, similarities, and differences. Those appropriate relations, whatever they turn out to be, are taken by physicalists as sufficient for establishing the physical ground for any putative claim to objectivity. Thus the objective facts, truths, etc. are either strictly physical or appropriately related to strictly physical facts, truths, etc.

Of course, the relevant sense of 'objectivity' here is in need of clarification. At least three distinct senses have been discussed and can be labelled as follows: inter-subjectivity, freedom from bias, and factuality. The first, inter-subjectivity, involves some sort of consensus or agreement, perhaps obtained as a result of conformity to certain standards, within a community. An objective truth is one that commands the assent of one's peers. Such an inter-subjective notion of objectivity ought clearly to be distinguished from the second, which requires that subjects abstract from specific interests: an objective understanding or judgement regarding some phenomenon is one not specifically shaped by individual or local group interests. Clearly, such a notion of objectivity is a graded one, which varies with context. And it is one central component in any plausible reconstruction of scientific objectivity (Scheffler 1967: 1).

Third, and very different from the first two 'epistemic' conceptions of objectivity, is a concept that focuses upon facts that are taken to be independent of epistemic or other cognitive matters. That is, they exist independently of human activity, interest, knowledge, etc. The objective facts, on this view, are those that are invariant across any change of perspective or interest (Post 1987: 67). Objective facts are those that exist whether we do or not; they are the putative objects of our knowledge, but they are in no way dependent upon that knowledge. And objective facts are those that make claims true according to some 'non-epistemic' conception of truth (Hooker 1987: 309).

Now, it is this third sense that is pertinent to the physicalist's claim that physics enjoys a privileged status with regard to objectivity. Perhaps the clearest expression of this privilege has been Quine's contention that since the physical truths do not determine the truths regarding translation, there are no objective truths regarding such matters: or, as he has often put it, there is no 'fact of the matter' (Quine 1979). The alleged failure of claims in linguistics and psychology to satisfy physicalist criteria of objectivity has led Quine to relegate those disciplines to a second-class status in our system of knowledge. The objective truths are truths describing what is real as opposed to what is fabricated or projected in a given situation. And, for the physicalist, what is real and true in a situation is grounded in what is real and true according to physics.

Two issues ought to be mentioned regarding this second aspect of the privilege of physics. First, one might ask whether this standard of objectivity cuts much ice if the objectivity of physics itself can be drawn into question. What good does it do to take physics as the measure of what is objectively real and true if we have no reason to believe that physics is objective in the required sense?

That something satisfies physicalist standards of objectivity may have no significance if physics is, say, a non-objective construction or if physics itself does not satisfy physicalist standards of objectivity. Second, objectivity in the required sense may well be, at best, an *internal* attribute, an attribute that must always be relativized to a language, a theory, or a system. If so, this would not necessarily impugn the physicalist's view of physics as being privileged with regard to objectivity, but it would considerably broaden our understanding of the nature and scope of that privilege. Both of these issues are important for a proper understanding of physicalism and will be discussed in subsequent chapters.

The third and final area in which physics is accorded special status by physicalists concerns explanation. The key idea is that of a unified explanatory system in which different branches of knowledge are organized hierarchically with physics at the foundation, and in which the generalizations and phenomena studied at each level in the hierarchy are explainable in terms of the generalizations and phenomena at lower levels. In such a hierarchy, physics is seen as the ultimate repository of mechanisms, states, processes, etc. that can be appealed to in an explanation of why some phenomenon occurs or why some regularity holds. Phenomena and regularities at any given level are explainable by appeal to other phenomena and regularities at the same or lower levels (not necessarily physics), but *all such appeals* are ultimately grounded in phenomena and regularities at the physical level. I shall say that a phenomenon or regularity is *explanatorily grounded* in the physical basis when there is a chain of 'vertical explanations'[9] that eventually links it to phenomena and regularities in physics. The chains may, in some cases, be quite short (for example, chemical bonding) or they may be quite long (for example, psychological or social processes). But in all cases there are, according to the physicalist, explanatory chains that ultimately connect higher-level phenomena with the physical basis.[10]

The sense in which physics is privileged here is that it provides

---

[9] A 'vertical explanation' of some phenomenon is, roughly, an explanation of why it occurs in terms of lower-level phenomena. See below and Ch. 4 for further clarification.

[10] I do not rule out the possibility that, in some cases, such an explanation might make appeal to higher-level phenomena. The key requirement is that every phenomenon is linked to the physical domain by a chain of vertical explanations, however circuitous the route may be.

the basis for a unified explanatory system, where the unification consists in the existence of an ultimate set of explanatory mechanisms, processes, states, etc. and the explanatory grounding of all phenomena and regularities in this basic set. In such a system, every phenomenon has a vertical explanation that is grounded in physics and every explanation of phenomena or regularities at any level of the system must appeal to processes (etc.) that are either in the physical basis or are grounded in that basis.

It is easy to see that this explanatory privilege is linked closely to the ontological privilege discussed earlier. If everything is dependent upon, supervenient upon, and realized by physical phenomena, then those physical phenomena would seem to provide the ultimate constituents that can be appealed to in an explanation of why any given phenomenon occurs or of why any given regularity holds. Thus the physicalist ontological structure strongly suggests that every phenomenon ought to have an explanation of its occurrence couched in terms of entities that are either elements of the physical basis or are themselves ultimately explainable in terms of such elements. Vertical explanations, which make linkages between different levels in a hierarchical system of explanation, mirror the vertical linkages in a hierarchical ontological structure grounded in the physical domain. For the physicalist, therefore, it is expected that physically-based explanation makes comprehensible both what can and what actually does occur in nature.

To understand this explanatory privilege further, a number of comments are in order. First, what is being claimed is *not* that every phenomenon or regularity has an explanation in *physics*. Physics is the branch of knowledge concerned with, *inter alia*, studying the basic entities and processes which ground the physicalist system. Erected upon this physical ground level is a hierarchical system of theories and explanations which is structured in part by vertical linkages between levels in the hierarchy. These vertical connections deploy physical mechanisms, states, and processes (or mechanisms, states, and processes that are physically grounded) in ways that explain the occurrence of higher-level phenomena. The character of such explanations is in need of being more fully understood; but there is no reason to suppose that it is part of the work of *physicists* to develop them. To explain the realization by physical phenomena of some higher-level phenomenon requires an appeal to physics, but in explanatory ways that link the physical with the non-physical, and hence in ways that may have little or nothing to do with explanation in physics. And in the case of an intermediate vertical explanation of some non-physical phenomenon in terms of lower-level, physically-grounded phenomena, there might be no mention at all of entities or principles drawn from the physical domain.

Second, I am not claiming that vertical explanation, in terms of physical or physically-grounded phenomena, is the *only* sort of explanation. Indeed quite the opposite is being asserted, since the hierarchical explanatory structure being alluded to should be understood as involving all legitimate patterns of explanation. Vertical explanation is only one possible pattern present in the system, one which ultimately links all levels of the system to the physical level. For reasons that I shall discuss shortly, this sort of explanation is valued highly by the physicalist, but not to the exclusion of other sorts of explanation. Further, along with the diversity of types of explanation, it should be understood that I am making no assumption here that there is only one sort of vertical explanation. That is, the explanation of the realization of high-level phenomena may be achieved in quite different ways depending upon the type of phenomena to be explained. For example, how the realization of chemical bonding is explained is likely to be very different from how the realization of mental states and moral values are explained. There is no one form that this sort of explanation must take.

Finally, it should be kept in mind that this view of a unified system of explanation grounded in physics must take into account the fact that some phenomena within any system are fundamental and unexplained. There are at least three sorts of unexplainable phenomena that are possible within a system: fundamental phenomena in the basis, ultimate vertical facts regarding realization, and higher-level phenomena that admit of no lower-level explanation. The first is generally acknowledged and non-problematic for the physicalist. The second may seriously compromise the physicalist vision and must be explored thoroughly, while the third provides direct counter-examples to physicalism and must be avoided if the programme is to be successful.

Complicating the picture will be the harsh, practical realities of providing explanations, even when they are in principle possible.

The physicalist's vision of a unified explanatory system must be understood as one whose actualization may be forever prevented by practical obstacles of many sorts. Thus, as will become clear, the explanatory privilege of physics, in addition to being difficult to articulate precisely, will also be difficult to defend conclusively. Difficulty in distinguishing those phenomena that have no physicalist explanation from those that have explanations beyond our epistemic reach confounds efforts to assess the acceptability of the doctrine.

With regard to all three respects in which physics is alleged to have privilege (viz., ontology, objectivity, explanation), some cautionary comments are in order.[11] First, to say that physics is privileged in some respects *is not* to endow physics with unconditional privilege in all respects or in all contexts. The physicalist must recognize that appeals to physics may well be unimportant, irrelevant, or downright harmful in many contexts, while in other circumstances such appeals may play a limited and quite secondary role to other modes of thought or experience.

Second, in recognizing the privilege of physics in the respects characterized above, the physicalist is not, *ipso facto*, endorsing classical reductionism, the identity theory, or the idea that physical language has universal descriptive power. As will become clear in Chapter 2, all three of these views have deep problems that make them unsuitable for a proper statement of physicalist doctrine. Third, none of the traditional associations of physicalism with other programmes or ideas, beyond those I have explicitly cited, should be made at this stage of the inquiry. The problem of formulation is that of developing a rigorous statement of the relevant senses in which physics enjoys privilege without making unnecessary commitments to potentially problematic or otherwise undesirable ideas.

To understand more fully the nature of the intended privilege, I shall shortly discuss some of the underlying values that motivate the physicalist. And then I shall discuss some specific *exclusions* from the list of ideas and values central to the physicalist programme (for example, scientism, conventionalism in ethics). Only subsequently shall I address the problem of how best to express the central ideas and values. The current point is that one should

---

[11] The first two points, regarding the nature of the privilege accorded to physics, have been extensively discussed in Post (1987).

suspend any antecedent commitments to the specific form that such expression should take and to additional ideas often associated with physicalism.

Having now identified what I take to be the key ideas that inform the physicalist programme, it is time to ask *why* one should accept them. For substantive as well as expository purposes it is important to distinguish between *a posteriori* reasons and *a priori* reasons for accepting such ideas. The former include such considerations as that the ideas are supported by relevant evidence and that the programme, guided by those ideas, has been or is likely to be successful. I shall discuss such reasons in later chapters. The later type of reasons, on the other hand, involve certain values that the ideas, and the programme guided by them, promote. Such values are the rewards that a successful physicalist programme has to offer, and they provide a strong *prima facie* incentive for pursuing the programme. One very important reason for distinguishing the two sorts of reasons is that, while the *a posteriori* reasons bear upon the degree to which one is justified in being a physicalist, the *a priori* reasons express what one cares about as a physicalist. Failure properly to attend to such underlying values can have negative consequences, not only for the formulation of physicalist doctrine, but also for a proper understanding of the significance of the programme and for the way in which it is pursued. It is to these crucially important *a priori* reasons for being a physicalist that I now turn.

Physicalism is a programme for constructing a system of knowledge that exhibits a certain sort of structure. Such structure has pay-off. And the *a priori* reasons for being a physicalist concern that pay-off, of which there are three sorts that physicalism can deliver: specific problems solved within the system, global features of the system, and broader implications of the system in various areas of human concern.

To begin with, physicalists have always seen advantages in their approach resulting from solutions it would provide to specific problems concerning such phenomena as life, mind, values, meaning, and society. The physicalist programme, if successful, would show how each of these, in so far as they are real and objective, fits into the natural order; and it would provide a basis for resolving foundational questions concerning theories of each. For example, the issues mentioned earlier, raised by the positing of mental representations in psychological theorizing, are a paradigm of this sort of problem. An additional problem which would be solved if physicalism were successful is that of delineating the boundaries of objective matters of fact. A physicalist system provides a measure of all that is objectively real, true, and possible. Thus the distinction between knowledge that concerns matters of fact and 'knowledge' that involves a component of subjectivity, or relativity to the knowing subject, can be drawn within the system. Quine's concern about the lack of objective status of linguistics and mentalistic psychology is an example of an appeal to a physicalist measure of objectivity.[12]

The second sort of value that would be promoted if the physicalist programme were to be successful involves global features of the resulting system. There are, broadly speaking, three categories of such features that deeply interest physicalists: (1) ontological parsimony and unity, (2) theoretical simplicity, consistency, unity, and comprehensiveness, and (3) substantial increases in understanding. This latter category is the most significant.

There is a substantial consensus among physicalists that the programme is motivated by ontological concerns. Those concerns involve, *inter alia*, a certain kind of ontological parsimony, i.e. that of not exceeding the ontology of physical theory. But such parsimony must be carefully understood. It consists in a restriction to one of the number of independent ontological types of object, attribute, and kind. That is, nothing exceeds the physical ontology in the sense that each object, attribute, and kind of thing that occurs in nature is either an element of the physical ontology or is dependent upon, supervenient upon, and realized by physical objects, attributes, and kinds in that ontology. Mental attributes, for example, cannot be instantiated in nature unless they are appropriately related to physical objects and their attributes. So although one can distinguish between ontological types of attribute (for example, mental and physical), and although one can allow that mental attributes are instantiated in nature, as a physicalist one requires a relation of dependence of the one ontological kind of attribute upon the other. Ontological parsimony here is a parsimony of *independent* ontological categories.

In order to understand better what kind of parsimony this is, it

---

[12] See Hellman and Thompson (1977) and Post (1987; 1991) for discussion of other problems allegedly solvable within a physicalist system.

is important to see what it is not: it is neither a parsimony of kinds of entity, nor a parsimony of kinds of attribute, nor a parsimony of individuals in nature. That is, this sort of ontological parsimony does not put severe restrictions on the kinds or groupings of things in nature that can be appealed to for the purposes of theoretical inquiry. Nor does it entail that entities in nature must have only physical attributes. Hence this ontological parsimony does not put severe limits on either what one can say about objects or on the amount of diversity one can acknowledge. It only requires that the things and attributes one says something about and the diversity one does acknowledge be appropriately related to physical things and attributes. Finally, there is no suggestion in this parsimony that there is a reduction in the number of individuals in nature: physicalists must, like anyone else, allow that there are as many things as there are. Again, the only requirement is that each individual be appropriately related to entities in the physical domain.

The physicalist vision, then, is that of a system of ontologically diverse elements grounded in a physical ontology, whose elements are the only ones that exhibit *ontological* primacy and independence. It is both appropriate and important to ask at this point: why is ontological parsimony of this sort of value? After all, many have claimed that parsimony is, at best, an aesthetic feature of a system and that it has no value in itself. What is important, on such a view, is to get the number of independent and ultimate ontological kinds right, and there is no special premium to be placed on that number being a low one as opposed to a high one.

Now, it is certainly true that one wants to get the numbers right; but that does not mean that fewer is not better for some purposes. And although parsimony may have little intrinsic value as a feature of a system, it contributes substantially to other, more intrinsically valuable, features. For example, parsimony, as conceived above, contributes to a certain sort of ontological unification, which in turn contributes to the overall explanatory power and understanding provided by the system.[13] Thus to understand more deeply the *a priori* reasons for being a physicalist, consideration must be given to what *ontological unity* means for the physicalist.

First, it is important to distinguish between unification of the

---

[13] See Friedman (1974), Kitcher (1981), Salmon (1984), and Hooker (1987) for significant discussions of how unification relates to explanation and understanding.

physical basis of the system and unification of the system as a whole. The punch of physicalism depends to a large extent upon there being a substantial amount of unity within the ontological framework of physics. To build a system upon a highly disconnected, unstructured basis with a large number of independent, ultimate entities and attributes would be to purchase a unification of the overall system with a counterfeit coin. Thus the physicalist looks, within physics, for reductions in the number of independent and ultimate entities, processes, and properties, and for a characterization of how all complex physical phenomena arise out of the basic physical entities and processes. Unification at the level of physics is a must for the physicalist programme. And although there are those who would dispute it (Cartwright 1983; Hacking 1983), the major trend in physical theorizing is towards greater and greater unification. See, for example, Hawking (1988) and Davies and Brown (1988) for clear and unequivocal statements regarding the trend towards unification in physics.

Assuming unification of the physical ontology, in what does unification of the physicalist system as a whole consist? Essentially, it is the construction of a structure upon this physical foundation which reveals everything else in the system to be a manifestation of the foundational elements. The relations alluded to earlier (dependence, supervenience, and realization) make explicit what it means to say that something is a 'manifestation' of the objects and attributes in the basis. Thus the unification provided by the system consists in these ways of relating everything to a comprehensive, unified base class of physical elements. It is a way of embedding all the diverse, non-physical aspects of nature within a physically-grounded ontological structure.

Why is unification of this sort valuable? As with parsimony, one might hold that unity is not intrinsically valuable: after all, there is just as much unity in nature as there is, no more and no less. But this claim, true as it is, cuts no ice with respect to the idea that a more unified system provides us with something that a less unified system does not. Specifically, a unified picture of nature provides more and deeper understanding than does a view of nature that represents it as a disunified aggregate of isolated and disconnected facts. Why is this so?

Ontological unification leads to increases in understanding provided by a physicalist system because all phenomena are embedded within a physicalistically structured ontology that is grounded in the physical domain, and because such embedding provides a framework for developing explanations of the occurrence of any non-fundamental phenomenon. Thus ontological unification means that there is a decrease in the number of unexplainable phenomena in nature:[14] the unification of the physical basis reduces the number of unexplainable phenomena contained within it, and the grounding of the rest of the system upon this basis clears away any unexplainable phenomena at higher levels. In this way, the physicalist's ontological unification programme reduces 'the unexplainable' by locating all fundamental phenomena in the basis and by precluding ontologically emergent objects (for instance, ghosts) and attributes (for instance, vital forces) relative to that basis.

The preclusion of emergent attributes is of special importance, since the grounding in the physical basis of all relations between higher-level objects and events means that all patterns of connection and influence are ultimately manifestations of physical connection and influence. Not only is every phenomenon grounded in the physical realm, but the interactions between phenomena in nature, to the extent that such interactions exist, are grounded in that realm as well. Understanding is thus enhanced because there are no unexplainable higher-level influences. Any unexplainable influences are relegated to the set of ultimate influences at the core of the physical ontology, whereas all higher-level influences are dependent upon, supervenient upon, and realized by those in that ontology. On the flip side, the embedding of all phenomena in a physicalist ontological structure reveals, and provides the basis for explaining, the *limits* of influence between natural phenomena. Any putative influences (for example, paranormal influences) that exceed physically-based limits are specious, and can be 'explained away' by appeal to the physical basis.

Finally, higher-level regularities can be explained by identifying lower-level, and ultimately physical, processes that realize their instantiations in nature. Since such regularities can exhibit exceptions from time to time, the sources of such exceptions can be revealed by a proper embedding of a particular higher-level system that exhibits the regularity in the larger, unified ontological struc-

---

[14] See Friedman (1974: 15, 18–19) and Kitcher (1981: 529–30) for versions of this idea.

ture of which it is a part. Exceptions can often be explained by
intrusions from outside the system of interest, sometimes they can
be explained by intrusions 'from beneath' (i.e. by features of the
physical realization of the phenomena in question), and sometimes
they can be explained by failures of other required conditions of
the system or its environment. Ontological unification provides a
framework for developing these sorts of explanation by locating
the phenomena of interest in a comprehensive, physically-based
structure. In sum, within a physicalistically unified ontological
system, since all objects, attributes, events, etc. are a part of a
structure that is grounded in the physical basis, all sorts of explanation can draw upon aspects of how a given phenomenon to be
explained is embedded in that structure.

The next group of valued features promoted by the physicalist
programme involves certain global improvements in systems of
theoretical knowledge. Classical definitional and derivational
reductionism was viewed by its proponents as a way of promoting
such systemic features as simplicity, consistency, theoretical unity,
and comprehensiveness. In so far as these features are prized by
the physicalist, it is important to keep in mind that classical
reductionism is just one way of achieving them.

Like ontological parsimony, simplicity of a theoretical system is
not an intrinsically valuable feature. Rather, it is important only
in so far as it contributes to other features of interest. Simplicity
is also notoriously difficult to define and measure. Thus in what
follows I shall focus upon the deeper and more intrinsically valuable features of theoretical systems that are of interest to the
physicalist. I shall assume that, *ceteris paribus*, simplicity contributes to such important systemic features as theoretical unity
and explanatory power. It therefore ultimately contributes to the
amount of understanding yielded by the system as well as, perhaps, to its pragmatic value.

Consistency of a theoretical system is, of course, an important
feature. It is desirable that our theories and explanations should
be capable of all being true at the same time. This is important,
not only because we want them all to be true, but also because it
is frequently important that they can be combined and integrated
in various contexts. Consistency is of value to the physicalist because
it is a minimal requirement for such integration. But there is much
more that can be said regarding how theories and explanations

relate to each other; with regard to theoretical systems, there is more structure and integrity besides consistency that is of value to the physicalist.

As I understand it, physicalism promotes unification of theoretical systems by requiring that they be vertically structured in certain ways. More specifically, physicalist *theoretical* unification consists in the grounding of all theories and explanations in the physical basis in the sense that any object, property, or relation posited or otherwise appealed to for the purposes of description or explanation is a physically-grounded object, property, or relation. Thus all such theoretical and explanatory structures are dependent upon, supervenient upon, realized by, and ultimately explainable in terms of physical (or physically-based) theoretical structures. The idea is that of a system of theories and explanations that are explanatorily linked to the physical basis either directly or indirectly via successive vertical links. Unification is achieved because there is a single ground relative to which all theoretical and explanatory structures are related in this way.

Further, theoretical unification for the physicalist consists in the potential for integration of different theoretical domains, a potential that is underwritten by physical substructure. In so far as theories of apparently distinct subject matters can be jointly deployed to relate those subject matters, there is a physical basis for such deployment. Our theories and explanations hang together in certain ways constrained and determined by the physical base of the system, and appeals to that base can make comprehensible these ways of hanging together. Thus the nature and limits of theoretical integration are based upon underlying physical structures and processes.

Examples of such theoretical integration include: theories of individual psychology with theories of economic, social, or political processes; multidisciplinary research programmes such as cognitive science; technological and social enterprises such as city planning; clinical enterprises such as medicine and psychiatry. In all such cases, theories from different domains are brought together for the purpose of theoretical or empirical research, or for more practical purposes; in all such cases, the integrative enterprise is underwritten by a physicalist framework. There are no ungrounded connections between theoretical domains. Such physical grounding can be exploited to improve theories and to deepen understanding of how different domains relate. But regardless of whether it is exploited, it exists and provides the basis for making such integration intelligible by circumscribing its nature and limits.

To sum up, physicalism underwrites both the occurrence of all natural phenomena and the connections between such phenomena. Theoretically this is expressed by a system of explanatory structures that is grounded in a physical base. Theoretical unity of the system consists in the ultimate connection of all theories and explanations to the physical base and to each other via the base. The value of this sort of unity is that it contributes to the development of more powerful theoretical systems, that it promotes understanding, and that it contributes to the solution of difficult social and technological problems by underwriting the integration of disparate theories in various contexts.

A fourth global feature of value to the physicalist is *comprehensiveness*, both with respect to physics and with respect to the system as a whole. Physics is comprehensive in the sense that everything is ontologically and explanatorily grounded in it, as characterized above. It provides a base for the dependence, supervenience, and realization of all phenomena and for the grounding of all theoretical and explanatory structures. As a consequence, the physicalist system as a whole is comprehensive in the sense that *everything* has a place in it. Nothing is left out of account. It is important to keep in mind, however, that such comprehensiveness does not require that all phenomena are in the domain of physics, but only that they are grounded in that domain. Comprehensiveness, like the other global features just reviewed (viz., simplicity, consistency, theoretical unity), is important because it contributes to the degree of understanding provided by a physicalist system. It is to this third general sort of feature of physicalist systems that I now turn.

The physicalist's deepest aim is to develop a system of knowledge that enlightens, that promotes understanding of ourselves and of the world around us. By building a system that involves ontological unity of the sort characterized above and that involves simplicity, consistency, theoretical unification, and comprehensiveness as just described, this aim is promoted. To sum up several of the points made above: understanding is increased because the system reveals why things happen and how they hang together. The physicalist programme is one of building a system in which all objects and attributes are embedded in a certain sort of onto-logical structure and in which all theories and explanations are embedded in a certain sort of theoretical structure. As a conse-quence, the physicalist system reveals how everything is related to the physical domain, and it provides a framework for compre-hending the nature and limits of the relations between all things and between all theoretical domains. The grasping of such connec-tions and limits is a paradigmatic instance of what it is to un-derstand the world.

Such ontological and theoretical structures, grounded as they are in a unified physical base, also reveal the considerable power and importance of physics. To see all things as a manifestation of the physical realm is to appreciate the significance of actually developing a physics that works. The constraints on such theoriz-ing are enormous, since physics, according to the physicalist anyway, must be adequate to the task of unifying everything and grounding all explanation and theorizing about nature. As a result, the actual development of such a physical theory would in itself constitute an enormous cognitive achievement.

As I have suggested above, a key element of physicalist under-standing is what I refer to as 'vertical explanation'. Such explana-tion purports to show how and why objects, properties, and relations are realized by lower-level, and ultimately physical, ob-jects, properties, and relations. Such vertical explanation is, per-haps, an expression of what has been called 'the Democritean tendency'. But unlike earlier versions,[15] I do not see such patterns of explanation as being restricted to a single sort of relation be-tween realized and realizing entities (for example, identity, fusion, structured wholes, composition of causal powers). How particular sorts of objects and attributes are realized in the physical world is a question that must be dealt with case by case. Vertical explana-tion must therefore be open to a variety of ways in which the many diverse phenomena that occur in nature arise out of the physical basis.

In addition to making intelligible how non-physical phenomena can occur in nature, vertical explanations also have valuable im-plications regarding our understanding and exploitation of the

---

[15] E.g. those of Democritus and Epicurus, or more recently, of Oppenheim and Putnam (1958) and Causey (1977).

patterns of vertical influence within various systems (for example, top-down and bottom-up influence)[16] and of the processes that underlie causal relations at high levels. The better we understand hierarchical structure, the more capable we become of grasping and manipulating causal processes. Provision of such vertical explanations constitutes the physicalist's ultimate payback on all promissory notes concerning the denial of the existence of ghosts and ontologically emergent properties and relations. Vertical explanation reveals both why certain non-physical objects and attributes are realized *and* why certain other putative objects and attributes are not.

It is important that physicalism should be viewed primarily as a constructional *programme* and not simply as a metaphysical *doctrine*. Only from this perspective can its full significance be grasped. Physicalism is a programme, on a grand scale, for building a comprehensive and highly diverse system. What matters for attaining the understanding promised by the programme is not abstract description of a physicalist system (via abstract metaphysical, semantic, and epistemological theses), but painstaking, detailed working out of the various parts of the desired structure. If there is no development of specific vertical explanations, there will be no actual understanding of the sort just described; thus the programme of physicalism will not have attained its primary objective (i.e. increasing our understanding of ourselves and the world). It should be further noted that there is much detailed work involved in developing the various vertical explanations required for full construction of a physicalist system. Such work requires considerable expertise in specific areas; it is evident therefore that physicalism is hardly a one-person philosophical programme.

The upshot of all of this unification and structuring of physicalist systems is a global increase in understanding attained in the ways to which I have alluded. The motivation for pursuing such understanding is not properly conceived of as 'fear of ghosts' or 'fear of darkness', as some have suggested (Haugeland 1984*a*). It is rather an expression of the desire for light and the recognition that such understanding can be of enormous consequence in various domains of human activity, as I shall now briefly discuss.

---

[16] Note that physicalism is in no way committed to the idea that because the physical realm provides the ontological ground for everything, there is only one direction of vertical influence: up!

A different sort of value that a physicalist system promises to deliver involves its impact upon human experience and action. There are four types of impact that make physicalism a potentially valuable programme. First, it is an important aspect of the physicalist programme that it tends to guide and improve research in various areas of science and philosophy. The guidance consists in providing a stimulus to ask certain questions (for example, concerning realization) and to seek certain sorts of answers (for example, in terms of attributes, processes, and mechanisms of realization that are grounded in physics). The guidance also consists in the application of constraints upon research, constraints that are based upon physicalist structure (for instance, if a certain type of hypothesized influence has no evident physical ground it will be viewed with suspicion).

Physicalism will also tend to lead to the improvement of theories as they become integrated into the system and grounded in the ways described earlier. An understanding of lower-level processes can provide new insights into how various phenomena are realized and how certain systems work (for example, the human visual system can only be understood if lower-level processes are taken into account). Such understanding can lead to explanation of anomalies, suggest refinements of a theory by identifying new factors and connections, suggest ways of extending a theory to new domains of application, as well as circumscribe the limits of such refinements and extensions. Finally, since theoretical integration is ultimately based upon a physical substructure that connects apparently disparate domains, the physicalist programme will improve the chances of integrating various specialized domains of inquiry now being pursued, sometimes in virtual isolation from each other. For example, current efforts to integrate domains (for instance, linguistics, psychology, neuroscience) in cognitive science research are underwritten by physicalist assumptions about how various types of phenomena are related and realized.[17]

A second pragmatic feature of a physicalist system is that, in so far as such a system is a comprehensive, integrated structure grounded in physics, it will tend to promote efficiency with respect

---

[17] I think it is important for physicalists and non-physicalists alike to reflect upon what such inquiry would be like if no physicalist assumptions were being made. When push comes to shove, such research activity cannot be made intelligible in a non-physicalist environment, in my opinion.

to human capacity for intervening in and influencing the course of events. As mentioned above, patterns of influence in particular systems can be comprehended and exploited as we gain increased understanding of the vertical structure of such systems. Thus within clinical psychiatry a physicalist framework enables comprehension of both how drug therapies can influence thought processes and how 'talking therapies' can influence the physiology of the brain. Further, the development of a general physicalist framework will tend to reveal how different domains of human interest are related to each other. Such improved understanding will tend to increase the capacity to influence one domain by tinkering with others (for example, influencing mental processes by manipulating the physical environment). And such understanding will allow us to comprehend and influence complex phenomena involving a confluence of different sorts of factors from different theoretical domains. For example, again in clinical psychiatry, biological, psychological, and social phenomena must be integrated if the phenomena of mental disorder are to be understood and responded to appropriately. Effective intervention requires a sufficiently broad framework for identifying and integrating relevant factors and for identifying and exploiting possible pathways of influence. A physicalist system provides such a broad framework.

These first two pragmatic values (improvement of research and improvement of capacity to intervene) feed into the last two areas of broader significance of physicalism: significance for social organization and significance for individual experience and activity. The latter are perhaps the most important areas of interest for the physicalist. After all, increased understanding and capacity to intervene have their deepest significance for us as they bear upon how we view ourselves and how we live our lives. Such interests have often been behind physicalist thought in the past.[18]

All the features of a physicalist system, discussed in the past several pages, can be brought to bear upon questions of deepest individual and social concern. The physicalist view of things has the potential to shape our view of ourselves, each other, and the world, and to inform the character and conduct of our lives by

---

[18] One can be reminded of this by a review of Lucretius's most eloquent exposition of Greek atomism. In recent years, the more technical side of physicalism has tended to obscure these deeper concerns. Post (1987), however, is one recent physicalist who has not lost sight of what physicalism is about in these areas.

providing answers to such questions as: What is objectively real, true, and possible in the world? What is my connection to the natural order? How am I influenced by the world around me? How can I influence the world? What are my origins and what is my destiny? What is the ultimate destiny of the human species? What are the origins and fate of the universe? What is the place of values (moral, aesthetic, personal, religious) in a world of fact? However, although this list could be readily extended, it is not my intention to suggest that physicalism provides the answers to all the problems that face an individual or a group. What it does do is provide one sort of framework for approaching those problems and for dealing with them. Given the various features of a success-fully completed physicalist system, there is considerable potential for solutions that are superior to the sorts of solutions that would be provided in various non-physicalist frameworks.

This completes my discussion of the *a priori* reasons for being a physicalist. I hope it is clear that there is real content to the claim that physicalism is a programme for developing a system with structure that yields pay-off. My contention is that if one values the pay-off, one ought to be motivated to entertain and pursue the physicalist programme. Of course, the promise of pay-off is not the only relevant consideration in deciding which horse to bet on. The remainder of this book should be seen as, in part, an extended discussion of many of the other relevant factors.

1.3. GRATUITOUS ASSOCIATIONS

Having outlined the key ideas and basic values of physicalism, I shall now consider what I call 'gratuitous associations', ideas and values that have been associated with some versions of physicalism, past and present, but which are not part of the core content of the programme. Combining an understanding of such core ideas and values and the gratuitous associations will put us in a good posi-tion to define and tackle the problem of giving more rigorous expression to physicalist doctrine.

Gratuitous associations to the physicalist programme arise from at least four sources: background philosophical frameworks and assumptions that are not necessary accompaniments to physicalism; background social or political agendas that are likewise unnecessary;

false beliefs about how best to express physicalist ideas and values; and false beliefs about the implications of such ideas and values. On these sorts of grounds, I shall compile a list of ideas and values about which the physicalist can withhold judgement while embracing the ideas and values described in the last section.

But first we must consider an important strategic question concerning how the gratuitous associations are to be identified and contrasted with the key ideas and values. The question is crucial because if there is disagreement over what is central to the programme and what is additional there will inevitably be disagreement over what a proper formulation of physicalist doctrine must express. This is especially relevant to my development of physicalism, since I consider explanatory concerns to be a core component of physicalist thought in a way that many recent physicalists apparently do not. Thus there is some pressure to justify my demarcation of the boundary between what is at the core and what is extra.

I should begin by conceding that there are no analytic or other *a priori* grounds for claiming that a certain set of ideas and values is essential to the physicalist programme. Since physicalism is a programme for research, the values, goals, and ideas that define it are set by those who are going to pursue it. As a consequence, there could be several physicalist programmes each defined by a different set of goals, ideas, and values. But this concession does not mean that anything counts as a physicalist programme. And it surely does not mean that different physicalist programmes are of comparable worth and are equally likely to succeed. Much hangs on how one conceives of the core content of the programme. But, to repeat, there is no reason to suppose that identifying this content is a matter of conceptual or linguistic analysis or of 'first philosophy'.

Further, I do not think that a *purely* descriptive approach is appropriate. Describing the ideas and values of past and present physicalists and looking for a consensus concerning what is gratuitous and what is essential raises rather than resolves the normative question of what ought to be counted as central and what additional. In studying foundational issues concerning physicalism, my aim is to study the foundations of a programme worthy of pursuit, not merely a programme with historical roots and with a historically-based identity. Proponents of physicalism clearly have not spoken with one voice and they continue not to do so. And not all extant versions of physicalism are equally worthy of being pursued. Thus although the study of the past is an important part of developing a characterization of the physicalist programme, the enterprise is not a purely descriptive one.

A third strategy would be simply to *stipulate* what the core content of physicalism is and what the gratuitous associations are. I resist this sort of high-handed approach since it is not good method nor does it properly capture the programmatic nature of physicalism. Rather, I view the drawing of the distinction between core content and gratuitous associations as a proposal subject to assessment and negotiation. As mentioned above, there is a range of physicalist programmes that are conceivable and that vary along such crucial dimensions as worth and viability. I do not believe that the proposal made in this book is the only one worthy of consideration, although I shall contend that it is superior to other contenders in the field, since it is a viable programme of greater *a priori* worth. None the less, I remain open to the need for on-going assessment concerning how best to formulate and pursue physicalism. A time may come when it is appropriate to reconsider the way in which the core content gratuitous association distinction was initially drawn. This, I take it, *is* in line with the programmatic nature of physicalism.

Having said what gratuitous associations are and how I shall identify them, I want to underscore the importance to this project of doing so. First, the distinction between core content and gratuitous associations is essential for proper formulation of physicalist doctrine: a formulation should correctly and fully express the core content, without incorporating unnecessary details. Second, this distinction is crucial for careful and accurate assessment of the programme. With the distinction in hand one can spot real, as opposed to specious, objections and sources of trouble, and one can diagnose and re-frame misguided disputes over the prospects of the programme.

Finally, the distinction will make the broader significance of the programme more readily comprehensible. What physicalism implies in philosophy, science, society, and an individual's life needs to be sorted out. Accomplishing this task is a condition for successfully fitting physicalism into these areas of human concern, and it depends upon a clear demarcation of the content of the programme. As I shall argue in Chapter 7, it should be recognized that, despite the stringent constraints it imposes, physicalism is likely to be consistent with a wide range of variation in the solution of many philosophical, scientific, social, and individual problems. The ideas and values of physicalism are compatible with a broad-ranging pluralism on many fronts. The identification of gratuitous associations will help identify the boundaries of that range, and it will help clarify the issues involved in integrating physicalism into various contexts.

What follows is a partial list of what I take to be gratuitous associations. First, regarding the philosophical spirit with which physicalism is proposed, the programme should not be construed as a species of so-called 'metaphysical realism'.[19] Such a construal has often been made by Putnam (1979: 611 ff.; 1983: 208) in recent years; although physicalism can be advanced in such a way, it need not be, as Putnam implicitly acknowledges. Second, one need not construe physicalism as committed to specific theories of truth and reference, despite the fact that some recent physicalists have formulated their physicalism in such terms (for example, Devitt 1984). Both of these points should be seen as relaxations of unnecessary and problematic philosophical baggage. Consequently, the approach I shall take below is to develop physicalism with as little initial commitment to such auxiliary philosophical stances as possible, introducing such views only when absolutely necessary.

Regarding the specific form that physicalist doctrine takes, there need be no commitment to theses requiring classical definitional or derivational reductionism, micro-reduction, eliminativism, or the identity of properties, states, processes, etc. All such views suffer either from failing to express the key ideas and values they are supposed to express or from requiring overly restrictive unifying relations. In Chapter 2, I shall consider in detail the various reasons for not being committed to these doctrines.

Turning to the key ideas at the centre of physicalist thought,

---

[19] 'Metaphysical realism' is roughly the view that there is a theory-, language-, and mind-independent reality possessed of a 'ready-made' structure awaiting our descriptions, conceptualizations, and theories. Unlike Putnam (1979; 1983), I do not include either the correspondence theory of truth or the idea that there is exactly one true theory of the world under this label, although these views are also gratuitous associations to physicalism. See Ch. 6 for further discussion of the relations between physicalism and realism.

there is good reason to avoid such views as that everything is completely physical, physics is the one true theory, physics describes and explains everything, the only patterns of explanation and the only legitimate methods of inquiry are those of physics, and everything is best understood from the perspective of theoretical physics. None of these views is implied by the core ideas and values I have identified, and there are substantial independent reasons for not endorsing them, as we shall see. Thus in placing physics at the foundation of a unified ontological and theoretical system, I am not committing myself to any claims asserting the absolute primacy of physics with regard to description, method, and explanation.

With respect to conceptualization of the ontological base of the physicalist system, there are a number of traditional associations that must be set aside. Such associations arose, often appropriately, from the character of physical theory at specific times in the past. But as physical theory evolves, so does our understanding of the physical domain. Thus, physicalists ought not be constrained by views of the physical ontology which involve, for example, the concept of matter, separable systems, 'particularism' (Teller 1986), deterministic systems, a space-time structure with four independent dimensions, or the value definiteness of physical magnitudes. What the physical ontology consists in is, of course, *the* central problem to be solved for proper development of physicalist doctrine. Recognition of the evolving nature of physical theory and the consequent rejection of specific *a priori* conceptions of the physical are critical aspects of that problem.

Finally, there are a number of views concerning matters of particular human interest which past thinkers in the physicalist tradition have embraced or which critics have attributed to the programme: for example, scientism, anti-religion, the devaluation of human beings in the scheme of things, the denial of subjectivity, conventionalism about values. Such specific views ought to be treated as gratuitous associations because they are not evidently implied by the core ideas and values identified above and there are no good reasons to count them among those motivating ideas and values of the programme.

Physicalism ought to be initially conceived and formulated without specific commitment to stands regarding such important issues, especially since the significance of physicalism is much harder to track than has usually been supposed. It has often been claimed that some aspect of reality (for example, moral values, intentionality, subjectivity) cannot be fitted into a physicalist framework. However, the grounds for such allegations have invariably fallen well short of being compelling. To include such claims, or their denials, within the core content of the programme, although a possibility, does not sit well with those who are motivated by increases in understanding as the key value of the programme. One's purpose as a physicalist is to understand these and other putative aspects of existence, not to legislate for or against them. It could turn out, of course, that one or another of these views is ultimately acceptable, but the grounds for thinking so will go well beyond purely physicalist ideas and values.

Regarding all the so-called 'gratuitous associations' just reviewed, the key point is that, in developing *physicalism*, the best method is to accumulate as few antecedent commitments as possible. The brief taxonomy above is intended to provide examples of the sorts of commitment regarding philosophical framework, formulations of the doctrine, and other agendas that physicalists ought not take too hasty a stand on either because physicalist ideas and values can be expressed in a number of different ways, *or* because the programme can have value in a wide range of larger frameworks, *or* because what the programme implies for many areas is unclear or not highly specific.

1.4. CRITERIA FOR ASSESSING PHYSICALISM

Given the discussion in the last two sections concerning the core ideas and values of the physicalist programme and the gratuitous associations, the objective of this section is to characterize a framework for formulating physicalist doctrine and for assessing the physicalist programme. Recall that the doctrine has three components: theses, presuppositions, and metatheses. The formulation of specific physicalist theses is the task of identifying appropriate physical base classes and appropriate unifying relations which suffice for effectively expressing the core physicalist ideas and values. It is this task with which I shall be concerned in the next several chapters, along with formulation of critical presuppositions and metatheses.

Subsequently, I shall conduct a partial interim assessment of the programme with respect to three dimensions of evaluation. First, the *adequacy* of a formulation of physicalist theses consists in its successfully expressing the central content of the programme (i.e. the core ideas and values). Thus an adequate formulation must have the property of being such that if the programme based upon that formulation were to be successful, then the key ideas would be fully expressed and the system that is constructed would exhibit the key features valued by the physicalist. It is one job of the physicalist philosopher to articulate theses that satisfy this criterion.

It is not, however, the job of the physicalist philosopher to build the relevant system single-handedly, since many sorts of specialized expertise are required. Nor must the theses of physicalism characterize the details of the system itself: it is important to distinguish between theses which call for a certain structure and the details of that structure. Thus an adequate formulation of physicalism calls for a certain type of structure without characterizing its fine details. For example, a formulation might call for the realization of all attributes by physically-based attributes without specifying how such realization is to be accomplished in particular cases. Further, the subjunctive conditional form of the criterion means that formulations of physicalism are not being assessed in terms of how well they motivate anyone actually to build a physicalist system. There may be many good reasons for not pursuing the physicalist project under various circumstances, but the assessment of adequacy of a formulation has nothing to do with such matters. Adequacy of a formulation concerns the features of a system that would result if it were constructed in accordance with that formulation.

A *minimally adequate formulation*, on the other hand, is a formulation of physicalism that includes no gratuitous associations and is the logically weakest adequate expression of the physicalist position. In this book, I shall focus upon the adequacy of the various formulations of physicalism. Questions of minimal adequacy will be considered only in passing because the issues of main interest arise regardless of whether a formulation is minimal.[20]

---

[20] See Lewis (1983) and Post (1987) for discussion of 'minimal formulations' of physicalism.

Once the question of the adequacy of a formulation is settled, the question of its acceptability arises. By an *acceptable* formulation I mean one which is adequate, defensible against the objections of critics, and plausible on the basis of relevant evidence. There are numerous objections that can be, and often have been, directed against physicalist doctrine. The working out of the physicalist programme requires providing sufficient responses to each. In addition, I note that the issue of what counts as relevant evidence for physicalism will be found in Chapter 5 to be much more problematic than is usually supposed. This finding seriously complicates the assessment of the acceptability of physicalist doctrine.

Finally, the assessment of the actual working-out of the programme based on a specific formulation of physicalist doctrine is of critical importance. By the *success* of the programme I mean the formulation of an adequate version of physicalist doctrine, the assessment of that formulation as acceptable, and the actual construction of a physicalist system of knowledge in accordance with the doctrine. Needless to say, success is a most stringent measure of how well one is doing. Identifying and evaluating the severity of various obstacles to success is part of the ongoing assessment of the programme.

I should observe that, as conceived here, successful working-out of the programme does not necessarily involve the incorporation of physicalism within *particular* larger philosophical, scientific, social, or individual frameworks. The implications of physicalism for such areas of concern will be traced, whenever possible, within the programme. Where necessary for the working out of the programme, the adoption of a solution to some specific problem will be made. But physicalism simply does not dictate solutions to all problems in these other areas, and the degrees of freedom remaining will have to be reduced in other ways. Consequently, the working out of the physicalist programme is not sufficient for the complete development of total philosophical, scientific, social, or individual frameworks.

2

Review of Past Formulations

Any robust materialist position should affirm . . . that what is
material determines all there is in the world.

Jaegwon Kim (1984: 162)

Determination principles are global . . . This saves us from
worrying about just how much of the environment of a system needs to be considered in order to determine the higher-
level predicates . . .

Geoffrey Hellman (1985: 610)

The principal objective of the present chapter is to examine a wide
range of past formulations of physicalist theses and to assess them
by the criteria just set forth. I shall argue that each fails to satisfy
the criterion of adequacy, and I shall diagnose these failures with
an eye to developing a clearer conception of what an adequate
formulation of physicalism should be like.

My primary objective in what follows is to pave the way for the
development of an adequate and acceptable formulation of
physicalist doctrine. Since achieving that goal involves identifying
a set of principles that are jointly sufficient for expressing physicalist
ideas and values and that are not vulnerable to disabling objections or countervailing evidence, it is essential to review the existing
catalogue of principles and to identify their strengths and weaknesses in these respects. I fully recognize that not everyone has had
my objectives in mind, particularly not the objective of expressing all the ideas and values mentioned in Chapter 1. Thus some
of the proposals I shall discuss were intended to characterize, for
example, specific types of inter-theoretic reduction that occur in
the sciences, and they were not specifically designed to express the
crucial relations involved in a grand physicalist architectural scheme.
Other proposals, though intended to capture aspects of the physicalist programme, were not designed to capture all the aspects I
have outlined above. When I criticize these views, especially with

respect to adequacy, I do so primarily from the perspective of what will serve my purposes; no unfair criticism is intended. With this caveat in mind, I turn to a review and assessment of extant principles that have been employed to express physicalist ideas and values.

2.1. REDUCTIONIST APPROACHES

It is both natural and important to begin with an examination of the programme of classical reductionism, since most recent developments bearing upon how best to formulate physicalism have been keyed in to the various reasons why classical reductionism appears to have failed as a formulation of physicalism. Such reasons generally involve putative counter-examples to reductionist theses demonstrating their unacceptability. What has not been observed is that the theses of classical reductionism also fail to provide *adequate* expression of core physicalist ideas and values.

As I shall conceive of it,[1] classical reductionism (CR) was a programme for a certain sort of unification within science that was guided by the following theses:

(DT) Every term in the languages of the special sciences is definable in the language of physics.[2]

(DL) Every law of the special sciences is derivable from the laws of physics and the definitions.

Such theses are purely linguistic in the sense that they involve quantification over linguistic entities only (for example, terms, law statements). Together, they were thought by their proponents to characterize a pair of structural relations that unify a theoretical system with respect to both ontology and explanation. Since it was also thought that such unification would contribute significantly to theoretical explanation and understanding, the goals of CR were well attuned to many of the core ideas and values of physicalism.

---

[1] More rigorous formulations of reductionism have been developed, but such refinements do not spare the approach from the criticisms that follow. See Hooker (1987) and Spector (1978).

[2] Criteria of definability can, of course, vary. The criterion for DT is nomological coextensiveness.

Buttressing this programme were two auxiliary philosophical assumptions that underwrote its deep significance: the D-N model of explanation and the thesis that all substantive metaphysical claims can be phrased in linguistic terms (Carnap 1967). As I shall argue below, the failure of CR adequately to express physicalist ideas and values is based largely upon the unacceptability of these two supporting doctrines. Thus not only is the programme of CR unlikely to succeed, but also, even if it were successful, it would not necessarily secure the unifications and increases in understanding that the physicalist seeks. In the light of this conclusion, the problem of formulation for the physicalist is set: to find an alternative way of expressing physicalist ideas and values that does not succumb to the problems besetting CR.

I begin by showing how poorly CR fares with regard to the ontological component of physicalism: i.e. that everything has a place in the physicalist ontology. As outlined above, this means that everything is dependent upon, supervenient upon, and realized by the individuals and attributes in the physical basis. In judging the adequacy of CR, it is useful to distinguish among ontological categories. For present purposes, the idea that everything has a place in the physicalist ontology is the idea that the ontology of physics bears the required relations to all individuals, attributes, and classes. This has seemed to many to be expressible by DT.

Now, this view that the physicalist ontological position can be expressed by a claim concerning the definability of terms in the physical vocabulary presupposes the following equivalence:

(R) Each entity is physical[3] if and only if each entity is such that there is a term that refers to/expresses it and that term is definable in physical terms.

R can be resolved into three theses to reflect the various ontological categories of interest:

(RI) Each individual is physical if and only if each individual is such that there is a term that refers to it and that term is definable in physical terms.

---

[3] I shall mean by 'each entity is physical' that each is either a member of the physical domain or is grounded in the physical domain in the required ways (i.e. dependence, supervenience, realization).

(RA) Each attribute is physical if and only if each attribute is such that there is a term that expresses it and that term is definable in physical terms.

(RC) Each class is physical[4] if and only if each class is such that there is a term that refers to it and that term is definable in physical terms.

To demonstrate the inadequacy of linguistic reductionism as an expression of the physicalist ontological position, I shall show that the above equivalences fail to hold in both directions.[5] The following two claims will be established:

(LR) Everything (individuals, attributes, classes) could be physical and yet there not be a physically definable term referring to/expressing each thing.

(RL) Everything could be referred to/expressed by a physically definable term and yet not everything would be physical.

The strategy behind existing arguments for LR has been to show that the ontology of *physics* cannot be specified by a system of definitions, and hence that the physicalist ontological position cannot possibly be correctly expressed by a definitional thesis: R and its component theses do not hold even for the ontology of physics, let alone for the ontology of every discipline. Towards this conclusion, Hellman and Thompson argue as follows:

When it is contemplated, moreover, that no matter how sophisticated the list [of basic physical predicates] and the 'defining machinery', there are bound to be entities composed of 'randomly selected' parts of other entities which elude description in the physical language, then it is evident that something is wrong with the whole approach. (Hellman and Thompson 1975: 553.)

The argument concerning individuals appears to be that the defining power of the physical language is not rich enough, no matter how strong it is made, to build expressions that would be satisfied

---

[4] By a 'physical' class I mean a class of physical individuals or a class of classes of physical individuals, etc.

[5] Note that I am not claiming that DT logically presupposes R, since DT can be true and R false. I am claiming that the view that DT gives full expression to the physicalist ontological position presupposes R and that, indeed, R is false. Thus linguistic reductionism as expressed by DT is not adequate as an expression of physicalism.

by entities composed of 'randomly selected parts of other entities'. The force of this argument, however, is elusive. Why should randomly selecting parts of other entities inevitably lead to the composition of entities which 'elude description in the physical language'? Hellman and Thompson's statement of the argument leaves unstated the assumptions upon which it depends. The following claims can be plausibly introduced to make the argument more explicit: (1) the physical language, P, is finitary; (2) the intended interpretation of P has an infinite domain (either denumerable or non-denumerable); and (3) the members of this domain can have either finitely or infinitely many parts.

Given these assumptions, there is a straightforward argument to show that combining randomly selected parts of entities in the physical domain inevitably leads to the composition of more entities than there are descriptions in P. The argument turns on the difference in cardinality between the class of terms in a finitary language and the class of entities constructible from the parts of the members of an infinite domain of individuals. What is troubling about Hellman and Thompson's statement of the argument is that they chose to focus upon random selection of parts, something which leads to the desired conclusion only in the presence of assumptions like those stated above. None the less, their strategy of exhibiting the inadequacies of the physical language for describing the physical ontology constitutes a significant development in physicalist thought. Granted the above three assumptions, their argument delivers a counter-example to RI by showing that LR is true for objects in the physical domain.

Another argument, along the same lines but more explicitly stated, has been offered by Boyd:

Briefly, the problem about definability arises because there is a continuum of possible physical states (if the true laws of physics are anything like those we now accept) but only countably many possible 'definitions' in the vocabulary of fundamental physics. (Boyd unpublished: 15.)

The problem here is straightforward: given the character of current physical laws, the cardinality of the set of all physical states is greater than the cardinality of the set of all physical definitions; thus there must be some physical states for which there are not associated physical definitions. LR, therefore, is true with respect to states, and we have a direct counter-example to RA. In light of this counter-example, it follows that the physical language is not adequate for describing the full ontology of *physical* states and that DT does not express the physicalist ontological position.

One of the premises upon which this argument is based needs clarification, however. What, in particular, is the feature of 'the true laws of physics' that leads to the indicated conclusion? Presumably it is that such laws express continuous, monotonic real-valued functions. Thus if for each value of such a function there is a distinct physical state, then there are uncountably many physical states, one for each real number in the interval of the reals for which the function is defined. And if the physical language is finitary, then the set of physical descriptions available is countable and the argument goes through readily.

In reply, one might try to reason as follows:

> Of course, it is true that, taken collectively, the class of all physical states outruns the class of all physical descriptions. But for any given physical state, a physical description can be constructed. The problem is not that there is a physical state that is not physically definable; it is that we cannot define them all at the same time. Hence the linguistic construal of the physicalist ontological claim is immune from the objection (i.e. LR is not shown to be true by the cardinality argument).

What is wrong with this reply is that, even if its premises are true, they are beside the point and do not entail that no counter-example to RA exists. Why does it fail?

Note first that R and related theses all have the following form:

$$(x)Px \longleftrightarrow (x)\ (\exists t)\ (Rtx\ \&\ (\exists p)Dpt),$$

where Px means x is physical, Rtx means x is referred to by t, Dpt means t is definable by p; and where t ranges over terms in some class of finitary languages, and p ranges over physical definitions formulable in the language of physics. The objection is supposed to show that a linguistic construal of the physicalist's ontological view fails of its purpose because the equivalence, RA, is false if the universal quantifiers range over a non-denumerable domain and the existential quantifiers range over denumerable domains. Physics and its intended interpretation provide such an interpretation for RA. Hence the ontology of physical states is not definable. If the ontology of physical states is not definable, then a thesis regarding definitions cannot express what the physicalist wants to say about the world. The fact that meta-linguistically the terms of a language can be reinterpreted does not alter the force of this argument: LR is true under the relevant interpretation, RA is false under that interpretation, and therefore DT does not express the full content of the physicalist ontological position.

The above argument has general applicability to other ontological categories (for example, properties, relations, events, objects). It applies to any entity that is characterizable along the continuum. Since it is plausible that such entities, in each of the general ontological categories, exist in physics, numerous counterexamples to R are in the offing. With regard to classes, it is a commonplace that there are more classes than there are terms in a finitary language. As a consequence, any purely linguistic thesis designed to express the ontological claim that every class is a class of physical objects (or a class of classes of physical objects, etc.) must fail.

To sum up with respect to this first group of cardinality arguments against DT: the central difficulty is that DT, conceived of as an expression of physicalist ontological claims, presupposes R, and R is false if the language adverted to by DT is finitary and the cardinality of the class of entities to be specified is uncountable. Since this latter assumption is not controversial, I conclude that, if we are working with finitary, first-order languages, the linguistic construal of physicalist ontological claims is inadequate. It appears that the only line of reply left open to proponents of such linguistic construals is to explore whether the introduction of an infinitary or a higher-order language would save R and hence the whole approach. I shall return to this issue below.

A second class of cardinality arguments for LR and against R should also be considered. These differ from the first kind in that, rather than arguing that there are more entities of a certain sort than there are terms, they argue that there are entities that are not describable in a finitary, first-order language. A very concise argument to this effect is offered by Earman in a discussion of the relevance of infinitary languages to problems with physicalism:

To provide some motivation for focusing on such languages, suppose that we want to express the 'state description' of the world as a sentence. Such a sentence may need to have an uncountable number of conjuncts, each of which specifies, say, the values of certain physical fields at some space-time point on a given 'time-slice'. And, to characterize the relevant features of such state descriptions, we may need an infinite string of quantifiers . . . (Earman 1975: 565.)

Thus the idea of a total state description of the world at a time requires that we work in a language rich enough to provide descriptions of every space-time point and to combine such descriptions to form conjunctions of uncountable length. Since finitary languages do not meet these specifications, we have an example of a single state that is not describable within such languages. Therefore another counter-example to R *vis-à-vis* states is plausible.

It might be replied that it is easy to find finite expressions that refer to the total state of the world at some time: for example, 'the total state of the world right now'. But such a reply raises serious concerns about the limits of acceptable physicalistic descriptions and definitions. Earman's objection is that it is not possible to find finite expressions that are physicalistically acceptable and which represent a full state description of the world. If the only descriptions satisfying these conditions must involve uncountable conjunctions that characterize what is (or could be) going on at each space-time point, then the resources of a finitary language will not suffice. The reply does not effectively meet this objection since the suggested description is not physicalistically acceptable. It makes use of an indexical expression, and it lacks appropriate content. Such a description provides no way of discriminating different possible total physical states of the world, and it provides no resources for characterizing theoretically relevant features of the present state. Thus it can play no significant role in a physicalistic system.[6]

A related line of attack against the left–right direction of R is based upon the so-called 'multiple realizability' of some kinds of states. The functionalist view of the mind, for example, holds that mental states are definable in terms of their causal role in the mental life of individuals. Since a given causal role can be played by an indefinite number of different physical states, even if each instance of a functional state is identified with, or realized by, an instance of some physical state, there is no *describable* physical state

---

[6] In correspondence, Post has suggested that there may well be theoretically significant descriptions of the sort I am here discussing: e.g. 'the distribution of mass–energy in the space-like hypersurface at *t*'.

that is nomologically correlated with that functional state. Therefore predicates which express functional states are not definable in
physical terms, and R is shown to be false.

To facilitate discussion of this objection, I assume that functional states are not 'type identical' to physical states. The brand
of physicalism countenanced here allows that there are non-physical
states, but it requires that actual and nomologically possible realizations of such states are physical. For the purposes of the above
argument, to say that functional states are physical means that
such states can only have physical realizations. I assume further
that the class of all nomologically possible realizations of some
functional state, F, is a well-defined class, although not one that
is particularly easy to list. And it is an assumption of the argument
that this class is of infinite cardinality, an assumption with which
there are no grounds to quarrel. Finally, let it be assumed that
each nomologically possible physical realization of F has a physical description.

The key issue raised by the argument, then, is whether there is,
corresponding to F, a single physical state that has a physical
definition. Now, it may be supposed that, allowing certain constructive operations for taking states and forming new states out
of them (for instance disjunction, quantification), there is a single
physical state corresponding to the class of physical realizations
for F. Such a state would be nomologically correlated with F since
all and only the nomologically possible realizations of F would be
instances of it. And if we allow such a construction, the question
is whether that state can be expressed by a physically definable
predicate. If so, then R, and hence DT, are unassailed. If not, then
there is another counter-example to R: i.e. a physical state which
is not expressed by a physically definable term. To explore this
question, I shall consider two cases corresponding to different
ways of constructing physical states: first-order disjunctive physical states and second-order physical states.

In the first case, the correlated state is the physical state, P,
which is P_1 or P_2 or P_3 or ..., where the '...' continues until
the class of all possible physical realizations of F is exhausted. The
corresponding *physical* predicate expressing this state would be
of the form 'P_1 v P_2 v P_3 v ...'. If the class of possible realizations of F is finite, then there is no problem: However, if, as is
assumed in the argument, the class is infinite, then no disjunctive

expression[7] in a finitary, first-order language would express the physical state in question. Thus again, if physicalist formulations are restricted to finitary languages, then counter-examples to R exist and linguistic construals of ontological claims are inadequate. This time the counter-example depends upon there being a physical state that is constructible out of an infinite class of other states. If such a construction is acceptable, then the counter-example follows from the difference of cardinality of that class and the finite length of expressions in finitary languages.

The second case (i.e. the correlated physical state is a second-order state) appears to be more promising. A second-order physical state is one which is characterized in terms of a quantification over first-order physical states. If there were a condition which all and only the members of the class of physical realizations of F share, then it would be possible to specify a second-order physical state, P, corresponding to the class as follows: P is the second-order physical state of being in some first-order physical state which . . . , where '. . .' stands for some condition which a physical state must satisfy to count as a realization of F.[8]

The strategy, then, is to have in the physical language a stock of terms expressing second- and higher-order physical states. Such terms efficiently bypass the need for individually specifying each realization of higher-order states, while allowing us to refer to such states using terms that are physically definable. Notice that no guarantee is made that there will be enough terms for all such states. The current concern, however, is with individual states that are difficult to define physically. But here, if cardinality problems don't threaten, then other equally serious problems do. Thus the problem facing the proponent of this approach is to find, for each higher-order state, a condition which any physical state must satisfy to count as a realization of that state. To find such a condition is to sail between the Scylla of finitely enumerating an infinite class and the Charybdis of saying only that each physical state in the class is a realization of the higher-order state in question. The latter alternative, though expressible, is clearly not worth

---

[7] Any other sort of term expressing this state is, in effect, subsumed under the next case, since conditions of applicability of the term must be provided if it is to be definable.

[8] E.g. the condition might be a certain causal role. See Putnam (1970), Field (1975; 1986; unpublished).

saying: for example, for F, the nomologically correlated physical state is the second-order physical state of being in some first-order physical state which is a realization of F. What is wrong here is, first, that no *physical* definition of the state appears to have been given. Of course, I have not provided a clear specification of what the demands on a legitimate physical definition involve. But the case in hand, making direct allusions to the non-physical state F, seems clearly unacceptable.

Second, the trivial specification of the second-order physical state in question is not responsive to the physicalist's ontological concerns regarding realization of non-physical attributes. It is entirely mysterious what makes each member of the class of realizations of some attribute, F, a *realization of F*, and it is a mystery that creates space for ontologically emergent properties (i.e. properties not in fact realized by physical properties). To eliminate the mystery and to fill the space, the proponent of this approach must place some non-trivial conditions on membership in the class of realizations of higher-order physical attributes that are in turn realizations of certain non-physical attributes.

The problem is not with higher-order physical attributes *per se*. Rather, it is with how to physically define such attributes when they are supposed to involve realizations of non-physical attributes. A legitimate physical definition must specify how the non-physical attribute is realized by the physical attributes in the class of its realizations. That is, the definition must unify that class *vis-à-vis* a class of realizations of the target non-physical attribute. Only in this way can trivial physicalization of non-physical attributes be precluded, where trivial physicalization of an attribute, N, results from the identification of some physical attribute or other that happens to be instantiated at the same time as N is instantiated. In such a case, whether or not N is realized by physical attributes, there will inevitably be a class or physical attributes that constitute a higher-order physical attribute that is correlated with N. This is no small difficulty for the approach.

Although there have been discussions (Boyd 1980; Field 1986; Van Gulick 1985) concerning how to avoid problems regarding an acceptable physicalist definition (for example, in terms of causal role and causal powers), and hence concerning how to specify the conditions on higher-order physical states that involve realizations of non-physical attributes, all such discussions have been too vague or too restricted to provide a general solution. As a result, the appeal to higher-order physical states does not clearly constitute a satisfactory approach for meeting the objection to R that there are terms, expressing non-physical attributes having only physical realizations, that are not physically definable. As a minimum, the approach needs to be supplemented by a general account of the conditions employed in permissible physicalistic definitions of higher-order terms. Such an account must be responsive to the problem, facing all physicalists who allow that non-physical attributes are realized by physical attributes, of showing how such realization can obtain within a physicalist framework that precludes mysterious emergent attributes. And although I am sympathetic to the efforts of those physicalists who have seen this need and have attempted to respond in the way now under consideration, I do not believe that the middle ground between the impossible and the trivial has yet been found.

In summary, the first class of cardinality arguments discussed above showed that no first-order, finitary language is adequate for describing all individuals, attributes, or classes in the physical ontology. As a result, R, RA, RI, and RC all fail in the left–right direction. Therefore, if physicalist linguistic theses (for instance, DT) are restricted to such a language, then they will fail to adequately express physicalist ontological concerns because of the limited expressive power of finitary languages. Further, consideration of the second class of objections to the left–right direction of R suggested that there are individual states that are not definable in a finitary, first-order language. This provides additional support for the failure of R for such languages. Finally, I argued that higher-order finitary languages also encounter difficulties with regard to expressing certain states, although this sort of approach has not been completely closed off. For now, *modulo* finitary first-order languages, I conclude that the physicalist ontological position cannot be adequately expressed by theses calling for definitions.

Turning briefly now to the issue of whether the employment of infinitary languages would save the linguistic approach underwriting DT, it is likely that the approach would *not* be helped in this way. First, relative to whatever language was introduced, there would inevitably be entities that would fail to receive a description on the sort of cardinality grounds cited above against the use of finitary languages. For any given infinitary language, there would be some physical entity (individual, attribute, or class), though generable via operations for entity construction, that would not be describable using the resources of the language. As Hellman and Thompson emphasized, the moral to be drawn is that an adequate formulation of physicalism cannot tie itself too closely to the defining power of the physical language, no matter how much it is augmented.

Second, as I shall discuss more fully below, even if enumerative definitions of infinite classes could be obtained by infinitary disjunction, the issues regarding realization of non-physical attributes would not be addressed. Specifically, conditions on *permissible* disjunctive definitions need to be imposed to guarantee that each disjunct is indeed a *realization* of the attribute being defined: an infinitary, disjunctive physical definition can always be trivially rigged without necessarily precluding emergent attributes. Thus proponents of the move to infinitary languages to save CR must stipulate appropriate conditions on permissible definitions.[9] In my view, it is not at all clear that such stipulations can be made while remaining within the purely linguistic framework of CR. This problem leads us to the next class of objections against R.

Although the failure of the left–right direction of R is sufficient for its rejection, it is crucial to see that there are also problems besetting the right–left direction (i.e. if each thing is such that there is a term that refers to it and that term is definable in physical terms, then each thing is physical). Counter-examples plausibly exist, and hence every individual (or attribute) could be referred to by a term definable in physical terms and yet it not be the case that every individual (or attribute) is physical.

Again, Hellman and Thompson have suggested an argument for demonstrating the existence of a counter-example. Against the claim that a strong form of reductionism (for example, DT) entails ontological physicalism they argue as follows:

To see this, consider a very simple theory, E, containing just two non-logical one-place predicates, P and Q, and the following non-logical axioms:

1. $(\exists x) (\exists y) (x \neq y \ \& \ (z) (z = x \lor z = y))$
2. $(\exists x) (Px \ \& \ (y) (Py \rightarrow y = x))$
3. $(\exists x) (Qx \ \& \ (y) (Qy \rightarrow y = x))$
4. $(x) (Px \lor Qx)$

---

[9] Lewis (1983: 358) expresses confidence in the possibility of infinitary definitions. However, he does not appear to be taking into account the problem of what a legitimate physicalistic definition involves.

That is, E asserts that there are exactly two objects and that exactly one object is a P and exactly one object is a Q and everything is a P or a Q. Now, in E, the following is provable: (x) $(Qx \leftrightarrow \neg Px)$. In other words, Q is definable in terms of P. Yet, this doesn't guarantee that all objects are, or are exhausted by, P-type things. In fact, in every model of E, there are two disjoint subsets of entities, one P-type, the other Q-type. (Hellman and Thompson 1975: 557.)

This line of argument apparently provides a direct counter-example to the right–left direction of RI, since it exhibits a situation in which every object is referred to by a predicate that is either a P-term or a term definable by P-terms, and yet not everything is a P-type thing. If this argument is successful, it reveals one of the deepest flaws of the linguistic approach: its failure to guarantee the physical nature of something picked out by a physically definable term. But despite the directness of this argument, it is not completely satisfying since there are some pressing and necessary clarifications. How much does the argument depend upon there being only two predicates in the formal system? And are negative definitions acceptable for the physicalist?

In a footnote, Hellman and Thompson claim that nothing depends upon there being only two predicates:

If use is made of certain relative terms, clearly within physicalist vocabulary as conceived by traditional reductionist positions, e.g. predicates of location, then parallel arguments can be constructed for theories containing any finite number of predicates. (Hellman and Thompson 1975: 557.)

But predicates of location, like negative definitions, are suspect: the locatability of all phenomena is allowed by some dualist positions. What the dualist denies and the physicalist asserts is that the phenomena themselves are physical. And being physical is not merely a matter of being located.

It is, of course, Hellman and Thompson's point that definitions couched in terms of negation and location are not sufficient for pinning down physical entities. But the point might be turned upon them as follows. A defender of the linguistic approach might contend that not just any definition is physicalistically acceptable. In particular, definitions that are negative or purely locational are not acceptable, whereas the definitions called for by DT and R must satisfy strict criteria for physicalist legitimacy. Hence it might be concluded that Hellman and Thompson have failed to establish what they set out to because their argument depends upon physicalistically illicit definitions.

The question that arises again is: What are the criteria that physicalist definitions must satisfy? Although it should be allowed that Hellman and Thompson have not provided a conclusive counter-example to RI, they have revealed a serious problem for the proponent of the linguistic approach to formulating ontological physicalism: viz., while staying within the resources of such a purely linguistic approach, to restrict physicalist definitions so as to guarantee the entailment from definability to a strictly physical ontology. The burden of this demand is, I think, more than such proponents can carry. It is a commonplace in mathematical logic, for example, that tightly constrained mappings between distinct domains are possible and that terms picking out individuals, properties, and relations in one domain are definable using terms that pick out individuals (etc.) in the other. No deep ontological morals, beyond isomorphism of abstract structure, follow from such mappings and definitions. And there is no way to frame more powerful *formal* restrictions that would guarantee the desired entailment. Therefore I shall work on the assumption that there is no way to both stay within the linguistic approach to physicalism and restrict the definitions in the required way.

In the case of attributes, there are also difficulties. The physical definability of a predicate expressing an attribute does not guarantee that the attribute is physical. It is, I believe, clear, although not uncontroversial, that the nomological coextensiveness of two predicates is *not* a sufficient condition for identification of the attributes expressed by those predicates. So even if all predicates are physically definable, there is no entailment of the ontological thesis that every attribute is a physical attribute: such wholesale definability is compatible with property dualism. The general point is that it is quite possible to relate two classes of attributes, subject to strict criteria, while preserving the ontological distinctness of the members of the two classes.

As outlined in Chapter 1, the physicalist freely allows for the possibility of non-physical attributes, but with the condition that such attributes satisfy the ontological requirements of physical dependence, supervenience, and realization. However, the nomological coextensiveness of predicates expressing non-physical and physical attributes is compatible with true ontological emergence:

since definitions need only be constructed to satisfy the criteria of
definability, there is no guarantee that such definitions cannot be
rigged to satisfy those criteria in the case of non-physical attributes
that are not identical to, composed of, or otherwise realized by
physical attributes (i.e. 'trivial physicalization' rears its ugly head
here too). Although additional constraints on the definitions might
ensure that no emergent attributes are physically definable, it is
again unlikely that purely formal or semantic constraints could do
this job. Thus the linguistic approach underlying DT does not
appear to have the resources required for fending off counter-
examples to RA. And thus the physicalist ontological position *vis-
à-vis* attributes is not adequately expressed by such an approach.
The moral here is that whether an attribute is physical and whether
an attribute is realized by physical attributes do not depend upon
formal or semantic relations between linguistic items.

To this point, I have argued for the general failure of the
equivalence between claims about the definability of terms and
ontological claims concerning the physical nature of all individuals, attributes, and classes of individuals.[10] Thus DT, the first
thesis of CR, is an inadequate expression of physicalist ontological
concerns. As is well known, there are reasons to believe that even
if DT did express the full content of the physicalist ontological
position, it is likely to be false. I shall briefly indicate some of the
main reasons for believing this since they place certain constraints
upon the acceptability of formulations of physicalist principles.

At least one source of objection to DT is to be found in those
discussions in the philosophy of mind concerning 'functionalism'.
Important components of those discussions were so-called 'multiple realization arguments' that were supposed to prove the un-
doing of 'physicalism' (i.e. the identity theory). Mental states,
according to the functionalist, are identical to certain sorts of
functional states, and functional states are abstract with respect to
their realizations, in a way that allows for many different possibilities. Thus functionalism is allegedly compatible with a wide
range of different sorts of entities being capable of mental states:
angels, humans, animals, computers, extra-terrestrials, etc. The

---

[10] Note that it follows trivially from the possibility of non-physical individuals
having physical definitions that there could be classes of non-physical individuals
that have physical definitions.

radically different natures of these entities being manifest, multiple realization has seemed a certainty. And the multiple realization of mental states is alleged to hold not only across types of entity. Both within types and within individuals of a given type, multiple realization of mental states is a likelihood according to the functionalist. (Boyd 1980; Fodor 1975; Putnam 1975, chapter 21.)

Although I have no serious quarrels with multiple realization arguments regarding mental states, I do not believe that they depend upon being a functionalist. Indeed, properly understood, functionalism is a theory of the nature of mental states and, especially, a theory of how mental states are realizable in various sorts of medium. Thus functionalism, to its credit, *purports* to tell a story about how mental states are realized by, *inter alia*, physical systems: it provides a response to the physicalist's concerns about realization. However, functionalism is only one such story, and it is a sadly incomplete and defective one. I say this without having firmly in hand a more complete and clearly less defective alternative: nobody has that. *But let it not be supposed for a moment that the attack on DT depends upon the outcome of these debates in the philosophy of mind. The failure of DT is a widespread one extending into virtually all domains of science and beyond.*

This is not the appropriate place to review the various arguments for the failure of DT in biology, psychology, the social sciences, physics, and non-scientific domains.[11] I take those arguments to have established beyond serious dispute that the physical definability of all terms is a false thesis as well as a false ideal, as argued above. The conclusion to be drawn from all this is that DT is both inadequate and unacceptable as a formulation of ontological physicalism. Its failure is tied to the shortcomings of the definitional powers of the physicalist language: specifically, its inability to define all physical and non-physical objects and attributes *and* its inability to express in definitions the ideas of realization and non-emergence. Theses of physicalism had, therefore, better not be too closely tied to strong assumptions about such powers.

I now turn to a discussion of how well CR fares with regard to the explanatory component of physicalism. As conceived within

---

[11] E.g. arguments regarding biology (Hull 1974; Kitcher 1984; Kincaid 1987) and social science (Currie 1984; Little 1991) have been presented.

the classical reductionist programme, this explanatory component involved development of an explanatorily unified system in which the laws of the special sciences are all derivable from, and hence explainable in terms of, the laws of physics and the definitions mandated by DT. Such unification was supposed to entail that all phenomena subsumable under the laws of the special sciences would, when redescribed in the language of physics, also be derivable from, and hence explainable in terms of, the laws of physics. A further consequence of such unification, the ultimate objective of the CR programme, was believed to be an increase in understanding resulting from a reduction of the total number of mysterious or unexplained phenomena, a reduction of the number of fundamental explanatory laws, and an increase in the power of those basic explanatory laws as a consequence of their increased generality. As is evident, these objectives of CR are well in tune with the objectives of physicalism as I characterized them in Chapter 1. Unfortunately, despite appearances, this approach is neither adequate with respect to the explanatory goals of physicalism nor acceptable.

That such an approach is inadequate as an expression of the explanatory concerns of physicalism is due largely to the fact that it depends crucially upon the shortcomings of the DN model of explanation. That is, the approach incorrectly presupposes that derivation from a set of laws is sufficient for an explanation in terms of those laws. It follows that the 'success' of the classical reductionist program (i.e. the construction of a system in accordance with DT and DL) would not necessarily provide the required explanations since there would be no guarantee that all non-physical laws would be explained by physical laws even though they would be derivable from them. Thus standard asymmetry and common cause counter-examples to the DN model (Salmon 1984) suffice to show that mere derivability does not guarantee the sort of explanation desired by the reductive physicalist.

Further, there would be no guarantee that the definitions alone, or in conjunction with the laws of physics, would provide proper vertical explanations of the instantiation of higher-level attributes, since those definitions are constrained only by the criterion of nomological coextensiveness. It is not required that they express, for example, principles of composition that explain why certain non-physical attributes are realized by certain physical attributes.

Since nomological coextensiveness is not sufficient for explanation, the derivation of the occurrence of non-physical attributes in no way guarantees explanation of their realization.

Finally, the success of the reductionist programme would not guarantee that the phenomena subsumed by the reduced non-physical laws would be explained. Not only ought we not to assume that derivation guarantees explanation, but also we ought not to assume that explanation is transitive. Even if the reduced higher-level laws are explained as a result of reductive derivation, the phenomena they explain need not be. An important instance of this problem has been discussed by Fodor (1978), who argues that the classical reduction of psychology to neuropsychology (or physics) can result in a loss of explanatory power since the neurophysiological (or physical) translations of psychological laws, based upon the definitions called for by DT, might fail to provide explanations of the phenomena subsumable under the psychological laws. The reason for this failure, according to Fodor, is that definitions constrained only by nomological coextensiveness need not preserve features of the defined predicates that contribute essentially to their roles in psychological explanation. For example, this is plausibly the case for predicates that express the content of a mental state and neurophysiological (or physical) predicates that are nomologically coextensive with them.

If this argument is sound, then it provides a counter-example to the claim that the derivability of a class of laws from another class entails that the phenomena subsumed under the members of the first class are explained by members of the second, and hence the laws in the second class gain in explanatory power as a result. This point, in conjunction with the ideas that the derived laws themselves are not *a fortiori* explained and that the definitions do not necessarily provide vertical explanations of higher-level phenomena, substantially undermines the classical reductionist programme with respect to the explanatory goals of physicalism.

It might be replied that these arguments have been unfairly premised upon the claim that CR presupposes that derivation entails explanation. Indeed it might be argued that although classical reductionism is closely linked to the DN model of explanation, that model is not committed to such a blatantly false presupposition. Thus CR can be saved from the above objections regarding explanation by *appropriate* deployment of the DN model. In particular, only explanatory derivations will count as legitimate reductive derivations of the sort called for by DL. My reply to this must here be brief. First, I do not see how such a strategy can be extended to cover the problem regarding the transitivity of explanation. It is not at all evident how the explanatory role of certain types of predicate (for example, predicates involving semantic content) is to be handled within the DN framework. Second, it is not clear what sort of constraints on derivations can be implemented to save CR. What constraints, for example, can be placed upon definitions to guarantee that they will yield vertical explanations? And what constraints upon derivations can be imposed to guarantee explanation of derived laws? To appeal to *ontological* considerations, as some approaches considered below do, would be to abandon the primarily linguistic nature of the CR programme. And, it is most implausible that further formal or semantic constraints are sufficient to distinguish explanatory from non-explanatory derivations. Finally, there is a substantial recent literature, with which I concur, outlining considerations showing that the DN model of explanation is itself not salvageable (Salmon 1984; Van Fraasen 1980; Kitcher and Salmon 1987). Thus I conclude that CR is inadequate with respect to the explanatory concerns of physicalism.

Like DT, DL is also unacceptable, and its failure to describe correctly relations between higher-level sciences and physics (or between higher-level sciences and lower-level sciences that are grounded in physics) is extremely widespread. Since the arguments for these failures have been amply produced and ably discussed elsewhere, I shall not review them here. It is important to keep in mind that it seems to be the rule, rather than the exception, that the laws of special sciences such as psychology, biology, and economics are not derivable from lower-level disciplines (Darden and Maull 1977; Dupré 1983; Hull 1974; Kincaid 1987; Little 1991; Wimsatt 1978). This does not mean that lower-level mechanisms and processes that underlie higher-level regularities are not identified and studied. It means that strict derivations of higher-level regularities are not, in general, to be had. Such failure is not restricted to biology, psychology, and the social sciences. There are reasons to believe that there are similar failures both in the macroscopic domains of physics itself and in various non-scientific domains.

Now, I do not want it to be thought that too much hangs upon the correctness of these charges of *unacceptability* regarding CR. The *inadequacy* of that doctrine is sufficient to establish the need to develop an alternative. It certainly should not be concluded that lower-level explanation of higher-level regularities are not to be sought and are not of importance to the physicalist. As emphasized above, vertical explanation is of the essence of the physicalist programme. But if the charges of inadequacy and unacceptability are correct, they put a restriction upon *how* explanatory objectives can be met. Specifically, definitional and derivational reduction cannot be counted on to play an essential role in securing those objectives.

To round out the discussion of CR, I shall consider how effectively it responds to the physicalist's concerns regarding objectivity. Recall that those concerns involved identification of the conditions of all objective matters of fact, all objective sameness and difference, all objective truth, and all genuine possibility in the natural order. The physicalist takes certain relations to physics as providing such conditions. According to CR, the theses DT and DL are sufficient to respond to these physicalist concerns by characterizing the required relations.

Unfortunately for this approach, the conditions of objectivity it can provide must inevitably be linguistic in character, either definitions of terms or derivations of sentences. This sort of condition must, for reasons discussed above, fall short of full coverage with respect to all matters of fact, all truths, all similarities and differences, and all possibilities. Not all matters of fact (etc.) are physically definable, nor need they have descriptions derivable from physical laws. As a result, CR simply lacks the resources for capturing the conditions of objectivity desired by the physicalist.

In summary, the situation regarding how effectively CR functions as an expression of physicalism is disastrous. CR is inadequate with respect to all three areas of physicalist concern. In addition to the inadequacy of DT and DL, they are very likely false as well. The two features that most clearly contribute to this poor showing with respect to adequacy are the purely linguistic character of the theses DT and DL and the apparently strong reliance of CR upon the DN model of explanation. Too tight an association between the theses of physicalism and the defining power of physical language is to be avoided, and the shortcomings of a strictly inferential approach to explanation must be overcome. Both problematic areas point to the necessity for introducing theses concerned with objects and attributes themselves in order to supplement, if not to replace outright, theses restricted to linguistic entities.

One natural way of responding to this situation is simply to append to CR the following claims concerning the objects and attributes in the domains of the languages presupposed by DT and DL:

(IC) The individuals in the domains of the special sciences are composed of individuals in the domain of physics.

(AI) The attributes of individuals in the domains of the special sciences are identical to physical attributes.

I shall refer to the view consisting of DT, DL, IC, and AI as 'CR''.

In IC, the nature of the composition relation is intentionally left open at this point. Mereological sums of entities (Oppenheim and Putnam 1958) as well as so-called 'structured wholes' (Causey 1977) have featured in important versions of this approach. As an ontological thesis, IC may well have a place in a viable formulation of physicalist doctrine. All depends upon how the notion of composition is developed. If one opts for mereological fusion, as many have, then the physical dependence and 'exhaustion'[12] of all individuals may well be secured. But much is lost if fusion is the only compositional relation. Many higher-level objects are complex structured aggregates of lower-level objects, while others are not evidently 'composed' of physical objects in any straightforward way at all (for instance, social institutions). Even if such objects are exhausted by the fusion of some class of physical objects, no account of their realization is given by a specification of the members of the relevant physical classes and the relation of mereological fusion. Thus with respect to physicalist concerns regarding *how* higher-level objects are realized by lower-level objects, fusion obviously falls short.[13]

On the other hand, to require that all objects are structured

---

[12] See Hellman and Thompson (1975) and Post (1987) for discussions of 'physical exhaustion'.

[13] In fairness to Hellman and Thompson and Post, their respective versions of the principle of physical exhaustion, based as they are upon fusion relations, are not intended to provide detailed accounts of how various high-level objects are realized by physical objects.

wholes composed of basic objects via a severely restricted class of structuring relations (for example, a class consisting of basic physical forces and processes) is almost certainly doomed to leave some objects out, since many objects (for example, social and cultural objects) are not simply complex physical objects. If such objects are countenanced within a physicalist system (as they should be if they are real), then they cannot be ontologically emergent. And the full physicalist account must show why they are not emergent by revealing how they are realized by physically-based objects. The likely limitations of the structured whole approach with respect to how higher-level objects are realized and the loss of detail inherent in the fusion approach mean that neither is fully responsive to this aspect of the physicalist agenda.

Further, CR′ falls to other objections that are as compelling as those against CR. Thus, the attribute identity thesis, AI, is very likely false, foundering upon multiple realization arguments of the sort that undermined DT. Over the past thirty years or so, including the heyday of the identity theory in the philosophy of mind, it has become increasingly clear just how hard attribute identity claims are to assess. No one should be terribly confident about how to individuate attributes or about how to decide when a given attribute identity statement is true or false. I believe that, as a consequence, physicalists are well advised not to tie their programme to theses like AI. On the view I sympathize with most, the identification of attributes across domains is a very infrequent occurrence. A more important relation is that of attributes in one domain being realized by attributes in others.

Causey (1977) has vigorously argued for attribute identity as an essential component of successful unification programmes in science. Thus his characterization of the unity of science involves nomological correlations between predicates in higher and lower branches. And he claims that in order for explanatory and ontological unification to be effected, those bridge laws must be construed as expressing attribute identities. Otherwise the correlations will constitute mysterious unexplained regularities that offset the benefits of successful reduction. I sympathize with Causey's understanding of the demands upon successful unification. The vertical relations between higher- and lower-level attributes must not be mysterious. However, I do not believe that identity is the only, nor the most promising, relation available to the physicalist in effecting ontological and explanatory unifications of a system. Thus in addition to the likely falsehood of any thesis that calls for identification of attributes, such theses are too restrictive with respect to how attributes can relate to each other in a *successful* unification programme.

The alternative to identifying attributes is to explain the nomological associations expressed by the bridge laws. I suspect that, in his discussion of such explanation, Causey has not properly distinguished between vertical and causal explanation. As a consequence, he has not seen that there is a false dichotomy that drives his argument (i.e. identify the attributes *or* causally explain the correlation). Vertical explanation is explanation of why a given physical attribute realizes the non-physical attribute that it does. It is not a matter of explaining the causal generation of one attribute as effect by another as cause, or of explaining both correlated attributes in terms of a common cause. Thus with the clarification of the *realization* relation between attributes in a physicalist system, a third possibility for handling the problem of mysterious correlations between non-physical and physical attributes opens up.

In addition to the shortcomings of AI, CR' does not clearly respond to the range of difficulties plaguing CR with respect to explanation. What is called for is a more thoroughgoing adaptation of CR' to the explanatory demands of physicalism. In particular, the position needs to be developed in a way that reveals how explanations of the right sort result from successful unification in accordance with its various theses. How, for example, do the theses AI and IC supplement DT and DL to guarantee explanations of laws, without incurring the liabilities of an inferential approach to explanation? And can the position effectively respond to Fodor's objection to CR? I shall bypass consideration of these questions, since in a real sense they are moot: IC, AI, DT, and DL are in all probability false. CR', however effectively it can establish its adequacy as a formulation of physicalism, almost certainly must fail to be acceptable for the reasons discussed earlier concerning CR. Like CR, it leaves us with the problem of how to formulate the doctrine of physicalism in a way that adequately expresses and promotes physicalist goals while retaining significant likelihood of success.

There are two other modifications of CR that it will be instructive to consider. First, Michael Friedman has suggested a thesis that he calls 'weak reducibility' (WR) as follows:

Let 'F_1x', 'F_2x', . . . , 'F_nx' be the primitive predicates of the theory to be reduced. Let a physical realization be a mapping B which associates each 'F_ix' with a set of open sentences containing only physical predicates, B('F_ix') = ['$A_{1i}x$', '$A_{2i}x$', . . .]. For any sentence containing only predicates from among 'F_1x', 'F_2x', . . . , 'F_nx', we can define truth under the realization B and satisfaction under B—they are defined just like satisfaction and truth, except that the clause for atomic formulas now reads: A sequence s satisfies 'F_ix' under B iff there exists an '$A_{ji}x$' in B('F_ix') such that s satisfies '$A_{ji}x$'. Let us now define weak reducibility . . . ; a theory is weakly reducible to physics if there is a physical realization B of its primitive predicates such that for each predicate 'F_ix' and each space-time point q, 'F_ix' is true of q just in case some '$A_{ji}x$' in B('F_ix') is true of q (i.e. 'F_ix' is not coextensive with any single physical predicate, but rather with a 'disjunction'—possibly infinite—of physical predicates) and in every model of physics the theory comes out true under B. (Friedman 1975: 358–9.)

Here, the non-physical truths are semantic consequences of the physical truths. And each non-physical predicate is nomologically correlated with a (possibly infinite) disjunctive physical predicate formed from the set of physical predicates making up the 'physical realization' for that non-physical predicate. For the sake of discussion, I assume that we are working with infinitary languages that allow for such infinitary definitions. And I assume that derivations of specific non-physical truths from physical truths and the definitions are in general possible.[14] Further, I assume for the sake of argument that such a thesis is true (i.e. it correctly describes a relation between all theories and physics). Does weak reducibility, so conceived, provide an adequate expression of physicalist doctrine? Does its well-motivated response to the problem of multiple realization provide a remedy to the defects of CR?

The answer, of course, is clearly not. First, weak reducibility does not avoid the general difficulties associated with the derivational approach to explanation: Friedman has generalized classical derivational reduction without providing additional criteria that would sort out explanatory from non-explanatory derivations. With respect to non-physical truths concerning specific matters of fact,

---

[14] This is a generous assumption due to the failure of completeness for infinitary languages. However, the points I wish to make are best made granting the possibility of such derivations.

derivations need not be explanatory. The same is true with respect to the derivation of laws, assuming they are possible via the infinitary definitions.

Second, WR highlights the ontological and explanatory concerns mentioned earlier regarding the physical realization of non-physical attributes. Couched in the language of predicates, it must be asked why a given predicate ought, or ought not, to be included in the physical realization associated with a given non-physical predicate. The (possibly infinite) enumeration of a set of physical predicates that are token associates of a given non-physical predicate does not begin to respond to this question. Since the realization of non-physical attributes by physical attributes must be explained in a physicalist system if WR is to have a chance of being an adequate formulation of physicalism, it must be supplemented by an account of the conditions of legitimate physical realizations. Without such an account, trivial physicalization is a real possibility, mysterious physical/non-physical connections could abound, and the possibility of ontologically emergent attributes (i.e. attributes not realized by physical attributes) would be a serious one. Thus WR neither calls for vertical explanation of the sort in which the physicalist is interested, nor does it rule out the possibility of ontologically emergent attributes.

Third, WR, being an essentially linguistic thesis, does not provide sufficient resources for a comprehensive doctrine concerning either the physical ontology or objective matters of fact. And, as argued earlier, even the employment of infinitary resources is not adequate to cover the full richness of the physical ontology of individuals, attributes, and classes. Thus I conclude that, although WR may be, as Friedman intends, a plausible candidate for explicating the thesis of physical *truth* determination (sententially construed), it does not provide adequate resources to handle all physicalist concerns about ontology, objectivity, and explanation.

The final modification of CR that I shall consider is one which employs the idea of higher-order physical attributes and associated definitions. The idea was introduced by Putnam (1970) and subsequently put to work by Field (1975; 1986) in his work on physicalist accounts of reference. Field explains it as follows:

. . . given any property ψ of physical states of a given order, we can define a higher-order state (also called a functional state) in either of two ways.

First, there is the state of *being in some state or other, s, such that* ψ*(s)*; second, there is the state of *being in some physical state or other, s, such that* ψ*(s)*. The former sort of functional state is 'topic-neutral', but the latter sort is reasonably called a higher-order physical state. Calling the state s for which ψ(s) the *realization* of the functional state, a functional state is physical if it can be instantiated only by having a physical realization. (Of course, the realization can vary from one instantiation to the next: that is the whole point of functionalism). (Field 1986: 111, fn. 33.)

On this approach, a hierarchy of *physical* attributes exists based upon the idea that attributes at higher levels are realized by attributes at lower levels satisfying certain conditions. As a result, the thesis, AI, can be broadened by construing it as associating all attributes capable of instantiation in nature with some attribute in the hierarchy of physical attributes. As Field suggests, each attribute is physical at least in the sense that it can have *only* physical realizations. Similarly, the thesis DT can be construed as calling for the physical definability of all non-physical terms, where the definitions include higher-order physical definitions associated with higher-order physical attributes. In this way the multiple realization problems that plagued both DT and AI are avoided. Finally, the approach can be viewed as attempting to retain the derivability of non-physical laws and truths from the laws and truths of physics and the definitions. Thus this version of physicalism, HR, involves the theses IC, AI, DT, and DL, where these theses are understood to cover higher-order physical definitions and attributes. HR purports to express the ontological and explanatory concerns of physicalism in a way that constitutes a sophisticated attempt to retain the spirit, and much of the letter, of classical reductionism.[15]

However, despite its sophistication, I believe HR must overcome some obstacles before it clearly meets the demand of adequacy required of a formulation of physicalism. With respect to ontology, although the view appears to respond to the problem of realization, it suffers from a certain lack of clarity regarding the notion of 'any property ψ of physical states of a given order' that is required for defining higher-order physical attributes. Without any stipulation regarding what sorts of properties are permissible

---

[15] In a recent paper (unpublished), Field has elaborated his views about physicalism in a way that clarifies his earlier writings and that more or less confirms the formulation in the text.

here, the road will be open to any attribute being associated with a higher-order physical attribute, without necessarily being *realized* by physical attributes in the sense required by the physicalist. For example, the property of *being a physical state that is instantiated when a ghost is present* would, if it were a legitimate property for generating higher-order physical states, allow us to define a physical state that would be associated with the state of a ghost's being present. This association does not at all guarantee that this latter state is realized by (i.e. constituted by) the physical states that are subsumed by the associated higher-order physical state. And this points the way to what is not a small difficulty for the approach: to put limits upon the sorts of properties that can play a role in generating the required higher-order physical states. Clearly, refinements are called for if the approach is to deal effectively with the non-emergence clause of physicalist doctrine.[16]

Further, there are potential shortcomings of HR regarding how it handles explanatory concerns. First, it is vulnerable to the general difficulty posed for all derivational theses that presuppose that derivation entails explanation of the laws of the special sciences. If no such presupposition lies behind HR, then clarification is needed concerning how it suffices for physicalist explanatory purposes. How, if at all, are higher-level laws to be explained within the system? Second, HR is probably vulnerable to Fodor's objection that the derivation of the consequences of higher-level laws may not explain those consequences, since the explanatory structures employed in the higher-level laws may not be preserved in the lower-level physical laws. I say 'probably vulnerable' since it is (remotely) possible that the higher-order physical definitions of a non-physical term might be explanatorily equivalent to that term. If so, then perhaps certain physical theoretical structures preserve higher-level explanation. This is unclear, and depends upon the sorts of properties that can be appealed to in higher-order physical property formation.[17]

---

[16] As I understand him, Field would now require that the relevant properties be sufficient for explaining the realization of the higher-order attribute. This, I believe, means that he is in substantial agreement with my views about the necessity for realization theories as will be discussed in Ch. 4.

[17] In his recent paper, Field requires that the explanatory structures found in the special sciences be preserved under translation into physics. Such a requirement is problematic, and is not clearly necessary for an adequate formulation of physicalism. See Chapter 4 for further discussion.

Finally, if we consider each of the physical realizations of some higher-level functional state, the question arises as to whether HR calls for proper explanations of why each realization is in fact a realization of the given functional state. Whether it does so depends upon whether it can adequately respond to the issues raised earlier regarding the nature of the defining properties. If we just stipulate that the physical states are realizations of the functional state, there is no explanation of such realizations. If we simply quantify over arbitrary properties of physical states, we have no guarantees that such properties, even if non-trivial, will be explanatory. On the other hand, if we make precise stipulations regarding which sorts of properties of physical states are relevant (for example, their causal roles (Putnam 1970; Field 1975), their causal powers (Boyd unpublished)), we are left with at least the question of whether such stipulations will not be too restrictive. Are appeals to causal roles and causal powers, for example, sufficient for explaining the realization of all non-physical attributes? This does not seem likely.

If the proponent of HR is effectively to defend the position, a specification of the range of properties of physical states, relevant to the realization and the vertical explanation of higher-level states, must be made in a way that is general enough to cover all attributes, yet restrictive enough to remain within physicalist boundaries, and substantial enough to be explanatory. This is, indeed, a demand that any adequate formulation of physicalism must meet if it is to be responsive to physicalist concerns about realization and explanation. I believe that Field is in substantial agreement with the legitimacy of this demand.

However effectively HR can cope with these difficulties, I want to conclude my discussion by saying that I am sympathetic to the approach. It is moved by what I take to be a full-blooded physicalist concern with the problems of realization and explanation of attributes. And it has introduced resources that take us some distance along the path towards an adequate formulation of physicalist doctrine. None the less, I think that some of its defects are probably irremediable (viz., the problems in how it handles explanation within a physicalist system) and therefore it must be modified.

Despite the alignment of goals between the above reductive approaches and physicalism, none of those approaches evidently succeeds in providing adequate and acceptable formulations of physicalist ideas and values. In the light of such failure, what I shall refer to as 'the problem of retrenchment' arises: to formulate a version of physicalism that is adequate and that does not fall foul of the difficulties that afflicted the reductionist approaches. Thus the problem is to develop a formulation that, *inter alia*, gives full expression to key physicalist ideas and values, that has no commitment to false views about the nature of explanation or the descriptive power of the language of physics, and that can accommodate the multiple realizability of many phenomena. In the next section, I shall consider various attempts to provide such a formulation of physicalism.

2.2. NON-REDUCTIONIST APPROACHES

In recent years, a number of proposals have been made that can reasonably be construed as responses to the problem of retrenchment. Unfortunately, there is considerable variation among the proposals offered, and none of them has clearly succeeded in providing an adequate formulation of physicalist principles. Such proposals do tend to be in agreement with respect to the rejection of theses concerning definition, derivation, nomological coextensiveness of properties, and type identity. As a result, I shall refer to each as a version of 'non-reductive physicalism' (NRP). In what follows, I shall review the most important of these proposals and show that none of them is adequate by the criteria I have set. It will become evident that most versions avoid the problems of reductionism at the expense of important physicalist ideas and values. That is, they trade off adequacy for acceptability. To my mind this is not the proper course for physicalists to follow.

I begin with what will be refered to as 'purely ontological' formulations. Examples include such theses as 'Everything is completely physical', 'Every fact is a physical fact', and 'All phenomena are physical phenomena'. Such formulations are aimed at capturing the ontological content of physicalism and are therefore mainly concerned with what exists. Although the above examples are surely too vague to be useful in developing a clearly articulated statement of physicalist doctrine, some version of this sort of thesis is required.

There are a variety of different, more specific statements of this kind. As versions of NRP, they all avoid type-identity claims and claims concerning nomological coextensiveness of physical and non-physical attributes. It is clear that such versions of physicalism are motivated by the avoidance of multiple realization difficulties. On the other hand, a primary potential *source* of difficulty for such formulations centres around the choice of ontological categories upon which to focus (for example, objects, events, attributes). Although such selections are necessary, each has the potential for masking controversial or problematic philosophical assumptions. In addition, such selections must, of course, be made judiciously so that the theses alluding to them satisfy the criteria of adequacy and acceptability.

Consider the following purely ontological statement of non-reductive physicalism:

(PO) Every object is either a physical object or an object composed of physical objects.

This is a thesis concerning particulars, not types or kinds of object, and its point is to establish the exhaustiveness of the physical ontology. Its full significance, however, depends on what is meant by an 'object' and what it means for an object to be 'composed' of other objects. I shall bypass the details of the various versions of this thesis, since for the most part the inadequacy of PO as a formulation of physicalism does not depend upon them. In general, I shall assume that objects are to be understood as particulars with reasonably well-defined space-time boundaries.[18] Further, PO allows for the existence of levels corresponding to the increasing complexity of the structure of objects in nature. But the modes of composition of objects need to be specified, if such a hierarchy is to be clearly articulated and if PO is to have significant content.

There are a number of problems with this sort of view, some perhaps remediable, though others show it to be inadequate and probably unacceptable as a formulation of physicalism. First, if the modes of composition are limited to those identified by physics (for example, composition of objects due to fundamental physical forces), then PO will almost certainly fail to cover all individuals in nature. Such modes are likely too restrictive, since, as I have remarked above, it seems evident that there are many objects in

---

[18] This is, of course, a problematic assumption, as I shall discuss below.

nature that are not complex physical objects built up from basic physical objects in accordance with such physical modes of composition (for instance, social institutions). An alternative construal of composition in terms of mereological fusion appears to fare no better, and for essentially the same reason: there are objects whose character is not captured by the fusion of physical entities. The idea that all objects are *token identical to* fusions of physical objects may have deep problems with Leibniz's Law, since it is, at the least, controversial that social institutions (for example, banks, universities) share all attributes with the fusions of the physical objects that occupy the same space-time regions as they do.[19]

If one responds by proposing that there are other modes of composition of objects beside fusion and the modes identified by physics, there emerges a quite general concern regarding PO: can the modes of composition be circumscribed in a way that is physicalistically justified (i.e. in a way that reveals how higher-level objects are constituted by lower-level physical objects) and that applies to the entire world of individuals? Failure to respond to this problem would undermine the physicalist's interest in precluding emergent objects from the natural order; therefore this concern must be dealt with by any adequate version of physicalism.

A second problem concerns the boundaries of objects. There is, of course, a concern about how to cope with 'quantum indeterminacy' and the 'smeared out' character of objects. And if macroscopic objects (for example, tables and chairs) are to be identified with some complex physical object, there may be a certain amount of fuzziness, if not indeterminacy, regarding with which physical object a given chair, for instance, is to be identified (cf. Quine

---

[19] I remind the reader that some proponents of versions of ontological physicalism do not intend that their formulations should capture the details of how higher-level objects are realized by physical objects. As Post has explained in correspondence, his concern with respect to theses like PO is to express purely ontological claims that do not encroach upon matters that are 'best construed as belonging to the epistemic, the explanatory, the methodological, and/or the metaphilosophical com ponents of non-reductive physicalism'. I certainly intend no unfair criticism. However, I contend that matters concerning composition, realization, and non-emergence are of critical *ontological* importance, and that the purely ontological component of physicalism must be responsive to them. The point of my discussion in the text is to underscore the idea that any adequate formulation of physicalism must make provisions for the realization and non-emergence of *all* objects in nature. Purely ontological theses such as PO are hard pressed with regard to this demand, although I am not saying the problem is intractable.

1981*b*). A number of stances might be taken with respect to this problem. One might argue for the elimination of chairs from the ontology of nature; but this is a view not to be taken seriously. One might agree to stipulations of which, from among the candidates, is the one to be identified with the chair—after all, there are many objects that will do; but such arbitrariness is philosophically suspect and may, in addition, be inconsistent with the physicalist programme.[20] Or one might abandon PO in favour of a less problematic thesis (for example, by giving up identity as the relation between macroscopic and physical objects or by rejecting objects as the pertinent ontological category for physicalist purposes). Again, any version of physicalism that includes PO or a relative must respond to these difficulties.

Although the first two objections may be tractable for the supporter of PO, the next few are not. Even if one assumes that PO suffices to express the idea that all objects in nature are ontologically dependent upon and realized by physical objects, it does not make strong enough demands upon the relations between physical and non-physical attributes. For example, since being a physical object does not preclude possession of non-physical properties and relations, PO is quite compatible with property pluralism. In my view this is not in itself a problem, assuming certain constraints are satisfied. But what is a problem is that PO is quite consistent with a failure of the supervenience of non-physical attributes upon physical attributes. Thus 'fixing' all of the physical properties and relations of objects in the world need not guarantee that the non-physical properties and relations of objects have been fixed.

In addition to falling short of basic metaphysical demands, this failure of supervenience also means that PO fails to express physicalist concerns regarding objectivity. PO is compatible with a failure of the key physicalist idea that the physical facts and truths determine all the facts and truths about nature. Thus this form of physicalism cannot play any serious role in characterizing the realm of objective matters of fact. A further difficulty, signalled by the failure of supervenience, is that PO is quite compatible with the existence of ontologically emergent attributes: attributes that do not supervene upon physical attributes are not realized by physical attributes either. Even if PO rules out ghosts, it does not

[20] See below, Ch. 6, for discussion of physicalist indeterminacy.

rule out the emergence of mysterious forces and properties. As a consequence, PO does not provide an adequate statement of even the ontological component of physicalism.

Finally, it will be evident that PO, being a purely ontological thesis of such limited content, does not give any expression to physicalist concerns regarding explanation. Even if PO is true and even if a system of knowledge were constructed in accordance with it, no requirement is imposed upon the system to provide explanations of the realization of non-physical properties and relations. Therefore such a system need not exhibit the sorts of ontological and theoretical unity that the physicalist seeks, as described above in Chapter 1. This is true of any statement of physicalism not specifically taking such matters into account. Thus in what follows it will be important to see exactly what additional sorts of theses are required to accommodate physicalist ideas and values: purely ontological theses like PO fall way short of the mark.

Next, there are those who would deny such ontological theses any place in a proper expression of physicalist thought. A second sort of candidate for providing a formulation of physicalism is what I shall call a 'purely supervenience formulation'. Given the variety of different sorts of supervenience relation that can be employed, there are many forms such a formulation of physicalism can take. I shall consider two that have been advocated in the physicalist literature.

To begin, it has been suggested that physicalism might be properly expressed in terms of a relation of *weak supervenience* of a family of properties A upon a family of properties B:

(WS1) Necessarily, for any x and y, if x and y share all properties in B then x and y share all properties in A.[21]

Thus physicalist ideas and values might be expressed by the following principle:

(WSP1) The family of non-physical properties weakly$_1$ supervenes upon the family of physical properties.

---

[21] This is Kim's formulation in Kim (1987), although he does not endorse WSP1 as a formulation of physicalism. WS1 is stronger than the relation Seager (1988) believes is sufficient for expressing physicalism about the mental; the points made in the text apply more forcefully against that claim.

This is a claim to the effect that in each possible world[22] individuals that share their physical properties will share their non-physical properties. However, as an adequate expression of physicalism, it fails.

First, in the event that two individuals differ in some non-physical respect, the relation of weak supervenience places no restrictions upon the types of physical properties or relations that must also distinguish them. Thus there is no requirement that the physical respects in which such individuals differ should be at all *relevant* to the respects in which they differ non-physically: *any* physical difference will do. WSP1 therefore characterizes a very loose connection between the physical and the non-physical domains, a connection that is too loose to capture what the physicalist wants to say about the world with respect to realization and non-emergence.

Second, and contrary to what is often thought, WSP1 does not imply the following principle:[23]

(WSP2) The family of non-physical properties weakly$_2$ supervenes upon the family of physical properties,

where the relation of weak supervenience in question is as follows:

(WS2) Necessarily for any object x and any property F in A, if x has F, then there exists a property G in B such that x has G, and if any y has G, it has F.

This is so because, unlike WSP2, WSP1 does not entail that there are any intra-world connective generalizations of the following form:

(CG1) For any x, if x has P, then x has N

where P is some possibly quite complex physical property or relation of individuals and N is some non-physical property or relation. Since the physicalist wants to express the idea that an individual possesses a non-physical property *in virtue of* its physical properties and relations, a *minimal* requirement is that each instance of a non-physical property be associated with instances of physical properties and relations that are materially sufficient

---

[22] For the present purposes, let it be understood that WSP1 is to be rendered as a claim about all physically possible worlds.

[23] Post (1991) was, I believe, the first to observe this.

conditions for it. Without such a connection, there is little force to the idea that the non-physical property occurs in virtue of the physical properties and relations of the individual. Further, there is little hope of explaining the instantiation of the non-physical property in terms of physical properties and relations: such explanation turns upon the idea that certain relevant physical properties and relations of the individual make it the case that the individual possesses the non-physical property in question. WSP1 fails to capture these ideas.

There is room for considerable misunderstanding on this point which I would like to head off. I understand 'the physical properties and relations of an individual' in a very broad sense, as follows. They include, at a minimum, the obvious, theoretically defined properties and relations of physics plus various Boolean compounds of those properties and relations, subject to certain restrictions to be discussed in Chapter 3. In addition, an individual possesses various relational physical properties defined in terms of some of its physical properties and relations *and* the properties and relations of other objects elsewhere in the universe. One clear example of such a relational physical property is that of being the tallest human. But this sort of example is not the only sort. The relational physical properties of an individual form an extremely large class, most members of which are theoretically uninteresting (for instance, the property of having a length of six metres while Mars is in retrograde motion). None the less, the class of physical properties and relations of an individual includes all relational physical properties that are based upon both an individual's properties and relations as well as the properties and relations of other objects. Clearly such relational properties can be quite messy. But they can also be quite important. Why?

I maintain that connective generalizations of the form CG1 (and CG2 below) are minimal requirements for capturing the idea that a given non-physical attribute of an individual is instantiated *in virtue of* the physical properties and relations of the individual. In particular, such a metaphysical connection requires that if a certain non-physical attribute is instantiated in virtue of a certain class of relevant physical conditions, then if the members of that class are instantiated again, the non-physical attribute will be instantiated again. If this is not so, then some relevant attribute was left out of the class as initially conceived. Without some such replicability, the notions of *relevance* and *in virtue of* lack force, or so I maintain.

Those who object to this demand often cite examples that show how a certain *restricted* class of physical properties and relations of an individual do not provide, either materially or nomologically, sufficient conditions for some non-physical attribute of the individual. Examples of such restricted classes are: the intrinsic physical properties of an individual and the causal properties and relations of an individual. The most forceful presentation of such an argument is given by Post,[24] who argues both that the truth of a certain belief (namely the belief that the rings of Saturn contain a certain molecule, called 'Ammon'[25]) depends, *inter alia*, upon whether or not a certain state of affairs obtains elsewhere in the universe outside of the individual having the belief (i.e. it depends on whether Ammon exists in the rings of Saturn) *and* that there are *no causal physical relations* between the individual having the belief and the relevant state of affairs. Since it is assumed, by those endorsing the view Post is attacking, that any connective generalization must involve at least some causal physical properties and relations which connect the believer with this state of affairs, the conclusion is that there can be no connective generalizations relating the physical conditions that make it the case that a true belief is realized in an individual and the belief that is realized in that individual.

The reason, of course, is that the restriction (viz., that there are some causal physical relations between the believer and the state of affairs that makes the belief true) screens off the relevant physical properties and relations. Once all the relevant properties and relations are included (for example, the property of being in a certain physical state that realizes the belief that Ammon is in the rings of Saturn,[26] while Ammon is in fact in the rings of Saturn),

---

[24] Post (1991), pp. 114 ff. In his discussion Post is mounting an objection against those physicalists who require that connective generalizations involve at least one causal relation. Such a requirement is unnecessary, in my opinion.

[25] See Kim (1987) for the original Ammon example that Post has revised.

[26] There may be questions concerning this way of putting things, since in at least some views of the determination of the content of mental states one cannot have a belief *about* Ammon without causally interacting with Ammon. But the point of the example and my reply concerns the possibility of connective generalizations when none of the relevant conditions to the realization of a certain attribute are causally/physically related to the bearer of the attribute.

connective generalizations are likewise established.[27] The effectiveness of my response here depends upon the broad sense of
'physical properties and relations of an individual' outlined above.
However, I see no reason why the physicalist must be saddled
with a more restrictive notion; as a consequence, I think the requirement that there be connective generalizations relating non-
physical attributes with the physical attributes that realize them
stands. But keep in mind that the requirement is a minimal one
carrying with it *neither* the suggestion that such connective generalizations, by themselves, establish that the non-physical attribute
is realized by the physical attributes that are involved *nor* the
suggestion that the connective generalizations provide explanations of the realization of the non-physical attribute *nor* the
suggestion that the connective generalizations are statable in some
theoretical language. The requirement is simply a minimal requirement on what is involved in saying that it is in virtue of one state
of affairs that another is realized.

Returning now to the review of physicalist formulations, suppose we amend the purely supervenience formulation so that both
WSP1 and WSP2 are offered as an expression of physicalism. Such
a formulation founders none the less, as others have pointed out
(for example, Kim 1984; 1987). Even if there are intra-world
connective generalizations of the sort entailed by WSP2, and even
if the modal force of WSP2 means that the claim holds for all
possible worlds of interest to the physicalist, WSP2 does not guarantee any cross-world stability of the connective generalizations.
Thus there need be no generalizations of the following form:

(CG2) Necessarily, for any x, if x has P then x has N.[28]

At best, weak supervenience relations require only contingent
connections between physical and non-physical attributes. They

---

[27] Post adds that 'even if there is some relation between Ammon and me which
is expressible in the language of physics—a relation that is thus in the broad sense
a physical relation—it is not likely to be a relation that does any work in determining that my belief is true, if Millikan's biosemantics is right' (Post 1991: 116).
My response to this is that, whether or not Millikan's theory is correct, there is a
true connective generalization that obtains when all the relevant facts are included
in the class of physical facts that make it the case that a true belief is instantiated.
[28] Again, the physical property P may be quite complex, may involve relations
to the environment, and may involve relational properties of the broad variety
recently discussed.

underwrite no counter-factual truths regarding what non-physical properties an individual would have if it were to have certain physical properties. In the light of this, WSP2 fails to satisfy the minimal physicalist demands upon realization and explanation for pretty much the same reasons as WSP1 failed. There is no required isolation of all *relevant* physical attributes that are nomologically sufficient for the realization of given non-physical attributes. As a consequence, there is room for emergent properties (i.e. properties that are not realized in virtue of physical properties) and there is no guaranteed basis for developing vertical explanations concerning realization. Thus a weak supervenience approach to the formulation of physicalism fails to be adequate.

A second approach to formulating physicalism purely in terms of supervenience involves an appeal to a *global supervenience* relation. The following is a typical formulation:

(GS1) The world could not have been different in any respect without having been different in some strictly physical respect. (Haugeland 1984*a*: 1.)

Haugeland suggests that a suitably refined version of this claim suffices to express what the physicalist wants to say. However, there are several reasons why such a formulation as this does not give adequate expression to physicalist doctrine.

To begin with, GS1 does not rule out the possibility of non-physical individuals and therefore it is entirely compatible with, for example, parallelist substance dualism.[29] Although not unaware of this objection, Haugeland (1984*a*: 9–10) insists that GS1 suffices to express the ontological ideas of the physicalist because all alternative ontologies are 'weird' and, in any event, *no* formulation of physicalism can preclude a wide range of 'kooky' metaphysical doctrines. His first reason is a clear-cut confusion of acceptability and adequacy of the formulation. It is not true that because all of the competitor's can be ruled out, GS1 suffices to express the physicalist's views regarding ontology. Indeed, GS1 falls short of capturing even the most basic ontological ideas of physicalism. His second reason, although more subtle, does not militate against requiring ontological theses. However, it does raise

[29] Of course, one could play around with the operative conception of *the world* in GS1, but this would be tantamount to introducing an ontological thesis in addition to the supervenience claim.

deep issues regarding the relation between physicalism and realism, issues that I shall take up in Chapter 6.

Further, GS1 is compatible with the existence of worlds in which there are instances of non-physical attributes but *no* instances of physical objects and attributes: this is because GS1 merely requires that if there is some non-physical difference between two worlds, then there must be some physical difference between them. Since any difference will do, GS1 is compatible with the existence of at least one totally non-physical world which *ipso facto* differs from any physical world in some physical respect. Thus even the ontological dependence of the non-physical upon the physical is not guaranteed by GS1.

Since GS1 is also compatible with the failure of both WSP1 and WSP2, it guarantees no stable associations between physical and non-physical attributes[30] of either the form CG1 or the form CG2. As a consequence, GS1, like weak supervenience formulations of physicalism, places only very limited constraints upon the relations between physical and non-physical aspects of things. And it is quite clear that GS1 makes no provisions for the isolation of physical attributes that are *relevant* to the realization of non-physical attributes, let alone any demand that physical attributes constitute the instantiations of non-physical attributes. Thus although GS1 succeeds in ruling out the existence of pairs of worlds that are physically indiscernible but non-physically discernible, it is not fine-grained enough to capture the physicalist's concerns about realization and non-emergence: GS1 makes no provisions for how, if at all, non-physical phenomena are embedded within the physical structure of the world.

Further, a purely supervenience formulation such as GS1 falls far short of giving adequate expression to physicalist explanatory concerns. GS1 is a claim that is formulated solely in terms of total states of the world without specific reference to any parts, regions, or other components. Such a restriction means that this formulation does not call for any more specific understanding of the physical realization of the non-physical. This should be strongly resisted by any physicalist. One very important aspect

---

[30] Except on a global scale: total physical states of the world are stably associated with total non-physical states. Thus even the most adamant non-reductive physicalist allows that there are nomological associations between some physical and non-physical states of the world.

of physicalist system-building is the isolation of relevant under-lying physical (or physically-based) states, processes, mechanisms, relations, etc. that provide a basis for explaining how specific non-physical properties, states, etc. are realized. All such expla-nation is viewed as inessential to an adequate formulation of physicalism by a purely supervenience formulation based upon a global supervenience relation.

In addition, as a number of philosophers have put it, GS1 fails to satisfy certain necessary conditions for the *genuine dependency* of the non-physical upon the physical (cf. Currie 1984; DePaul 1987; Grimes 1988; and Kim 1987). For example, since global supervenience is not an asymmetrical relation, GS1 is compatible with the global supervenience of the physical upon the non-physical. Such a possibility means that GS1 simply fails to express the full thrust of the physicalist's idea that everything occurs *in virtue of* what goes on in the physical domain: genuine dependency is a decidedly asymmetrical relation. Now, I allow that this notion of genuine dependency is still vague to some extent. But I think that it is evident that it goes beyond mere covariation of attributes in different domains,[31] and that it involves a clear direction of influ-ence of attributes in one domain upon attributes in the other. Whenever one domain depends upon another (in the relevant sense), the direction of influence is from the latter to the former, and it is this notion of direction of influence (for instance, causation, realization) that GS1 fails to capture.

Note that the problem is not quickly resolved by requiring that global supervenience be asymmetric in the case of physicalism, since the asymmetry of global supervenience is *not* a necessary condition for genuine dependency relations or realization relations (i.e. a global supervenience relation can hold in both directions even while there is a definite direction of influence from one do-main to the other).[32] The bottom line is that global supervenience

[31] E.g. as called for by GS1 or by the more restrictive relations of nomological sufficiency or equivalence.

[32] See Post (unpublished) for discussion of this position. He cleverly shows that dependency relations of a certain sort can go in both directions between certain types of phenomena. But it should be noted that his examples are of dependency relations between A and B at different times (i.e. at time 1, A depends upon B, while at time 2, B depends upon A). The point of my claim in the text is that genuine dependency of the relevant sort must be asymmetric at a given time (i.e. the dependency of A upon B at time t means that B is not dependent upon A

is, in my opinion, just the wrong sort of relation for expressing many key physicalist ideas, even if it is useful in expressing some of what the physicalist wants to say.

Finally, I should note that Kim (1987) has argued that GS1 is defective since it leaves open and mysterious the question of why the non-physical globally supervenes upon the physical. He suggests that the explanatory gap is filled by supplementing GS1 with a version of what he calls 'strong supervenience', a thesis that implies generalizations of the form CG2. Post (1991: 109 ff.) replies that generalizations of that form are equally in need of explanation, and he concludes that GS1 is not defective for the reason that it is not explained in terms of more specific nomological connections between physical and non-physical attributes. In my opinion, Kim has deflected attention from the central physicalist concerns here. It is not that GS1 requires explanation so much as that the physicalist has significant metaphysical and explanatory concerns regarding the realization of attributes. Although GS1 expresses a certain amount of the modal force possessed by relationships between the physical and the non-physical, it completely abstracts away from the details of the relationships between specific physical and non-physical attributes. But it is exactly in those details that the physicalist has primary metaphysical interest: each attribute that is instantiated in nature must be constituted by some set of physical (or physically-based) attributes. Indeed, the physicalist has an interest in explanation and understanding of such realization relations. Post has, I believe, drawn the wrong conclusion in his reply. A better one is that both global supervenience and generalizations of the form CG2 require supplementation if physicalism is to be properly expressed. Below, I shall argue that Kim has failed to fill either the gap he believes he is filling or the gap he ought to fill with his appeal to *strong supervenience*.

I conclude that purely supervenience formulations, like purely

---

at *t*). GS1, however, is compatible with contemporaneous global supervenience of both A upon B and B upon A. Hence GS1 does not suffice to capture genuine dependency. Thus I differ with Post on his conclusion that global supervenience relations suffice for capturing the physicalist's conception of determination of the non-physical by the physical. Such a conception requires more than is expressed by determination theses such as GS1.

ontological formulations, are not adequate by the criteria set forth above. Neither is sufficient to guarantee the sorts of ontological and theoretical unification that is central to the physicalist programme. As a minimum, perhaps, they must be combined to yield an adequate formulation of physicalism. In exploring proposals that combine ontological and supervenience theses, I shall begin with what can fairly be called 'the consensus view' (CV). It is essentially a combination of the ontological and supervenience theses just discussed.

In his book on physicalism, John Post offers the following set of physicalist theses (Post 1987: 185):

(INV)   Everything whatever is either an ST sum of entities that satisfy a positive predicate on the list from physics, or else is a set that belongs to some rank R(a) in a set-theoretic hierarchy [grounded upon physical entities taken as ur-elements].[33]

(MND)   For any P-world W,[34] if x and y are N-discernible in W, then they are P-discernible in W.[35]

(TT*)   Any two P-worlds indiscernible as regards the physical properties and relations of the things in them are also indiscernible as regards their things' non-physical properties and relations.[36]

INV is a much more sharply formulated ontological thesis than PO above. MND is a version of the weak supervenience principle,

---

[33] This principle is a slightly modified statement of Hellman and Thompson's (1975) 'principle of physical exhaustion', as Post clearly acknowledges.

[34] For Post (1987: 173) a P-world (i.e. a physically possible world) is a triple <D,P,R>, where (*a*) D is a set both of space-time points and of things that can be said to occupy them, (*b*) P and R tell us (among other things) which of such points are occupied and by what (or to what degree of field intensity or of probability), and (*c*) this distribution of occupied/unoccupied over the space-time points is in accord with the laws of physics.

[35] In this formulation, x and y are P-(N-)discernible in W if there is some P-(N-)property/relation that x has and y lacks in W, where a P-property is a physical property and an N-property is any other property. See Post (1987: 176).

[36] In fact, Post (1987: 185) formulates his minimal set of physicalist principles as including the following principle instead of TT*:

(TT)   Given any two P-worlds, if the same P-sentences are true in both, then the same N-sentences are true in both.

He suggests that the two principles are equivalent and that TT involves semantic ascent that may be useful. However, TT does not imply TT*, even if TT* does imply TT. Thus, I think it is best to formulate Post's physicalism in terms of TT*.

WSP1, concerning individuals within a P-world.[37] TT* is a global supervenience principle explicitly restricted to P-worlds. Post takes these three principles to constitute the minimal set of physicalist principles and clearly indicates their affinity to other formulations in the literature which are only minor variants. As he puts it:

> ... family quarrels among physicalists over the final ultrarefined formulation of their principles should not be allowed to blind others, let alone themselves, to the profound implications the minimal principles have in whatever variation. (Post 1987: 189.)

It may appear that Post believes that the above principles should be taken as constituting an adequate formulation of physicalism. However, his view is correctly understood as a minimal formulation of the *metaphysical* component of physicalist doctrine; it was not Post's intent to be expressing the explanatory concerns of physicalism. Thus although I shall argue that the view just formulated is not adequate, only those objections bearing upon metaphysical concerns are criticisms of Post's actual position. In any event, it is clear that a consensus has formed around this sort of view, as is evident in the work of many.[38]

Equally clear, however, is that the sharpening of the ontological thesis and the clarification, in terms of P-worlds, of the modal force of the supervenience theses do not remedy all the defects that undermined the ontological and supervenience theses taken separately. Specifically, even if it is assumed that INV fully captures the ideas of ontological dependence and realization of objects and that INV, MND, and TT* together fully capture the physicalist's concerns about objectivity, combining the three theses does little

---

[37] Post (1991: 120, n. 22) states that the Ammon case constitutes a counterexample to MND for essentially the same reason that it constitutes a counterexample to connective generalizations of the forms CG1 and CG2: viz., that if the class of relevant physical properties and relations does not include relational properties of the sort discussed earlier, then two individuals could indeed differ in some non-physical respect while agreeing in all physical respects. Since, as I have argued, there is no reason to restrict the class of physical attributes in this way, I see no need for the retraction. But the arguments for or against the consensus view do not depend upon the outcome of this dispute.

[38] In addition to Post, Lewis (1983), Horgan (1984) and Teller (1985) endorse views of this sort. As I believe all will agree, Hellman and Thompson (1975; 1977) richly deserve credit for first introducing this approach to the formulation of physicalist principles.

to respond to the shortcomings of the previous proposals with respect to the realization and explanation of attributes.

First, conjoining MND (WSP1) with TT* (GS1) does not imply even the weak supervenience principle WSP2. Hence such a combination does not imply any contingent, intra-world generalizations of the form CG1. Second, even if one adds WSP2 to the consensus view, it will not imply any stronger, inter-world generalizations of the form CG2. Thus CV requires no necessary connections between specific physical and non-physical attributes short of those concerning total states of the world. As I have argued above, such failures mean that CV does not express the idea that for each non-physical attribute that is instantiated there are physical attributes, often complex and relational, that are *relevant to* and *sufficient for* the realization of such non-physical attributes. Such physical attributes are, in most cases,[39] distinct from the remainder of the physical attributes that are instantiated at the same time, attributes that can vary widely without affecting whether the non-physical attribute in question is realized. As a result, CV fails to require that there are relevant physical attributes that constitute the non-physical attribute. Thus it fails to rule out emergent phenomena.

Haugeland (1982; 1984b) has argued that it is not in general possible to isolate those physical phenomena that can be identified with, or deemed relevant to the realization of, non-physical phenomena. Hence physicalists are advised to drop from their views either sort of claim. Although I have no stake in claims of identity, I do think physicalists cannot give up on claims regarding relevance without sacrificing the most central ideas and values of their position. And I find none of Haugeland's arguments very compelling. In particular, he appears to assume that the distinction between what is relevant to realization and what is not must be circumscribed by some general principle (Haugeland 1984b: 65–6). It is rather more likely that such distinctions are to be drawn on a case-by-case basis.

---

[39] That there are attributes that are *holistic* in the sense that they are realized by the total state of the universe needs to be acknowledged; but there is no good reason to suppose that all attributes are holistically realized. Although many attributes are realized by relational physical attributes, this is not a reason to endorse only global supervenience theses, as is so often suggested. See below for discussion of this slide from local to global supervenience principles.

He further suggests (Haugeland 1984*b*: 66) that theses requiring the isolation of relevant base attributes cannot cope with the problem of context dependence: i.e. the problem posed by attributes of an individual that depend upon what is going on outside the individual. However, it is surely consistent with physicalism that some attributes depend upon what is going on at some spatio-temporal distance from the individual possessing them. And it is exceedingly plausible that for many, perhaps most, attributes only a reasonably well-circumscribed class of physical attributes, some internal to the individual and some, perhaps, external, are relevant to their realization, while all other physical attributes either are part of the supportive background conditions or are, as Haugeland calls them, 'innocent bystanders'. It is the job of what I shall call 'realization theories' to make such distinctions, and thus to show how a given attribute is realized by physical (or physically-based)[40] attributes.

A further difficulty for CV is that, because it does not call for the isolation of relevant physical attributes that suffice for and constitute the instantiation of particular non-physical attributes, it does not require, nor does it provide a solid metaphysical framework for, the vertical explanation of such instantiation. The failure to call for a distinction between relevant and irrelevant physical attributes is fatal for any effort to explain the realization of non-physical phenomena; relevance is the essence of such explanation. Therefore the failure to require that physical attributes *constitute* non-physical attributes undermines CV's capacity to provide for the required sort of explanations, in addition to undermining its capacity to rule out emergent phenomena.[41]

I should also observe that Post (1987), like Hellman and Thompson (1975; 1977), chooses to treat attributes in a set-theoretic fashion and conceives of emergence as concerned with predicates that express attributes.[42] Space precludes pausing to

---

[40] Although I have been framing the discussion in terms of physical attributes, I fully intend to allow that the realization of certain non-physical attributes can best be understood in terms of constitution by physically-based, non-physical attributes.

[41] I remind the reader that Post (1987: ch. 5) has much to say regarding the explanatory connections between the physical and other domains. My criticism of CV is not meant to suggest that he has ignored this important dimension of physicalism.

[42] That is, the issue of emergence concerns what Post calls 'the domain status' of attributes (i.e. what predicates express them), as opposed to their inventory status as set-theoretic entities that fall within the inventory of mathematical-physical entities.

justify fully my dissatisfaction with this approach, but I contend that the identification of attributes with functions defined high up in a set-theoretic hierarchy does little to clarify and constrain realization relations between physical and non-physical attributes, and hence does little to preclude ontologically emergent attributes. Thus the fact that attributes can be represented as functions defined over set-theoretic structures representing possible worlds is not an appropriate underpinning for the sort of physicalistic, ontological unification required by the approach. But the issues here involve us in fundamental philosophical differences that must be left unaddressed for now.

Along a different line, I note that Lewis (1983) and Horgan (1982; 1987) have each made appeals, in somewhat different ways, to a distinction between alien and non-alien attributes when formulating their versions of physicalism. Roughly, an attribute is alien to a world if it is neither instantiated by individuals in that world nor *analysable* in terms of attributes that are instantiated. Thus attributes alien to our world are not capable of instantiation in the very world with which physicalist doctrine is concerned. It is both Lewis's and Horgan's strategy to restrict the possible worlds that circumscribe the modality of physicalist theses to those worlds in which only non-alien attributes are instantiated. However, this move does not suffice to guarantee that there are no emergent attributes in the sense with which physicalists are concerned. Specifically, it does not preclude worlds in which there are emergent attributes (relative to the physical basis) and other attributes composed out of them, if such attributes are actually instantiated in this world.[43] Thus Lewis's and Horgan's respective versions of the consensus view do not spare it from the criticism that it is consistent with a violation of the physicalist's non-emergence clause. To achieve this it must also be required that all non-alien attributes are composed of, or perhaps analysable in terms of, *physical (or physically-based) attributes.*

To sum up, with respect to attributes, CV only requires that each non-physical attribute is instantiated or not given the total physical state of the world and that if two worlds differ in some non-physical respect then they must differ in some physical respect.

---

[43] Contrary to Hellman and Thompson (1977), physics need not preclude such emergent attributes if they were correlated with, but not realized by, physical attributes.

Since appeals to total world states do not pinpoint relevant attributes that constitute a given non-physical attribute and since the required physical differences between worlds are largely unconstrained by CV, it does not appear that CV effectively expresses the physicalist's concerns with non-emergence and, as a consequence, explanatory unification. Thus CV is inadequate by the criteria I have set.

It is not a sufficient reply here to claim that in the case of each instantiated non-physical attribute there will always be a distinction between relevant and irrelevant physical attributes and that it is always possible to provide explanations where they are possible even if they are not explicitly called for. The failure to include any thesis bearing upon such explanation, in addition to the failure effectively to preclude emergent attributes, means that CV is compatible with both the existence of emergent attributes and the existence of a system of knowledge that does not provide any vertical explanations. This, I have been contending, is not true to the spirit of physicalism. An adequate formulation of physicalism should explicitly require the non-existence of emergent attributes and the existence of physically-grounded vertical explanations of non-physical phenomena (i.e. vertical explanation in terms of physical (or physically-based) attributes.

I conclude that the consensus view, although popular among significant physicalist philosophers and although it captures an important component of what the physicalist wants to say, fails as an adequate formulation of physicalism. Those who would despair at this point because they perceive their options as being either classical reductionism or the consensus view are labouring under a false dichotomy. There is no reason to suppose that either the kind of restrictions on non-emergence or the kind of vertical explanation that the consensus view fails to call for can only be secured by the sort of unification thought to result from a classical reduction. And, as I argued above, CR itself does not provide for such restrictions or explanations. The burden of this chapter is to begin to explore and develop a view of what the serious alternatives look like. In further pursuing this end, I shall examine two more formulations of non-reductive physicalism.

A natural extension of the consensus view is to supplement it with a further supervenience claim appealing to what Kim refers to as the 'strong supervenience' of one family of properties, A, upon another family, B:

(SS1) Necessarily, for each x and each property F in A, if x has F, then there is a property G in B such that x has G, and necessarily if any y has G it has F.[44]

For physicalist purposes, A and B are families of non-physical and physical properties respectively. And the relevant notion of necessity, left open by SS1, should be construed in terms of physically possible worlds. Thus the following physicalist thesis can be added to CV:

(SSP) The family of non-physical properties strongly supervenes upon the family of physical properties.

SSP does indeed supplement the consensus view, since it implies, but is not implied by, both MND and TT*; and it neither implies nor is implied by INV. Further, SSP does *appear* to directly respond to the failure of the consensus view to isolate relevant physical factors regarding the realization of specific non-physical properties: for each instantiation of a non-physical property we are assured of the existence of physical property that is a nomological sufficient condition for that non-physical property (i.e. SSP implies generalizations of the form CG2). However, since different instantiations can be associated with different sufficient conditions, SSP is also quite responsive to the problem of multiple realization. And it avoids many of the other problems that undermined classical reductionism, as it makes no reference to the defining power of the physical language, the universal explanatory power of physics, or the deductive unification of the physicalist system of knowledge.

None the less, the consensus view supplemented by SSP is still not an adequate formulation of physicalism. To see this, I shall begin by clearing away a couple of objections that do not cause the formulation any difficulty. First, it might be alleged that non-qualitative properties reveal the principle SSP to be false.[45] The objection is supposed to be that two individuals could share all their qualitative physical properties and yet differ with regard to the non-qualitative properties of being the specific individuals they happen to be. Horgan (1982) has properly dealt with this by

---

[44] See Kim (1984; 1987) for discussion of strong supervenience and its relations to other supervenience claims. But, see Post (1984) for an argument that 'strong supervenience' is not really a *supervenience* thesis at all.

[45] The distinction between non-qualitative and qualitative properties is roughly the distinction between those properties that do and those that do not make essential reference to particulars.

pointing out that the families of properties involved in all physicalist theses should be construed as qualitative only. Second, strong supervenience is formulated in a way which presupposes that the bearers of physical and non-physical properties are identical, a presupposition that is problematic (for example, micro-physical properties are not in general properties of macroscopic objects). Space does not permit detailed discussion of this objection; I will assume that there are successful strategies for disarming it and that it does not matter to the discussion that follows which strategy is employed.[46]

What I take to be the first more serious objection concerns SSP's apparent commitment to strict localization of the relevant physical properties. Many (for example, Hellman 1985; Horgan 1987; Petrie 1987; Post 1987; Teller 1986) have contended that, since some non-physical properties are relational or context-dependent, physicalists are best advised to abandon theses like SSP and to opt for global supervenience claims like TT* and GS1. Teller offers the following in support of this move:

Local physicalism faces an immediate problem, for it is not clear how it can deal with relational properties. Clearly, two individuals can be exact physical duplicates or replicas and yet can differ as to properties such as being a Cadillac owner or being the largest planet in a planetary system. One can try to get supervenience of general relational properties on the physical by applying the idea of supervenience to all the relational and non-relational physical properties holding of or among all the things entering into the relevant supervening relations. But it would be messy to carry out this idea systematically because often one must deal with a non-specific number of relevant individuals entering into the relations, for example when the individuals are covered by an existential quantifier. So some authors appeal to the simple expedient of covering everything at once . . . Global physicalism automatically takes care of the problem of relational properties. (Teller 1986: 72.)

But the reasoning here does not support the conclusion. SSP is a metaphysical principle asserting a relation that either holds or fails to hold between families of properties in the world. Metaphysically,

[46] One strategy for disarming the objection is to take the individuals to be regions of space-time. Another is to develop an account of how macroscopic individuals are realized by microscopic individuals in various configurations and to tailor the supervenience thesis accordingly. See Kim (1988) for a version of this strategy.

there is no question of systematically carrying out any procedure in which, for example, we would have to identify all the relevant physical individuals and their properties/relations. And there is no messiness that anyone would ever have to deal with, given the acceptance of SSP taken as potentially involving 'wide' supervenience bases for non-physical properties and relations. Thus no reason is offered here for dealing with relational properties by adopting TT*, rather than the stronger SSP. From the point of view of explanation, the messiness involved in wide supervenience bases would have to be managed, but why think there is only one way to handle such messiness or that it is best dealt with by dodging it entirely?

Post also opposes localized supervenience principles as follows:

Such versions would be stronger than TT because they would require that each N-phenomenon x be determined by P-phenomena within some relatively small ST-region around x, whereas by TT the relevant P-phenomena need not be thus local but can be global. But the weaker TT seems preferable, if only to provide for the possibility (shall we say the likelihood) that there are non-physical truths about the world that are true only in virtue of the totality of the P-phenomena ... Also there are problems in specifying how large the 'relatively small region' must be, within which the P-phenomena determine x. Furthermore, if determination holds for every relatively small ST-region, it need not also hold either for the whole world or for all the truths about the whole world. (Post 1987: 188.)

Thus, according to Post, TT's (TT*'s) virtues include the flexibility of allowing for holistic as well as local determination of non-physical truths and attributes. While not precluding relatively local determinations, it also makes room for holistic ones, unlike SSP.

The importance of reviewing this line of objection to SSP is that it requires us to be clear on just what SSP involves. To begin with, it need not be construed as local in the sense that the supervening properties and the supervenience base properties must be instantiated in the same regions or must be intrinsic properties of individuals. Rather, SSP requires only that for each non-physical property (or relation) of an individual, there exists a physical property (or relation) of that individual that is a nomological sufficient condition for it. There is no stipulation that the supervened-upon property be localized in the individual or in some relatively small region surrounding the individual. Thus there is no problem of specifying the size of relevant regions. And if relations to the rest of the world are involved in the physical realization of a particular non-physical property, that is in no way ruled out by SSP.[47] Thus SSP, like TT*, can accommodate holistically realized or context-dependent properties. Post's and Teller's reasonings labour under a false dichotomy: either physicalists must embrace an implausible, strictly localized version of supervenience or they must opt for a global version. The point of the reply is that SSP is neither of these.

If the above comments suffice to disarm Teller's and Post's opposition to SSP based upon the problem of context dependence, then abandoning it in favour of the weaker TT* is not justified by their arguments and one can still hope that it provides the additional content needed for an adequate formulation of physicalism. Thus the consensus view, centred as it is around TT (or TT*), is consistent with, but does not call for, there being metaphysically and explanatorily relevant physical factors regarding the realization of non-physical phenomena. The promise of SSP is that it does call for isolation of such relevant factors and thus that it avoids the central defect of the consensus view. As a result, if SSP suffers from no other flaws, it should be conjoined with the consensus view to provide an adequate formulation of physicalism.

Unfortunately, there are other flaws. In a nutshell, supplementing CV with SSP does not effectively respond to the shortcomings of CV regarding the realization and explanation of attributes. With respect to realization, the problem is that although each instantiated non-physical attribute is associated with a nomological sufficient condition, SSP places no further constraints upon the character of these conditions. As initially conceived by Kim, the supervenience base property associated with a given non-physical property, N, is what he refers to as a 'B-maximal property' (BMP), a property that is the maximal conjunction of *all* the individual's physical properties. But such a BMP is, in fact, a supervenience base for all the individual's non-physical properties that are co-instantiated with N. It is not constituted by *only* physical properties that are

---

[47] As discussed above, the physical properties and relations called for by SSP can include, in addition to intrinsic properties of an individual, causal and non-causal relations, as well as relational properties defined in terms of conditions obtaining elsewhere in the universe.

relevant to the realization of N. Thus this way of conceiving of the supervenience base property is clearly not responsive to the problem of isolating *all and only the relevant* physical properties.[48] Kim (1984: 165) is aware of this shortcoming, but his only response is to affirm the importance of identifying a minimal supervenience base property that consists of properties that 'justify' us in saying that an individual has N in virtue of those properties or that its having N consists in having those properties. Thus, SSP can be qualified by requiring that the physical property associated with any given instantiation of a non-physical property is a 'minimal base property', or a 'least nomological sufficient condition', for that non-physical property, where such a minimal property satisfies the condition that 'any property weaker than it is not a supervenience base' (Kim 1984: 165), and hence is not a nomological sufficient condition for N.

Such repairs are not enough, however. The problem of isolating *only* relevant base properties is easily seen to be left untouched by the suggested requirement. Let B be a base property that consists of all and only the relevant properties bearing upon the realization of N, and let us suppose that it is indeed a least nomological sufficient condition. Let B′ be a base property composed of nomic equivalents of those properties that compose B but which do not themselves compose N. Then B′ will be a least nomological sufficient condition for N without being composed of properties that constitute N.[49] That is, being a least nomological sufficient condition does not guarantee relevance of the sort required by the physicalist. It does not guarantee that the properties composing the supervenience base for N are those base properties that constitute the realization of N. Thus although SSP implies generalizations of the form CG2, and hence underwrites counter-factual claims, such generalizations and counter-factuals fall short of expressing what the physicalist wants to say about the relation

---

[48] There is also a problem concerning whether BMPs, as conceived by Kim, are in fact *physical* properties, since BMPs are constituted by, *inter alia*, properties that are the negations of physical properties. Post (1984) and Hellman (1985) have both objected that, even if negation is a legitimate property-forming operation, it is unlikely that the negation of a physical property is always a physical property.

[49] Thus SSP falls short in part because it does not guarantee uniqueness of the supervenience base for a given instantiation of a non-physical property. Such uniqueness, of course, would not preclude multiple realization, which involves variation of supervenience bases for a given attribute across instantiations.

between physical and non-physical properties (i.e. that the latter are realized by the former).

As a consequence of this difficulty with BMPs, the view, as revised, is not responsive to the problem of emergence. Since SSP does nothing, beyond requiring nomological sufficiency, to constrain the kinds of physical properties that can be supervenience base properties, it is consistent with there being a non-physical property that is *not* instantiated *in virtue of* the physical properties that compose *any* of its minimal supervenience bases on a given occasion. The nomological sufficiency of the physical base property for a given N does not guarantee that N is instantiated *because* the base property constitutes it. Hence emergent attributes that are nomologically connected to certain physical properties are not precluded by SSP, and the key physicalist idea regarding realization of attributes is simply not expressed.

A general point can be gleaned from this difficulty: there is no way of staying within the resources of a strong supervenience approach to formulating physicalist principles while at the same time expressing the key physicalist idea regarding realization and non-emergence. Being a (least) nomological sufficient condition for a property is not a strong enough relation for guaranteeing that the target property is instantiated *in virtue of* base properties or that it consists of base properties. Interestingly, in responding to objections, Kim was using the right language for expressing what are the issues of most importance (viz., 'in virtue of', 'consists of'), but in doing so he was stepping outside his own approach.[50]

That SSP fails to satisfy the explanatory concerns of physicalism follows from a number of considerations. Quite evidently, since SSP is a metaphysical principle it makes no explicit demands upon the structure of systems of knowledge that are developed. The physicalist programme, however, is quite concerned with such structure and is not a purely metaphysical enterprise. Therefore an adequate formulation of physicalism must call for, or at least entail the existence of, explanations that isolate all and only those

---

[50] Note that, as with global supervenience, the strong supervenience relation is not asymmetric. And, as with global supervenience, the asymmetry of strong supervenience, even if it can be made plausible in the case of physicalism, is neither necessary nor sufficient for genuine dependency of the non-physical upon the physical. See Grimes (1988) for a lucid discussion of these issues. And see Kim (1984: 166–7) for heroic efforts to motivate the asymmetry of strong supervenience of the non-physical upon the physical.

physical properties that are relevant to the instantiation of specific non-physical properties and that show how the relevant properties constitute such non-physical properties. Since SSP does not call for such explanations, its only hope is that it entails the existence of such explanations. Unfortunately, it does not.

Although it might be claimed that SSP guarantees a metaphysical basis for developing the sorts of explanation that the physicalist desires, this is not the case for the reasons just discussed with regard to the problem of emergence. An explanation of the realization of a non-physical attribute must isolate all and only the relevant physical attributes and show how they combine to realize that attribute. But as I argued above, SSP does not put sufficiently strong constraints upon the physical properties that constitute a supervenience base for some non-physical property to guarantee that all and only the relevant physical properties are isolated. Thus further constraints must be placed upon supervenience base properties if SSP is to have a chance of satisfying the explanatory demands of physicalism.[51]

SSP also fails to entail the requisite sort of explanation because the nomic sufficiency of a physical attribute for some non-physical property does not account for *how* the physical property constitutes the non-physical property. Without such an account, no clear understanding is provided of why the non-physical attribute is instantiated and of how it is embedded in the physical fabric of the universe. Simply displaying an array of physically-based facts does not suffice to explain the realization of a non-physical attribute, even if such an array provides a nomologically sufficient condition for that attribute. SSP, in effect, abstracts away from the crucial details required for this sort of explanation. Such abstraction also undermines classical reduction and many of its variants, as was pointed out earlier in this chapter.

I should observe that there may well be many reasons why such explanations, even if possible, are not actually developed, reasons that have to do with the accessibility, representability, and comprehensibility of the physical base attributes. But these problems

---

[51] Although I am framing the discussion in terms of physical properties, remember that vertical explanations of the instantiation of a non-physical property need only be formulated in terms of *physically-based properties*. I am not suggesting that vertical explanations always appeal to physical properties in the narrow sense (i.e. properties in the domain of physics).

do not undermine the physicalist programme. As long as the instantiation of non-physical attributes are explainable in the way I have indicated, physicalism will be satisfied, even if the working out of the programme is not or cannot be successfully completed. SSP, however, does not entail that attributes are explainable even in this idealized sense (i.e. in the sense that abstracts away from problems of human access, etc.) Hence SSP does not capture this crucial dimension of physicalism.

I conclude that SSP fails to provide the necessary supplementation to the consensus view required for raising it to the level of an adequate formulation of physicalism. It fails to do this because it provides no guarantee that, for a given non-physical phenomenon, there are physical properties relevant to its realization which can be isolated and related to it in a way that rules out emergence and that constitutes the basis for an explanation of the appropriate sort. Hence there is no guarantee that the consensus view supplemented by SSP will yield the sort of understanding sought by the physicalist, an understanding of the physically-based 'vertical structure' of the natural order.

The final proposal I shall consider arises from some of Post's ideas for yet another sort of relation between physical and non-physical phenomena, a relation which may provide the required constraints upon realization and the required explanatory connections. In defending his principle TT against more localized supervenience theses, Post writes as follows:

... once in possession of TT or an equivalent, we can easily explicate the notion of the specific P-phenomena that are the ones to determine some given N-phenomena x. ... we merely define them as the phenomena that form the least set that suffices to determine x (or if there is more than one such set, their union) ... Local determination principles are probably too strong, but TT enables us anyway to speak of the specific P-phenomena that determine some given N-phenomenon ... Thus TT combines specificity with the ... allowance for holistic or ecological thinking ... (Post 1987: 188.)

Thus TT (or, better, TT*) incorporates both specificity and holism by allowing for the isolation of specific physical phenomena associated with each non-physical phenomenon. The isolated physical phenomena are 'the P-phenomena that form the least set that suffices to determine x', where x is some given non-physical phenomenon. The critical issue here is whether conjoining this idea with CV provides any further metaphysical or explanatory content that suffices to fill the gaps in CV. To determine this, I shall consider the following supplementary thesis:

(LSP) For each non-physical property N capable of instantiation in a P-world, there exists a set L of P-properties, such that L is the least set of P-properties that suffices to determine N.

In this formulation, a set of P-properties (i.e. physical properties) is the *least* set that suffices to determine N just in case it, but no proper subset, suffices to determine N. And a set of P-properties, L, determines N just in case, for any two P-worlds W and W', W and W' are alike with respect to the P-properties in L only if they are alike with respect to N. Thus LSP differs from SSP in at least two ways. It does not require that there be a single complex physical property that is a nomological sufficient condition for N, and it concerns both the occurrences and the non-occurrences of N. Further, LSP holds out promise for handling explanation in ways similar to SSP in that it appears to isolate explanatorily relevant physical properties bearing upon specific non-physical properties, and it is formulated in terms of a potentially explanatory relation between the physical and non-physical properties involved (viz., determination). Perhaps, explanation derives from appreciating that N-phenomena co-vary with P-phenomena in this way.

Now, does LSP suffice for filling all the gaps left open by CV? On the plus side, LSP does handle the problem of multiple realization and the problem of context. And it appears to handle the problem of isolating only the factors relevant to the instantiation of non-physical properties by requiring that L be the least set of P-properties sufficient for determining N. None the less, LSP falls short with respect to both concerns about emergence and concerns about explanation. First, although LSP *appears* to call for all and only the physical base properties that are relevant to the realization of some non-physical attribute N, it, like SSP, fails to do so. It can be seen to fail for essentially the same reason: a class of nomic equivalents that are not relevant to the constitution of N could satisfy the demands made by LSP. Thus irrelevant properties *vis-à-vis* realization are not precluded from L. Further, the relation of determination holding between L and N does not express the idea that the properties in L constitute or realize N. Systematic covariation does not imply or require any tighter relations between a given property and those properties in its determination base. Thus there are no resources within the approach for distinguishing those properties that are relevant to the realization of N and those that are not.

Therefore LSP is compatible with emergent attributes since it does not impose strong enough conditions on L and its relation to N: the relation of being the least class sufficient for determining N does not guarantee that N is *realized by* the members of L. This is because LSP, like SSP, abstracts away from the connections between the members of L and N that underwrite the determination relation.[52] Post believes that this is a virtue of LSP since it represents a form of determination that does not require connecting generalizations. I think it is a major defect because it means that LSP cannot satisfy the physicalist's concerns about emergence. The general point here is that determination, like nomological sufficiency, is too weak a relation to express what the physicalist wants to say about how non-physical attributes are related to the physical domain.

Second, LSP is defective with regard to providing expression of physicalist explanatory concerns. Quite evidently, LSP is an existential assertion that has no implications regarding whether explanations are possible or forthcoming in terms of the properties in L. The mere assertion of LSP in no way expresses the idea that non-physical phenomena admit of physicalist explanations or that such explanations ought to be developed. Since LSP lacks the resources for guaranteeing that the properties that satisfy the thesis are those that in fact realize a given non-physical property, it also does not guarantee the isolation of a class of properties that could play a role in the development of a vertical explanation of that property. Thus, even if it is assumed that the properties in L are available for use in explanation, those properties need not be relevant to the realization, and hence the vertical explanation, of N.

---

[52] As discussed earlier, I am not suggesting that we require clarification of these connections in order to explain determination, although that is perhaps a consequence of such clarification. Rather, the requirement is fundamental to the physicalist programme in so far as it precludes emergent attributes and leads to understanding of the realization of all non-physical attributes.

Further, although it can be allowed, assuming that the relevant properties are isolated, that the relation of determination does have some explanatory force, it is not an appropriate relation for explaining the realization of one property by others. Such vertical explanation involves showing how relevant physical (physically-based) properties combine to constitute the realized property. Thus even if one grants that some understanding is gained by seeing that a determination relation holds between the properties in L and some non-physical property N, there are major explanatory gaps that need to be filled. LSP does not involve other explanatorily useful connections that can be appealed to in specific cases in order to understand how or why N is realized by the properties in L. And since it is the establishment of such intricate vertical connections and consequent understanding that the physicalist is interested in, LSP is, therefore, too weak to fill the explanatory gap left by both CV and CV plus SSP. The sort of explanation provided by LSP is simply not the sort the physicalist requires (cf. Kincaid (1988) for discussion of these issues; see also Chapter 4 below).

It is no defence against this conclusion to claim that it is consistent with CV supplemented by LSP that connecting theories can be developed to clarify the intricacies of the relations between physical and non-physical properties, and thus to provide vertical explanations. At issue is whether an adequate formulation of physicalism can be provided without saying anything about such connections. Anyone who restricts their physicalism to a metaphysical formulation like CV plus LSP has forsaken a central dimension of the physicalist programme. However, it should be observed that Post's discussion of 'connective theories' (Post 1987: chapter 5) is a valuable contribution to the literature on such matters. But, as Post would agree,[53] it should be seen as part of a statement of physicalism, not an important additional physicalist project. Physicalism is ultimately concerned, not only with what exists and what counts as an objective matter of fact, but also with deeper understanding of how non-physical phenomena arise.

---

[53] Although Post focuses upon the metaphysical side of things when formulating his version of physicalism, he is in agreement that a full-blooded formulation of physicalism must take into account explanatory concerns. My principal objection against him is that his *metaphysical* formulation is inadequate because it does not successfully preclude emergent attributes.

Thus CV supplemented by LSP fails for exactly the same reasons as do all other formulations of non-reductive physicalism.

2.3. DIAGNOSIS AND CONCLUSIONS

In this last section, I shall offer a preliminary diagnosis of the failure of the above proposals to provide adequate formulations of physicalism. In brief, none of the formulations successfully expressed the physicalist's concerns about non-emergence and explanation. Most versions did not address the problem of isolating all and only the relevant physical attributes bearing upon the realization of non-physical. And none of the versions constrained the relations between physical and non-physical attributes in a way that rules out emergence. Thus all failed to address successfully the most significant *metaphysical* interest of physicalism: viz., the non-emergent realization of all that there is by what goes on in the physical domain.

Further, in one way or another, all versions failed to properly express the physicalist concern with regard to the following two questions:

1. In virtue of what physical[54] properties did the instantiation of a given non-physical property occur?
2. In virtue of what did such-and-such physical properties constitute the instantiation of such-and-such non-physical property?

That is, none of the above formulations succeeded in addressing the problem of how best to express physicalist concerns regarding either the isolation of explanatorily relevant physical properties bearing on the instantiation of non-physical properties or the identification of connections between physical and non-physical properties that explain why given non-physical properties get instantiated. In the absence of some way of addressing these concerns, a formulation of physicalism fails to engage the deepest and most important motivations of the physicalist programme: the drive to develop an understanding of how the non-physical aspects of

---

[54] Again, 'physical' here should be construed in a broad sense that includes physically-based properties as discussed earlier.

the world are embedded in the physical fabric that, according to the physicalist, exhausts, determines, and realizes all that there is. We can come to understand how such phenomena as mind, meaning, and values are possible in a physical world and are determined by the physical aspects of the world only if we have in our possession explanatory theories that isolate relevant physical properties and that relate them to those non-physical phenomena. Since all the formulations reviewed above do not call for or entail this sort of explanation, they fail to call for or entail the kind of understanding that the physicalist wants to have, among other kinds of course. If progress is to be made towards formulating physicalism adequately, steps must be taken to develop physicalist theses that call for and characterize the required kind of explanation and understanding. Nothing short of this will do, if we are to frame a significant physicalism that both informs us and drives research in important directions.

One primary reason for these failures is the attempt to formulate physicalist doctrine in terms of theses concerning supervenience of one sort or another. Although initially appealing as a way of responding to the failure of classical reductionism, it is now apparent that, although supervenience does not succumb to some of the problems that undermined definitional and derivational reductionism, it is simply too weak and too abstract a relation to capture the most significant metaphysical and explanatory concerns of the physicalist. Systematic covariation, though important and illuminating to some extent, does not capture the specific connections that relate the non-physical aspects of existence to the physical basis. Likewise, nomological sufficiency and correlation are also neutral with respect to the fine details of the connections between the properties that are nomologically related. For better or worse, it is these intricate connections with which the physicalist is most concerned, from both a metaphysical and an explanatory perspective. Thus physicalism must be formulated in terms that may include, but must go beyond, supervenience and nomological connection.

In the light of this diagnosis, let us assess where we stand with regard to the problem of retrenchment. Recall that the problem is to provide an *adequate* formulation of physicalism that does not encounter the same difficulties as classical reductionism. We can and should now add that the formulation does not encounter the same difficulties as do the formulations of non-reductive physicalism just reviewed. Thus the constraints upon an adequate formulation are beginning to mount up. Neither a purely ontological formulation nor a purely supervenience formulation, nor any of the formulations reviewed which combine these two sorts of thesis, suffice. To formulate physicalism properly, theses deploying more powerful conceptual resources are required (for example, theses concerning *realization* and *explanation*).

At this point it might be objected that the standard of adequacy which I am employing imposes such implausibly strong requirements that any formulation of physicalist theses that satisfies it will be obviously false. Thus the fact that it leads to the conclusion that extant formulations of physicalism are inadequate should not disturb us all that much. The objector might reason as follows. According to the standard of adequacy employed, we must frame physicalist theses that call for the *perspicuous* embedding of non-physical phenomena in the physical world. This means calling for the *isolation* of relevant physical or physically-based properties for each non-physical property and the *identification* of connections between the physical and the non-physical which reveal how the latter are realized by the former. But there are many reasons for thinking that this is too much to expect in general; to require it is to require that false theses be developed.

For example, it might be contended that some relevant physical properties are unavailable for the purposes of explanation because they are (in principle) inaccessible, *or* that certain non-physical properties are, for humans, *mysteries* (i.e. they are properties whose theory of realization is forever beyond our intellectual grasp), *or* that for certain non-physical properties there is no principled distinction between relevant and irrelevant physical base properties, *or*, finally, that even if such relevant physical properties exist and are in principle available to us and even if theories of realization are possible for every non-physical property, explanations may never be developed because of contingent technical or human limitations. For these reasons the objector may be led to conclude that it is folly to adopt my criterion of adequacy, to reject the consensus view (or some modest extension of it), and to require that theses concerning realization and explanation be explicitly written into a formulation of physicalism.

I find none of the above arguments very convincing. Briefly, this is because, first, the fact that certain physical properties are inaccessible to us does not imply that explanations of the non-physical properties that they realize do not exist. It is important to distinguish between theories and explanations on the one hand and our capacity to formulate them on the other. The inaccessibility of certain properties bears only upon the latter. Second, if it is the case that human cognitive capacity is not sufficient for grasping an explanation of the realization of some phenomenon (i.e. its realization is a mystery), we would not, strictly speaking, have a counter-example to any thesis that called for such an explanation. Rather, we would have a gap in our theoretical understanding, of which there are many others, no doubt. Third, if there are non-physical properties that are not determined by any restricted set of relevant physical base properties, then such a property might be viewed as being realized by the total state of nature. If so, then physicalism is unassailed if there is, none the less, an explanation of how that property is realized by the total physical domain. If no such explanation is possible, physicalism fails, but not because the property is holistically realized. Finally, that contingent human or technical factors would preclude the development of possible explanations should not be seen as at all threatening to the truth of any explanatory thesis calling for such explanations. Again we would, at worst, have a gap in our knowledge rather than a counter-example to physicalism. Thus I do not believe that the objector has made a case for the claim that my criterion of adequacy is too strong.

Further, however, I believe that the objector's strategy is misguided. My contention is that there are certain metaphysical and explanatory motives that are central to a proper understanding of the physicalist programme and that physicalist theses ought to express. This contention is partly justified by the history of physicalism as well as by the intrinsic value of developing the sort of explanation involved. If such motivations are impossible to satisfy (i.e. if they are 'utopian'), that is a fact of deep importance requiring understanding. But I know of no very good argument (surely not any of the stock objections to physicalism) that supports this sort of opposition to such motives. I am encouraged by the work of philosophers and scientists in a wide variety of areas (for example, work concerning mental representation, moral values, aesthetic values, knowledge) that shows promise in overcoming the obstacles often thought to impede explanatory progress along physicalist lines. Thus any retreat from central metaphysical and explanatory motives ought to be done with high visibility and with stronger justification than is currently available.

My conclusions are, first, that an adequate formulation of physicalism is much harder to provide than has sometimes been supposed, requiring as it does the proper taking into account of concerns about non-emergence and explanation. Second, opting for the consensus view, or some modest extension, as a replacement for the failed enterprise of classical reductionism compromises the deepest motivations of the physicalist programme. Given the current situation, therefore, physicalists are best advised to pursue the 'high road' in search of more adequate ways to formulate their doctrine.

3

Identification of the Physical Bases

Materialist metaphysicians want to side with physics, but not to take sides within physics.

David Lewis (1983: 364)

... the typical physicalism, for example, while prodigal in the platonistic instruments it supplies for endless generation of entities, admits of only one correct (even if yet unidentified) basis.

Nelson Goodman (1978: 95)

The objective of this and the next two chapters is to formulate an adequate and acceptable version of physicalist doctrine. In so doing, I shall work within the guidelines and constraints identified in Chapters 1 and 2, and I shall identify and respond to a number of objections to the doctrine. The task is thus to weave a path through the objections and constraints and to formulate a version that satisfies the criteria of adequacy and acceptability.

Towards this end, the present chapter will be devoted to the identification of the physical bases of the physicalist system. In pursuing this objective, I shall critically review a number of past, failed, strategies for characterizing the bases, and I shall suggest a set of constraints upon any viable approach to this problem. Hence the way will be paved for the development of a more effective strategy. I shall also identify and assess a set of presuppositions of any significant formulation of physicalist theses. Such presuppositions exist because the content of physicalist claims depends upon there being a principled and determinate set of physical bases of the right sort to warrant privileged status in a physicalist system *and* the existence of such bases depends upon the truth of the presuppositions.

The problem of identifying the physical bases is one of the deepest foundational issues facing physicalists. It has, with only a few exceptions (for example, Hempel 1969; Hellman 1985; Post 1987), been all but ignored by proponents of physicalism.

Opponents of physicalism, on the other hand, have often pressed the matter, but never in a conclusive way (for example, Chomsky 1980; Goodman 1978; Putnam 1979; Sussman 1981). Nowhere, to my knowledge, is there a detailed and comprehensive discussion of the issues involved, and my intention is that the present chapter should count as a step towards filling this gap.

3.1. BASES FOR ONTOLOGY, IDEOLOGY, AND DOCTRINE

In general, a physicalist thesis is a claim concerning a relation between two classes, one of which is designated the 'base class' and contains physical entities[1] as members. Thus any adequate formulation of physicalist doctrine requires an antecedent characterization of *the physical*. Further, since physicalist theses can and do vary with respect to the types of entities and relations involved (for example, linguistic and non-linguistic entities, reductive and non-reductive relations), it is necessary to specify what being physical consists in for each type of entity implicated in the formulation of physicalist theses. The purpose of the present section is to provide an account of the required physical bases.

As a preliminary, I shall briefly characterize a distinction, first introduced by Quine (1951) and subsequently clarified by Hellman and Thompson (1977) and Post (1987), between ideology and ontology. Quine writes as follows:

> Given a theory, one philosophically interesting aspect of it into which we can inquire is its ontology: what entities are the variables of quantification to range over if the theory is to hold true? Another no less important aspect into which we can inquire is its ideology . . . : what ideas can be expressed in it? (Quine 1951: 14.)

And again:

> The ideology of a theory is a question of what the symbols mean; the ontology of a theory is a question of what the assertions say or imply that there is. The ontology of a theory may indeed be considered to be implicit in its ideology; for the question of the range of the variables of quantification may be viewed as a question of the full meaning of the quantifiers. (Quine 1951: 15.)

---

[1] I use 'entity' as a completely generic ontological category subsuming, *inter alia*, objects, attributes, terms, and sentences. Specific physicalist theses will typically concern only one sort of entity.

Although there is much to be said regarding this distinction, the following comments must suffice for present purposes.

Ontology concerns what, *according to a theory*, there is. I shall construe this in terms of both what individuals and what attributes there are, given the truth of the theory. In addition, matters of ontology concern what ontological kinds there are (for example, *event*, *property*, and *individual*), and they concern qualifications of such kinds (for example, *mental*, *physical*, *concrete*, and *abstract*). Every entity satisfies one or more ontological kind predicate, and thus everything has some ontological status or other (cf. Hellman and Thompson (1977) for a more extensive discussion). For the physicalist, such matters of ontology are central to an understanding of what the physical basis for ontology consists in and of how other ontologies relate to it. Shortly I shall be concerned with the question of what individuals and attributes exist if physical theory is true. Eventually the questions of whether there are non-physical attributes and whether they undermine the physicalist programme will loom large.

Regarding ideology, the central concern is with 'what ideas can be expressed in' the language of a theory: what symbols are employed in framing the theory?, which symbols are fundamental and which are derivative?, what do the symbols mean? Roughly, matters of ideology concern the expressive power of a language, and thus they concern the structure of a language and its interpretation. The *ideological status* of an entity is determined by which predicates express it (for example, physical predicates, psychological predicates). Hence only attributes enjoy ideological status (Hellman and Thompson 1977). Since an attribute can be expressed by predicates drawn from multiple, ideologically distinct vocabularies, an attribute can enjoy multiple ideological status. With regard to physicalism, ideology bears crucially upon the problem of identifying the physical bases: it is with the language of physics that the effort to isolate the bases ultimately begins.

For expository purposes, I shall extend Quine's distinction to include what I shall call 'doctrine', which is concerned with what true sentences are formulable in a language. It is, of course, vital to distinguish between sentences that are accepted as true and sentences that are in fact true. Thus 'doctrine' admits of two meanings of which it is essential to keep track. Either way, such sentences are central to the formulation of a theory as well as to the formulation of explanations, laws, and statements of contingent matters of fact. I distinguish sharply between truths, theories, explanations, and laws on the one hand and true sentences, theory formulations, law sentences, and explanation formulations on the other. In matters of doctrine, it is only the members of the latter group that are central. As with ideology and ontology, doctrine is a crucial ingredient for developing the physical bases to which all else is related.

Given this tripartite distinction, the problem of characterizing the physical bases for ontology, ideology, and doctrine must now be confronted. In taking on this problem, I shall begin with a few general strategic points and a review of the most important approaches advocated by others. The point of the review will be to build a list of constraints that will help to develop a viable characterization of the bases, i.e. a characterization that will avoid critical objections as well as do the work required by the physicalist programme.

It is important to distinguish between *a priori* and *a posteriori* approaches to the task of isolating the physical bases. By an '*a priori* approach' I mean one which depends upon some form of conceptual or linguistic analysis or upon some other form of *a priori* argument purporting to provide definitive grounds for drawing the distinction between physical and non-physical entities, be they objects, attributes, terms, or whatever. By an '*a posteriori* approach' I mean one which relies upon appeals to the empirical study of nature (for instance, as conducted in the sciences) for grounds that distinguish the physical from the non-physical. This sort of approach is followed by those who hold that *the physical* is whatever is in the domain of physics, and it is this latter approach that seems to be favoured by most recent physicalists. However, it should be noted that most *a posteriori* approaches have deep difficulties that may reveal a need for some sort of *a priori* stipulation.

For example, it is often suggested, even by those favouring an *a posteriori* approach, that a basis containing entities possessed of consciousness or states with propositional content could not be a legitimate physical basis. That is, it strikes many that the physical base classes that ground a physicalist system could not contain what we had antecedently identified as mental entities. Along these lines, Rorty (1979: 20; see also Van Gulick 1985: 56, fn. 7) stipulates that the *mental* is the 'opposite' of the *physical* and thus that these ontological categories are disjoint. Other *a priori* constraints upon what can be included in the physical bases might preclude values, semantic attributes, and *abstracta*, or they might impose conditions such as impenetrability or having location. Whether such constraints reflect more than past or present bias is not something over which I shall now pause, although in most cases I suspect that they do not. As will become evident below, this question about *a priori* constraints is no mere trifle for the physicalist: the content, empirical status, and methodological significance of the theses all depend crucially upon how it is resolved. For now, I shall follow those who endorse an *a posteriori* approach, and later I shall return to a discussion of possible *a priori* constraints upon the physical.

What then is a good *a posteriori* strategy for isolating a set of physical bases for the physicalist programme? Most authors have attempted to circumscribe the physical bases by first isolating the physical ideology and doctrine, and then giving a derivative characterization of the physical ontology. Although there may be alternative strategies,[2] I shall employ the more standard one for the following reason. It looks as if the three bases must be interdependently characterized. I know of no plausible way of proceeding with independent characterizations. And, as we shall see, it does not appear likely that one can get very far in the attempt to characterize the physical ontology without relying heavily on what one takes physics and its referential vocabulary to be. Given this general *a posteriori* approach, it will be of considerable assistance to examine a number of prominent proposals that have been advanced along such lines.

For historical as well as conceptual reasons, it is appropriate to begin by looking at the distinction, drawn first by Meehl and Sellars (1956), between 'physical$_1$' and 'physical$_2$' terms as follows:

physical$_1$: terms employed in a coherent and adequate descriptive, explanatory account of the spatio-temporal order.

---

[2] E.g. to identify the physical ontology and then to characterize the physical ideology and doctrine in some derivative way; or to identify the bases independently of each other.

physical₂: terms used in the formulation of principles which suffice in principle for the explanation and prediction of inorganic processes.

This distinction forms the basis for distinguishing two very different types of physicalist proposal.

First, *physical₁* was intended to capture the full vocabulary of natural science, where any term of such science that is applicable to some region of the 'space-time causal order' (i.e. the natural order) falls within the *physical₁* category. Hence the terms of physics, chemistry, biology, psychology, and so on are all elements of the *physical₁* ideology.³ The *physical₁* doctrine, then, consists of the sentences that are formulated in *physical₁* terms and are true. And the *physical₁* ontology contains all entities and attributes in a structure characterized by the *physical₁* doctrine. Physicalism based upon this conception of the physical is a pretty mild doctrine, although its defenders do not consider it trivial. It is mild because the bases for ideology, ontology, and doctrine contain, rather than ground, everything in natural science. It is not trivial, however, since the doctrine would require anything not explicitly an object of study in the sciences to be appropriately related to the entities contained in the *physical₁* bases.

In my view, however, this form of physicalism is much too weak. It might be endorsed by someone who believed that the branches of natural science cannot be individuated in any principled way, and hence by someone who believed that physicalism ought *not* to be conceived of as a programme that accords a place of privilege to physics, as distinct from the rest of natural science. However, these contentions are controversial at best, and they threaten to undermine the various values and benefits of the physicalist programme that I have highlighted in Chapter 1. The strength and significance of physicalist doctrine is, *ceteris paribus*, inversely proportional to the extent of the bases. Those versions that include everything, or everything within natural science, lead to principles that are weaker and less significant than those that are narrowly construed in terms of the entities that fall explicitly within the domain of theoretical physics. What this weak form of physicalism does is pose a challenge to anyone who would

³ On a broader construal of *physical₁*, terms that are applicable to regions of space-time, whether they are scientific or not, are classed as *physical₁*. See Cornman (1971), Davidson (1970), and Malinas (1973) for examples of this construal.

advocate a more stringent conception of the physical and hence a stronger version of the physicalist programme: viz., to show that it is possible to develop a narrower conception of the physical bases which is cogent and which supports physicalist theses that accord privilege to *physics*.

Physical$_2$ represents an attempt to characterize the theoretical vocabulary of physics and chemistry as distinct from the vocabulary of the rest of natural science. *Physical$_2$* therefore represents a move toward a stronger version of physicalism based upon a narrower construal of the physical bases. Physics and chemistry are, according to this conception, the branches of science concerned with the explanation of all inorganic processes. And physicalism is the doctrine that grounds all else in bases derived from these sciences. Unfortunately, there are a number of problems which make this approach undesirable.

First, as a characterization of the terms of physics and chemistry, *physical$_2$* suffers from the defect of not being supported by an independent characterization of what 'inorganic' means. As a consequence, the major problem for the physicalist has simply been pushed back a step. If the key question facing the physicalist is 'What is a correct conception of *the physical* and what correctly falls under that concept?', then that question has been traded in for 'What is a correct conception of *the inorganic* and what correctly falls under that concept?', without any obvious gain in clarity or prospect of resolution. Now, this latter question may not be unanswerable, but I shall leave it to the proponents of *physical$_2$* to do the answering, since there are other, more serious, difficulties besetting the approach.

Physical$_2$ is unacceptable in a second way since it lumps together physics and chemistry. The physicalist, whose aim it is to accord to physics a place of privilege within the physicalist system, should be seeking a characterization of the terms of physics, not physics plus chemistry. Only if the bases are tightly associated with physics will they underwrite physicalist theses that accord physics the sorts of privilege identified earlier. *Physical$_2$*, though narrower than *physical$_1$*, provides too broad a conception of *the physical* to do the work the physicalist wants done.

A third objection[4] is that there are certain developments in science that could prove highly embarrassing for proponents of

---

[4] This objection was suggested by Ned Block in conversation.

physical₂. If our empirical study of mind evolves to the point that it is well established that certain highly complex machines are capable of mental activity (for instance, robots that can think), then it appears that purely inorganic processes would exhibit properties that, under most construals of physics and chemistry, could not be accounted for by those disciplines. Thus certain features of a class of inorganic objects would not be explicable within the confines of physics and chemistry. Consequently the *physical₂* vocabulary would not contain all terms required for the explanation of inorganic processes.

To respond to this objection with the claim that, because of this alleged possibility, physics and chemistry (hence *physical₂*) subsume psychology is exactly the kind of move the physicalist must try to avoid, since it is preservation of the boundaries between disciplines that gives the physicalist doctrine much of its bite,[5] and, as described, such a move would also be blatantly *ad hoc*. A better response, therefore, is the rejection of *physical₂* as a characterization of the physical ideology. It should be recognized that such categories as *organic* and *inorganic* may not prove to be metaphysically or epistemologically interesting ones from the point of view of physicalism. And the features of inorganic systems may not all be explained by the principles of physics and chemistry, as those disciplines are properly conceived.

A final objection to *physical₂* is suggested by Chomsky (1968: 83) and developed by Block as follows:

> Briefly, it is conceivable that there are physical laws that 'come into play' in brains of a certain size and complexity, but that nonetheless these laws are 'translatable' into physical language, and that, so translated, they are clearly physical laws (though irreducible to other physical laws). Arguably, in this situation, physicalism could be true—though not according to the account just mentioned (i.e. *physical₂*) of physical property. (Block 1980: n. 4.)

The suggestion here is that the notion of physical law need not be restricted to phenomena of a certain degree of organization and complexity. Especially, it need not be restricted to inorganic phenomena. Rather, there is reason to speculate that which physical laws are actually operating at a given stage of the evolution of the

---

[5] The issues raised here are complicated, however, since it should not be concluded that downward incorporation of the sort imagined is never appropriate. See below, Chs. 5 and 6.

universe depends upon the degree of organization and complexity to which it has evolved. And in brains of a certain size and complexity, *physical* laws may operate that do not operate at lower levels (for example, inorganic levels) of complexity.[6] If this speculation is correct, then $physical_2$, as a characterization of the physical, is clearly defective. Of course, the burden of the objector is twofold: to develop an alternative conception of the physical that is defensible and to establish the truth of the speculation just described, given that alternative conception. The objector must do both because the objection depends upon an antecedently understood conception of the physical. Assuming this burden can be carried, the objection, in addition to undermining $physical_2$, reveals a general constraint on any characterization of the physical: viz., that it be compatible with the idea that there are 'emergent physical laws'.[7]

Given the objections just reviewed, $physical_2$ appears to be an unacceptable account of the physical. However, I want to draw attention to some good points of the proposal: (1) it purports to provide a principled way of identifying the sciences from which the bases will be derived; (2) it is an attempt to isolate a vocabulary that is narrower than the $physical_1$ vocabulary and hence one that could serve in the development of bases for significant physicalist principles; (3) it attempts to provide a principle for controlling the admissible extensions of the physical bases; and hence (4) it underwrites the idea that the growth of physics is compatible with a non-trivial formulation of physicalist doctrine. Thus, despite its fatal flaws, $physical_2$ teaches us a lot about how to proceed.[8]

---

[6] Note that the point does not rest crucially upon the cosmological evolution of the universe or upon biological evolution. It can be made in terms of whether or not certain more or less specialized conditions are realized either by natural or intelligent processes. See Hacking (1983) for discussion of phenomena created in the physics laboratory.

[7] I.e. *physical* laws that come into play only at certain stages in the evolution of the universe or only under highly specialized conditions. Emergence here is of an innocuous sort *vis-à-vis* the physicalist programme.

[8] Many philosophers have formulated characterizations of the physical bases in ways derivative from $physical_2$ (e.g. Field 1975; Putnam 1983: 223; Van Gulick 1985). Thus the physical is often conceived of as including whatever falls within the domain of physics plus whatever is appropriately related to the individuals and attributes in that domain. And Rorty (1979) characterizes a series of conceptions of the physical, the $physical_n$ for indefinitely large n, where each element is an extension of the one that precedes it. Although these ideas have value for some purposes, they are clearly defective as conceptions of the physical bases. At best, they capture the idea of physically-based entities.

A third strategy for isolating a set of physical bases begins by attempting to read off the physical ideology, ontology, and doctrine from the activity and productions of physicists, i.e. those who are engaged in the conduct of physics. Thus the physical ideology consists in whatever terms are typically used by physicists and found in physics texts when essential elements of physical doctrine are being presented. A core physical ontology, then, is picked out and characterized by the physical ideology and doctrine, and the full physical ontology is generable from this core. This proposal amounts to leaving it to the physicists to determine what the physical ideology, ontology, and doctrine are. It is a proposal favoured by, perhaps, the majority of contemporary physicalists.[9] Unfortunately, it is also a proposal that is quite unacceptable for reasons I shall roughly characterize now and discuss more fully below.

First, 'leaving it to the physicists' is unhelpful unless it is known who the physicists are and, more deeply, what physics is. Otherwise, the question of what the physical consists in is begged. As a consequence, this approach may be no better than the previous approaches with respect to the problem of isolating a relatively narrow basis for the physicalist system, since the class of 'physicists' may be quite broadly construed. And it is quite vulnerable to the charge that 'physics' may be revised and extended freely and without significant constraint. The boundaries of physics and of the class of physicists must be shored up if this approach is to have any hope of success.

Second, the proposal fails to provide a principle for identifying physics and physicists that is clearly relevant to the metaphysical and epistemological concerns of the physicalist. It is therefore vulnerable to the objection that what counts as physics and who counts as a physicist are based upon arbitrary administrative decisions or other forms of socio-historical accident. For example, who is identified as a physicist may result from arbitrary, or otherwise physicalistically irrelevant, decisions having to do with how to organize and run a university. Such prospects seriously threaten the content and empirical status of physicalist theses.

Third, the proposal leaves open whether it is current physics,

---

[9] See e.g. Boyd (unpublished), Friedman (1975), Hellman and Thompson (1975), and Post (1987). I think it likely that each of these philosophers, though not being explicit on the point, would grant most of what I am about to say.

future physics, or some ideal physics that is to play the critical role in identifying the physical bases. If it is current physics, then there would appear to be no room for the growth of physics from the point of view of any particular formulation of physicalism. Hence current physics must be true if physicalism is to be true; and this makes it appear that physicalism is obviously false.[10] If it is some future physics that is intended, then no principle has been provided that constrains the evolution of physics so that *ad hoc* modifications of physical theory designed only to save the physicalist doctrine from embarrassment by counter-example can be identified and ruled out. In the absence of such a principle, physicalism has seemed to some to be a trivial doctrine (Chomsky 1968: 84) or a vacuous one (Feigl 1969; Hellman 1985; Hempel 1969). Further, as I shall argue below, those (Lewis 1983; Putnam 1983: 212) who advocate the idea that the physical is defined by modest extensions of current physics that are similar to it have not successfully avoided the difficulty here. Finally, if it is some ideal physics that is intended, then again a principle is required for putting some content into the proposal. I conclude that if the physical bases are to be identified in terms of physicists and physics (whether current, future, or ideal), then physicalists must be responsive to the problems just mentioned. It simply will not do to say that a physical term is a term written in a physics text or uttered by a physicist.

Strategically, the problem for the physicalist is that, unless there is some antecedently specifiable principle for identifying the physical bases, physicalism cannot be formulated in a significant way. Given the standard approach of beginning with the identification of the physical ideology and doctrine and then developing a view of the physical ontology in terms of them, a principle is first required for identifying physical terms and physical truths. Such a principle must function to isolate such terms and truths in metaphysically and epistemologically relevant ways *vis-à-vis* the physicalist programme. And it must be capable of accommodating changes in our conception of what the physical consists in as knowledge grows, without compromising either the content or the

---

[10] See Hellman (1985) for an example of someone who holds that the content of physicalist doctrine is tied to current physics, and thus that the theses of physicalism are likely to be false at every stage of the development of science. See further discussion of this issue below.

empirical status of physicalist doctrine. Given the shortcomings of the first three proposals for accomplishing these objectives (i.e. *physical*$_1$, *physical*$_2$, 'leaving it to the physicists'), it is now evident that this whole strategy depends crucially upon a principled identification of *physics*. The next set of proposals are specifically directed toward this problem.

A number of philosophers have suggested that physics is the branch of science whose goal it is to provide a comprehensive supervenience base of a certain sort. Consider the following formulations of this approach by Lewis and Quine respectively:

> Physics (ignoring latter-day failures of nerve) is the science that aspires to comprehensiveness, and particular physical theories may be put forward as fulfilling that aspiration. If so, we must again ask what it means to claim comprehensiveness. And again, the answer may be given by a supervenience formulation: no difference without physical difference as conceived by such and such grand theory. (Lewis 1983: 356–7.)

> One motivation of physics down through the centuries might be said to have been . . . : to say what counts as a physical difference, a physical trait, a physical state. The question can be put more explicitly thus: what minimum catalogue of states would be sufficient to justify us in saying that there is no change without a change in positions or states? (Quine 1979: 163–4.)

Thus physics is the branch of science aimed specifically at providing theories and catalogues of basic attributes that provide supervenience bases for all other phenomena and the theories that concern them. There are, however, several difficulties with this approach.

First, both Quine and Lewis appear to have confused the aims of physics with the aims of physicalism. Indeed, as *physicalists*, they are right in insisting that physics satisfy the constraint that it provide a supervenience basis for everything else. And they are certainly correct in claiming that physicists do aim at comprehensiveness of a sort. However, the idea that the aim of physics is to vindicate physicalist supervenience (cf. Healey (1978) for a discussion of this view) is untenable. There is, for example, little reason to suppose that physicists are *aiming* to provide a supervenience base for moral, aesthetic, or social phenomena: their focus is neither on supervenience nor on these sorts of high-level phenomena. It is physicalism that calls for such supervenience relations.

Further, without venturing too far into the problem of the modal status of physicalist theses, it should be noted now that if the aim of physics were to vindicate physicalism, then there would be no possible worlds in which our physics held and physicalist supervenience claims did not. As I understand Lewis and most other physicalists, this is not compatible with a correct understanding of the modal status of physicalist principles.[11] For example, anyone who believes that it is possible for our physics to hold in a dualistic world or in a world in which supervenience fails is committed to finding a characterization of physics according to which the truth of physics does not entail the truth of physicalism. Thus it appears that the comprehensiveness that physics strives to attain must be conceived in some other way. The idea that the aim of physics is to vindicate physicalism involves confusing the aims of physics with a constraint imposed upon physics by those who endorse physicalism.

In addition, the suggested characterization of physics is much too abstract to distinguish physics from other candidates for a comprehensive supervenience base. If Goodman's somewhat cryptic claim (Goodman 1979) that psychology (of a sort) aims at providing such a supervenience base is correct, then physics is not uniquely picked out by either Quine's or Lewis's characterization.[12] Notice that I am not here endorsing the view that psychology does have an equal claim to providing such a basis; that issue will be discussed below. The present point concerns the *aims* of various research programmes and whether or not a supervenience formulation suffices to uniquely single out physics. I suggest that such a formulation does not, since what a general supervenience condition leaves out of a characterization of physics is any reference to the typical questions that physicists address, the types of phenomena they study, the methods they employ, and the kinds of answers they propose. It is my view that an adequate characterization of a research programme or a scientific discipline must make reference to some or all of these matters. Otherwise the characterization will give little insight into what practitioners in the

---

[11] See below in Ch. 5 for discussion of the modal status of the theses.

[12] Neither Quine's allusion to positions and states nor Lewis's mention of a physical difference suffices to provide a non-question-begging way out of this difficulty.

discipline are up to, and it may (as in the present case) simply fail to distinguish one discipline from others.

Finally, a supervenience condition is so abstract that it allows that the boundaries of 'physics' could, if necessary, be expanded to include chemistry, biology, psychology, moral theory, and whatever else is required to satisfy the condition. As stated, it makes such inclusions a matter of course in physics, and not subject to any further constraint. Thus physicalism is guaranteed to be true if physicists will only do their job, and this could include doing chemistry, biology, etc. Clearly, this is not the intent of either Lewis[13] or Quine. However, I suggest that unless there is another characterization of physics at work they cannot avoid this difficulty; and if they cannot avoid the difficulty, then physicalism is trivialized and devoid of the appropriate content.[14]

An alternative approach involves the idea that physics is the branch of science that is completely general. As I have argued in Chapter 2, it won't do to cash this in as the claim that physics is the branch of science that provides a *description* of everything. Thus the generality of physics must be understood in some other way. Some possibilities that have been suggested are the following:

Whatever the means in physics, the end is the same: to frame theories so general as to apply everywhere everywhen. (Post 1987: 205.)

the physical sciences are unique in being universal; they apply to everything there is in a way economics and psychology or even colour statements do not. (Stroud 1987: 270.)

any event that falls within the universe of discourse of a special science will also fall within the universe of discourse of physics. (Fodor 1975: 13.)

No science is more pretentious than physics, for the physicist lays claim to the whole universe as his subject matter. (Davies and Brown 1988: 1.)

Now, whether or not the authors here intended to give characterizations of essential characteristics of physics or simply to formulate

---

[13] Lewis's characterization of physics is richer than I have suggested in the text since, in addition to the supervenience requirement, he discusses other features of the inventory of properties proposed by physics. My point is only that something more than a supervenience condition must be offered in characterizing the research programme of physics.

[14] Haugeland's (1982) suggestion that physics is the branch of science that constrains all possible world histories fails for reasons similar to those that undermine the supervenience formulation of the aims of physics.

contingent claims about physics, it is important to see that such formulations do not suffice for characterizing physics and its aims. As a result, such appeals to generality do not serve the physicalist's need for a principled way of identifying physics. This approach suffers from defects similar to those of the supervenience approach: i.e. it neither distinguishes physics from other programmes, *nor* precludes the idea that physics is whatever it takes to attain generality, *nor* provides constraints on revision of physical theory, *nor* provides details regarding what the physicist is up to (for instance, typical questions). Further, it should be noted that vague talk of universality leaves open what such universality involves. For example, does the idea that the physical sciences apply to everything mean that nothing can ever be in violation of physical laws, or that each law has specific applicability to every object and event and to every region of space-time? A characterization of what physics is should both have content and be free of such ambiguity. All the criticisms just mentioned point to a need to provide a more detailed characterization of what physics is.

In light of the shortcomings of the approaches just reviewed, a partial list of constraints upon an approach to identifying the physical bases can be compiled:

- Develop a relatively narrow conception of the bases.
- Develop the bases in terms of physics, as distinct from the rest of natural science.
- Permit evolution of our conception of the physical bases without compromising either the content or the empirical status of the theses.
- Circumscribe the range of permissible extensions of our conception of the bases.
- Employ no question-begging terms (for example, 'inorganic', 'physicist').
- Allow for compatibility with 'emergent' physical phenomena and laws.
- Characterize physics and the physical bases in metaphysically and epistemologically significant ways.
- Do not impose inappropriate *a priori* constraints upon the form or content of physical theory or the bases.
- Do not conflate the aims of physicalism with the aims of physics.
- Do not characterize physics in a way that is so abstract as not to distinguish it from other disciplines.

- Provide a characterization that is free of ambiguity or vagueness that compromises its content.

The central idea for an *a posteriori* approach is to pin down *physics* as a distinct branch of knowledge and to develop the physical bases for ideology, doctrine, and ontology in terms of it. The list of constraints reflects the various perils that can befall one who pursues this course. And although satisfaction of the constraints would not guarantee success, it would put one in a more powerful dialectical position *vis-à-vis* critics.

The approach I favour begins with a characterization of physics that can be briefly summarized as follows: physics is the branch of science concerned with identifying a basic class of objects and attributes and a class of principles that are sufficient for an account of space-time and of the composition, dynamics, and interactions of all occupants of space-time. The crucial features of these classes are that they are minimal with respect to the descriptive and explanatory purposes they serve, that the magnitudes are defined for all regions of space-time, and that each occupant of space-time satisfies the principles governing those magnitudes. It is both the types of phenomena they are introduced to explain (i.e. composition, dynamics, interactions) and their complete generality that distinguishes these magnitudes and principles from others, and hence that distinguishes physics from other branches of inquiry. For example, a psychologically-based system, while claiming priority for psychological properties and principles in conceptual or evidential matters, does not introduce such properties and principles to explain the dynamics, the composition, or the interactions of the occupants of space-time. And although biology is concerned with composition, dynamics, and interaction, it lacks the generality of physics. The minimality condition in effect keeps physics from being gratuitously identified with all of natural science or with physics plus chemistry.

Fundamental physics, as Davies and Brown put it earlier, lays claim to the whole universe as its subject matter. In addition to being concerned with such matters as how elementary particles and their properties arise, how they combine to form more complex objects and their properties, how they interact with each other, and how systems of such particles evolve over time, the physicist is concerned with the nature of space-time itself as well as with the origin and fate of the universe. Thus physics is concerned with developing a unified picture of all that there is, a picture that succeeds in portraying the fabric in which everything is woven and is connected with everything else. For example, physicists surely conceive of the elementary particles and their attributes as 'the basic building blocks' for all more complex chemical, biological, and other phenomena. Further, all dynamics and all interactions among phenomena are constrained by fundamental physical principles. But this does not mean that physics is concerned with all complex phenomena with respect to *all* of their properties. For example, social phenomena are not of interest to the physicist *as* social phenomena, although they are of interest as occupants of space-time (i.e. in so far as they involve causal processes or entities which 'take up space').

As a consequence, it is not incorrect to say that physics is a universal science, but its universality needs to be understood in terms of the kinds of questions that physical theories typically answer. Such questions might be expressed as follows:

- What are the fundamental constituents of *all* occupants of space-time?
- What are the fundamental processes that underlie *all* causation and *all* interaction between such occupants?
- What parameters are relevant to the dynamic unfolding of *all* systems in space-time and hence to *all* change?
- What is the nature of space-time itself, its origin (if it has one), and its destiny?

Both the types of questions asked (i.e. questions concerned with space-time and its occupants) and the required generality of the answers are characteristic of physics. It is these features that establish the ontological and epistemological *relevance* of physics to the physicalist programme, as a quick review of the central ideas and values discussed in Chapter 1 will reveal.

This way of characterizing physics is, in my view, superior to the Quine/Lewis approach discussed earlier because it focuses upon the specific questions and concerns of physics as opposed to merely providing an abstract global constraint. That constraint (i.e. that physics identify an adequate supervenience base for all natural phenomena), as I suggested above, need not distinguish physics from a radically different sort of 'universal' science (for instance, psychology). And such a constraint does not allow for the isolation of *physics* from any other subject matter that might be required to guarantee an adequate supervenience base. Thus the general constraint, taken alone, means that the truth of 'physics' entails the truth of physicalist supervenience principles, if only because 'physics' is simply whatever it takes to satisfy the constraint (for example, 'physics' could include sociology). The current proposal involves no such entailment. Physicalism constitutes a significant doctrine regarding the structure of nature and knowledge that is quite additional to the enterprise of physics.

Further, given its more substantial content, the proposal suffices to guarantee a relatively narrow conception of the physical bases. But it makes no commitment to any particular terms or theories or to any particular entities, processes, or magnitudes, with the one exception of a commitment to space-time. It follows that this characterization of physics is quite consistent with the growth of physics and an evolving conception of what falls within the physical domain. With the exception as noted, no further *a priori* constraints upon the form or content of physical theory are imposed. The main desideratum bearing upon what counts as a physical term, theory, object or attribute is whether something contributes essentially to addressing the questions that it is the aim of physics to answer. Since this view is compatible with 'emergent physical laws', contains no question-begging terms, and is evidently relevant to the ontological and epistemological concerns of the physicalist, it appears to have a strong *prima facie* case for meeting the constraints upon an approach to identifying physics and the physical bases. If it is a defensible approach, then it provides a strategy for identifying physics and the physical realm in a way that overcomes the obstacles encountered by earlier approaches. Hence, if it is defensible, it will underwrite the development of the physical bases and the formulation of significant physicalist theses.

At this point, I shall proceed on the assumption that a *prima facie* case has been made for the defensibility of my characterization of physics, although I shall also consider some objections. What theory is it that answers to this characterization? No one knows. But the best guess that can be made is that it is current physics or some suitable elaboration, given that portions of current physics are underdeveloped or tentative. And it is clear that one can accept current physics as the best available approximation to the physics of our world without giving up the idea that it could be revised tomorrow. Hence the best available approximations to the actual physical bases for ideology, ontology, and doctrine are to be understood in terms of current physics or some suitable elaboration. As it goes with physical theory, so it goes with the physical bases: what one takes to be in the bases is subject to revision as knowledge in physical theory evolves. It is this picture, in my opinion, that justifies the practice of most physicalists to look to current physics when characterizing the bases.[15]

Now, having buttressed the strategy for developing the physical bases with a conception of what physics is, it is time to turn to the task of sketching out how a characterization of those bases might be developed. My aim here is strictly that of offering only a sketch of such a development and of identifying its major presuppositions. I shall assume that physics is identifiable and that there is a formulation of fundamental physical theory that is available for present purposes. For concreteness, the theory should be conceived of as characterizing a space-time manifold of four dimensions in which various regions are occupied by objects or attributes.[16] How, then, does one proceed towards giving a full characterization of the physical bases?

I shall begin with the basis for ideology. Assuming a basic stock of non-logical terms that are employed in formulating the theory and which express the fundamental physical magnitudes and kinds of object, the central problem is that of specifying the definitional apparatus that can be legitimately employed in building complex predicates from members of the basic stock. It is quite beyond the scope of this project to extricate from a formulation of physical theory some stock of basic predicates and rigorously to characterize the full range of allowable constructions built up from the basic stock. But I shall indicate the directions in which this is likely to go.

---

[15] See Post (1987: ch. 3) for a valuable discussion of the distinction between 'the manifest physical universe' and 'a true physical universe beyond the manifest universe as conceived by today's physics'.

[16] But physicalism should not be taken as being permanently committed to this sort of theory. Issues such as what the number of dimensions of space-time is, whether space-time is a real entity, whether space-time theories are compatible with 'quantization', and the like are open and left to physics for resolution. As physics develops answers to these questions, adjustments in the characterization and development of the physical bases may have to be made.

If we allow that the physical basis for ideology is to be initially conceived as a class of predicates built up from the basic predicates in terms of standard first-order logical apparatus and if it is closed under such predicate-forming operations, then three questions must be addressed: (1) Is the class of physical predicates closed under negation? (2) Is higher-order apparatus being inappropriately excluded? and (3) Is infinitary apparatus being inappropriately excluded?.

As Post (1984; 1987) and Hellman (1985) have suggested, it is implausible that the class of *physical properties* is closed under negation. Specifically, complementation negation of a physical property is not going to preserve the physical character of the property even if one allows that such a property-forming operation is legitimate (for instance, *not being an electron* is a very dubious physical property). Thus in the present context, where the concern is with developing a conception of the physical ideology that will underwrite the development of the physical ontology, allowing negation in the generation of the physical ideology is highly suspect. On the other hand, physical description and explanation frequently require the use of negation; and, they do so without flirting with metaphysical disaster. Thus I shall proceed with a characterization of the physical ideology that allows for negation in the generation of the class of physical predicates. However, for certain purposes, constraints will have to be introduced to govern the use of predicates that are entirely negative. And certain presuppositions for the application of negative predicates will have to be satisfied (for example, that the negation is not complementation-negation), when the physical ontology is being characterized. Thus the answer to the first question is that the class of physical predicates is closed under negation, but the class of physical properties is not likely to be so.

With respect to the inclusion of higher-order and infinitary apparatus in the generation of the physical ideology, my stance is quite liberal. I know of no good reason for excluding such apparatus, and there may well be theoretical advantages for inclusion, especially in the case of higher-order quantification. The practical difficulties involved in wielding an infinitary language do not necessarily undercut those theoretical advantages, and context will determine which of these considerations is most important. For present purposes, theoretical advantages are of primary importance.

Allowing that the logical apparatus can be appropriately enriched in these ways, the physical ideology looks something like this: there is a basic stock of physical predicates from which is generated a class of physical predicates built up from members of this basic set in terms of the logical machinery decided upon. The members of this set constitute the physical ideology, and, subject to the constraints upon negation just mentioned, all the properties, states, processes, relations, and theoretical kinds expressed by these predicates are, ideologically speaking, *physical* properties, states, etc. That is, they are expressed by physical terms and are available to play a role in the physical description and explanation of natural phenomena. Again subject to the constraints upon negation, those properties, states, etc. are all, ontologically speaking, *physical* in character. On the other hand, they do not exhaust the class of all physical properties, states, etc., since, as argued in Chapter 2, the physical ontology is quite likely to outrun the physical ideology.

Given this characterization of the physical ideology, how is the physical doctrine to be specified? Again, I offer here only a rough specification: the physical doctrine consists of all true general and singular sentences formulable employing the terms of the physical ideology. This class of sentences is certainly not supposed to be finitely axiomatizable. In effect it constitutes the full physical theory of nature, and it includes all law-like and counter-factual truths, as well as all contingent truths concerning past, present, and future physical states. When physicalists say that the physical truth determines the whole truth about nature, it is this class of true sentences that is being referred to as 'the physical truth'.[17] Formulations of physical explanations are also drawn from this class.

Given the physical ideology and doctrine, there are at least two strategies for identifying the physical ontology. The first is to identify the physical ontology with the class of those entities which satisfy, or are expressed by, one or another of the terms in the physical ideology. Although this is a commonly endorsed approach (see, for example, Davidson 1970; Haugeland 1982), its difficulties for first-order, finitary languages are, as argued in Chapter

---

[17] If a truth is construed as a non-linguistic entity of some sort (e.g. a proposition), then a different construal of the physicalist doctrine ensues. On a linguistic construal, all the truths probably do not express all the facts of nature, whereas on a non-linguistic construal they do.

2, clearly insurmountable, and for more enriched languages involving higher-order and infinitary apparatus there is little reason to hope for much better. Therefore I shall not pursue this first strategy.

The alternative approach is to employ the physical ideology in order to identify a basic class of physical entities and to specify constructive apparatus sufficient for generation of the physical ontology from the basic class. The entities need not be 'occurrent'; they must only be capable of being instantiated in the space-time of this world. Thus the generated objects and attributes, though outrunning the physical ideology and doctrine, should be viewed both as constituting the intended interpretation of the language in which the physical doctrine is formulated and as encompassing all physical objects and attributes that are instantiated plus all that could be instantiated, given the generative apparatus and the laws of physics. Different versions of this strategy result from different specifications of the type of entities in the base class and from different apparatus for generating the full physical ontology.[18]

My preference here is to follow those physicalists (Carnap 1967; Quine 1975; Friedman 1975) who begin with a four-dimensional space-time consisting of points which can be uniquely specified by a quadruple of real numbers. By a 'region of space-time' I shall mean any set of points within this space-time framework; each region is uniquely specified by a set of quadruples of real numbers.[19] Given such a framework, the strategy for completely specifying the full physical ontology involves a number of steps: (1) to determine which ontological categories will be employed (for example, objects, events, states, attributes, kinds); (2) to determine for each such category how to work with it (for example, to answer such questions as 'What is an event?' and 'How are events to be individuated?'); (3) to determine, for each category, a 'core class' of physical instances; and (4) to determine, for each such category, a constructive apparatus for generating the full class of physical instances of the category, given the core class.

In what follows I shall only informally and partially specify the

---

[18] See Hellman and Thompson (1975) for a paradigmatic example of this procedure.

[19] I intend to bypass all issues concerning whether characterizations of space-time require coordinate systems or whether coordinate-free characterizations are preferable.

physical ontology in accordance with the above strategy.[20] I assume that a satisfactory specification of this ontology includes a treatment of concrete individuals and attributes. However, within these main categories, it is unclear which traditional subcategories must ultimately be included in a full specification of the physical ontology. I shall leave it open both whether and, if so, how events, processes, and states, for example, are to be developed. Rather, I shall focus on a minimal specification of the physical ontology upon which any more elaborate version can be based.

My outline for the development of the physical ontology proceeds as follows. First, as initial ingredients, are the ontology of space-time points and regions as described above and the class of physical predicates distilled from a scouting of contemporary physics as just outlined. The selected predicates express either physical magnitudes (with possibly infinite ranges of values) or kinds of object.[21] I assume that each value of a given magnitude is an attribute (i.e. a property or relation). Thus the core class of physical attributes consists of all attributes expressed by the primitive and complex attribute-predicates of the physical ideology.[22] And the core class of physical object kinds consists of all kinds expressed by the primitive and complex kind-predicates of the physical ideology.[23]

Second, more complex physical objects and attributes are generated from the core classes of objects and attributes in accord with appropriate object- and attribute-forming operations. In the case of attributes, these operations will include quantificational and Boolean operations, subject to restrictions regarding negation,

---

[20] The working-out of the details of the bases for each ontological category involves interesting and important issues. However, for present purposes such development would divert attention from the work of identifying and assessing presuppositions, theses, metatheses, and objections to physicalism.

[21] Concerns about the status of objects in fundamental physics have led some to forgo talk of material objects. Although I share these concerns, I now believe that, for the purposes of presenting physicalist doctrine, talk of physical objects is useful, especially in discussions of how higher-level objects are realized. If an object ontology must be discarded in favour of a space-time ontology over which physical attributes are distributed, the current formulation can be readily revised.

[22] Since the range of values that a physical magnitude can assume is possibly non-denumerably infinite, it should be understood that all attributes corresponding to distinct values of the magnitude are being included in the core class of attributes, whether or not there is a predicate for each value. Quantification over all values suffices here for expression of each of the attributes corresponding to those values.

[23] See Causey (1977) for discussion of attribute- and kind-predicates.

and they may be finitary or infinitary. In addition, they will include specific physical processes which result in the formation of complex attributes. In the case of objects they also include both Boolean operations, again subject to restrictions on negation, and physical processes. The results are two fully generated classes, one of physical attributes and one of physical objects.

Third, the members of these two classes are evaluated[24] at every space-time region of the actual world. The result is a complete distribution of all of the generated physical objects and attributes over the entire space-time ontology. Relative to this distribution two further generative processes are applied. All relational attributes of the sort alluded to in Chapter 2[25] are generated, as are all fusions of classes of the occupants of space-time regions. Roughly, a fusion of some class of physical objects permits the aggregation of the members of that class to form a new object, however spatio-temporally discontinuous those members are and however arbitrary the class of objects is.[26] The result of applying these two operations to the original distribution of physical objects and attributes is a total distribution of physical objects and attributes over all of space-time. Such a distribution constitutes the total current physical state of the actual world.

Finally, the physical ontological base is completed by a characterization of the class of all possible total distributions of physical objects and attributes: such a class consists of all alternative distributions of objects and attributes subject to the constraint that the laws of physics are satisfied. Thus the physical ontology should be understood as including a class of all possible total distributions of physical objects and attributes over the regions of space-time. One of these distributions will correspond to the current total distribution of such objects and attributes; the others will be alternative total distributions that define the nature and limits of what is physically possible in this world.

With these constructions in hand, the physical bases for ontology

[24] Note that 'evaluation' in this context does not have the connotation that measurements are actually made by some person or device. Nor does it mean that fundamental magnitudes must take on determinate values or that objects have determinate locations. Appropriate probability distributions suffice.

[25] I.e. attributes of an individual that are defined in terms of what is happening in other individuals separated from the first by some space-time distance.

[26] See Hellman and Thompson (1975) and Post (1987) for explanation of how fusions of classes of physical objects are to be understood.

involve three classes: the class of physical attributes consisting of all the fundamental physical properties and relations and every property and relation generable from them, the class of all physical objects consisting of all the fundamental objects and every object generable from them, and the class just described of all possible total distributions of objects and attributes over the regions of space-time. These three classes will play crucial roles in physicalist theses concerned with ontological matters.[27]

In the present project, there is no need to develop the physical bases in a more detailed way, since the characterization just offered is sufficient for the purposes of the next several chapters. Rather, I shall turn to a discussion of some presuppositions of the foregoing approach, presuppositions shared by virtually every extant version of physicalism. The three I shall focus on are as follows:

(P1) There are principled divisions among branches of science, and, in particular, between physics and the rest.

(P2) There are determinate physical bases to be developed, given that physics can be isolated.

(P3) Physics occupies a special place in science which justifies developing the physical bases in terms of its ideology, ontology, and doctrine.

For each, I shall discuss its content, why it is a presupposition, and the major objections that have been advanced against it.

However, before proceeding with a systematic discussion of P1–P3, I want to highlight the importance of such a discussion for the defence of the physicalist programme. Clarification of these presuppositions is important because they have been the target of numerous objections over the years, objections to which physicalists have not properly responded. The following is a characteristic form of argument implicit in the critical discussions of physicalism given by a number of writers:

---

[27] This ontological framework can be extended to include a set-theoretic hierarchy built upon the individuals of the ontological system. Hence, as Quine puts it, 'we suffer no shortages' when it comes to our need for classes of objects. See Hellman and Thompson (1975) and Post (1987) for such an extension. As I shall argue below, it is problematic whether to count such a hierarchy as part of the *physical* ontology. What is not problematic is that, if one is a realist about sets, then given the physical ontology of space-time regions and their contents, a hierarchy of sets grounded in a set of physical individuals exists.

1. The theses of physicalism depend for their content, significance, or truth upon the determinate and principled identification of the physical bases.
2. The determinate and principled identification of the physical bases depends upon the truth of presupposition x.
3. Presupposition x is false.
4. Thus there is no determinate and principled identification of the physical bases.
5. Thus the theses lack content, significance, or truth.

Since my approach (and that of virtually everyone else) assumes that both (1) and (2) are true when x is replaced by a name of any one of the three presuppositions, it is rather important to look closely at those presuppositions and to defend them against the many arguments that have been advanced. It is my intention in what follows to close off a number of lines of objection to the physicalist programme and, into the bargain, to make clear why (1) and (2) are true.

3.2. PRESUPPOSITION 1: DIVISIONS BETWEEN BRANCHES OF SCIENCE

The content of P1 is that the institution of science has a certain structure: viz., it is partitioned into more than one part on the basis of some principle, or set of principles, that have metaphysical and epistemological significance with regard to the physicalist programme. Such significance concerns, for example, identification of relatively basic objects and attributes that could play a role in the realization and vertical explanation of other objects and attributes. Or, as in the case of most efforts to identify physics, it concerns a constraint such as comprehensiveness relative to some domain of phenomena. The key idea is that P1 requires that physics be identified in terms of its subject matter and typical cognitive pursuits, since it is these that have potential relevance to the sorts of ideas and values central to physicalism. Since the standard strategy for identifying the physical bases requires only that there be such a division between physics and the rest of science, P1 likewise requires no more than this. Of course, although this is all that is needed, it is possible that more fine-grained divisions within the body of scientific knowledge can also be justified.

What kind of claim is P1? To be sure, it is not a superficially descriptive claim about the current divisions among branches of science. Thus it is compatible with P1 that current science is not cleanly carved up along the boundaries indicated by principles that are needed for the purposes of physicalism. Failure to satisfy such principles may represent the influence of contaminating factors such as budgetary constraints at universities, funding patterns by various agencies, and political considerations that lead individuals to join forces. It may also be due to intellectually significant factors like more or less transient patterns of collaboration among researchers and the sharing of important cognitive resources that bear upon research activity.[28] P1 requires the existence of principles of ideal structure that will underwrite physicalistically relevant divisions, while current, manifest divisions can be viewed either as approximations to this structure or as crosscutting it. Neither of these possibilities necessarily impugns physicalism.

However, the physicalist must take care here. An important question to ask is, 'What are the best ways of understanding the organization of scientific knowledge and activity?' It is a question that the physicalist must take seriously as part of the programme. And it is a question that must always be answered relative to specific purposes. The idea that there is such a thing as *the* structure of science is fatuous. The idea that physicalist purposes are the only important purposes is entirely untenable. And the idea that physicalist structure can serve all purposes cannot be seriously entertained. It is a basic truth that different purposes often require different ways of organizing a subject matter. Thus the physicalist must allow that there are other purposes requiring certain ways of organizing science, and that these other organizations may be very different from the organization required for physicalism. For example, to account for certain paths of development within a research programme it may be necessary to focus on patterns of collaboration, or on the sharing of resources, that may crosscut physicalistically significant boundaries. And cross-disciplinary research efforts, such as those in the cognitive and neurosciences, may be absolutely essential for fully understanding some phenomenon (for example, how the mind and brain work) even though such groupings may not directly correspond to principled divisions of the sort required by physicalism.

The crucial issue for physicalists is, ultimately, whether science

---

[28] See Giere (1988) for important discussion of such factors.

admits of being correctly carved up in ways relevant to physicalism, regardless of whether there are other correct ways of doing the carving. I shall contend that the arguments of sceptics, for the claim that no relevant principles for underwriting the required divisions exist, are inconclusive, and that there are some plausible candidate principles that deserve exploration. Whether science can be organized in the way physicalism requires is, at the very least, an open question. Thus P1 will emerge from the following discussion largely unscathed, and some approaches to justifying the claim that there are divisions among branches of science that do reflect metaphysical and epistemological distinctions of pertinence to physicalism will be identified. Again, I want to emphasize that P1 in no way precludes the idea that there are contaminating factors that can occlude physicalist divisions, or the idea that there are alternative ways of organizing science that are possible and necessary for certain purposes.

Why is P1 a presupposition of physicalism? Recall that the strategy for development of the physical bases requires that physics be identified, where by 'physics' is meant a branch of science that satisfies certain conditions and that is narrower than all of natural science. Given this identification, the physical bases for ideology, ontology, and doctrine can be developed. The characterization of physics at the heart of this approach is supposed to be metaphysically and epistemologically relevant to physicalist ideas and values (for example, by calling for identification of *fundamental magnitudes* and for comprehensive answers to certain questions). Thus it is explicitly assumed in this strategy that physics is a distinct branch of science, a branch distinguished from others on physicalistically relevant grounds.

In addition to explicit assumption, however, there are other reasons for holding that P1 is a presupposition of physicalist theses. To begin with, it is revealing to reflect upon the consequences of rejecting P1. Specifically, there are two related ways in which things go radically wrong for physicalism if P1 is denied, ways which correspond to two different ways of denying P1. On the one hand, it might be claimed that there simply are no distinct branches and that every part of science is just a part of one undifferentiated total science (i.e. 'physics'). On the other hand, it might be claimed that there are divisions within science, but that the principles forming the basis for the divisions are metaphysically and epistemologically irrelevant to the physicalist programme. For example, the divisions among branches of science might be alleged to be the exclusive products of socio-historical forces that have no significance for the physicalist enterprise.

If P1 is rejected for the first reason, then it appears to be an unavoidable consequence that the theses of physicalism are trivially true for natural science because, given the strategy for identifying the physical bases, every term of science falls within the basis for ideology, every truth of science falls within the basis for doctrine, and every object and attribute posited within science falls within the basis for ontology. Such a consequence is not at all in the spirit of the physicalist programme. If P1 is rejected for the second reason, then two untoward possibilities present themselves. First, the boundaries between disciplines, being based on irrelevant factors, need not reflect the metaphysical and epistemological primacy of physics (for instance, the objects and attributes in the basis for ontology need not be fundamental in the right ways). Thus the actual content and significance of the theses would depend upon factors irrelevant to physicalist ideas and values, and hence they would probably not be what the physicalist intends. Second, the boundaries, being sensitive to factors irrelevant to physicalist concerns, might be easily reshaped for irrelevant purposes or for *ad hoc* defence of the programme. Thus the boundaries between 'physics' and other branches of science may be too infirm and fluid for the theses to have significant empirical content. If physics cannot be isolated in relatively stable and appropriate ways, then it and the derivative physical bases may be subject to open-ended and unconstrained revision, thereby making physicalist theses trivially true and bereft of significant *a posteriori* status.

Strictly speaking, in the light of the above considerations, P1 is not a logical presupposition of the theses of physicalism: to show *that* it must be shown that the denial of P1 entails their denial. But the force of calling P1 'a presupposition' is still pretty strong if its denial leads to significant distortion or trivialization of the content of physicalist doctrine. I therefore conclude that such considerations, although not sufficient to establish that P1 is a logical presupposition, are sufficient to show that if the physicalist intends to express a significant doctrine concerning ontology, objectivity, and explanation, a doctrine whose fate is neither sensitive to irrelevant factors nor trivially guaranteed, then P1 had better be retained.

There are, however, objections to P1 that require a response. Since the prospects for the programme depend crucially on fending off these objections, it is curious that for the most part physicalists have chosen to remain silent regarding them. Critics of physicalism have advanced an assault on P1 that has been typically expressed as follows:

for many purposes, a simple division of Science into branches is very useful, but we have no sufficient reason for assigning deep significance to this classification. We conclude that Science is an enormous area of human research which is united by a common method. Its divisions are for convenience in describing results and do not represent a fundamental feature of Science. (Kemeny 1959: 182.)

The division of science into areas rests exclusively on differences in research procedures and direction of interest; *one must not regard it as a matter of principle. On the contrary, all the branches of science are in principle of one and the same nature, they are branches of the unitary science, physics.* (Hempel 1949: 382.)

The province that is actually regarded today as belonging to each science is very largely the result of historical accident; . . . Such considerations clearly justify the view that science is a single whole and that the divisions between its branches are largely conventional and devoid of ulterior significance. (Campbell 1953: 13–14.)

The assault on P1 here appears to involve both a positive and a negative thesis. The *positive thesis* is that the apparent divisions among branches of science are the product of socio-historical factors that are irrelevant to the ontological and epistemological concerns of the physicalist programme. As observed above, this claim suggests that, even if there are divisions among branches of science, such divisions do not give interesting and appropriate content to physicalist theses and that such divisions are vulnerable to irrelevant and arbitrary shifts. In so far as the theses of physicalism depend upon divisions in science, the positive thesis of the critics appears to imply that they lack the ontological and epistemological content that the physicalist supposes they have. To buttress this attack, a *negative thesis* is advanced to the effect that there are no acceptable ontologically or epistemologically relevant principles for distinguishing branches of science: all alleged principles of this type are defective. Hence there *could not* be principled divisions of the kind required by the physicalist. And thus P1, in the sense required for the isolation of the physicalist bases, must be rejected.

My response to this two-pronged attack will correspondingly proceed in two stages. First, I shall show that the positive thesis, even if true, does not entail the alleged consequences. And second, I shall argue that the negative thesis is essentially undefended and that there is a plausible principle for characterizing such divisions. I shall conclude from this that the negative thesis is very likely false, and thus that P1 is defensible.

The positive thesis, more specifically, is the claim that the structure of science (i.e. the divisions among and relations between branches) is the product of arbitrary socio-historical factors which concern human interests, the dynamics of group interaction, and the features of institutional structures that happen to exist at a given time. Many philosophers have concluded, on the basis of this thesis, that P1 is false. The issues before us here are whether the positive thesis is true and whether it entails that the apparent divisions among branches of science are arbitrary with respect to the physicalist programme.

To begin with, it is hard to deny that the above-mentioned factors play an influential role in determining what scientists do, how they are identified, and what lines of development are opened and closed in scientific research. As a result, the conduct of science, the products of scientific activity, and the relations between scientists are all affected by such factors. Thus, in at least this sense, I take the positive thesis to be true. However, this thesis does not reveal how various factors differentially contribute to scientific activity and to the structure of scientific knowledge. Part of what is at issue is whether it is *only* metaphysically and epistemologically irrelevant influences that shape science, or whether there are also significant contributions from 'human reason', 'scientific method', and the 'objective elements in experience' that go into forging various aspects of the institution of science.

As a consequence, there are two elaborations of the positive thesis that need to be recognized: (1) that social (etc.) forces shape science to some extent but there are also other relevant factors that are operative as well, and (2) that only social (etc.) factors shape science to the exclusion of more relevant factors. In what follows, I shall suggest that the existence of contributions from more significant factors of the sort just listed leads to the endorsement of (1), to the rejection of (2), and to an understanding of how there can be divisions in science that are non-arbitrary *vis-à-vis* the physicalist programme, despite the influence of socio-historical factors. Subject to this clarification I shall accept the positive thesis as true, but contend that it has no interesting consequences for the physicalist programme.

Everyone should agree that it simply does not follow from either version of the positive thesis that the structure of science must be, or in fact is, a 'socio-historical accident' with no ontological or epistemological significance. Science could, after all, be an accident that gets things right. Therefore it is a *non sequitur* to conclude, from the claim that socio-historical factors (partially or totally) influence the course and structure of science, that such structure is arbitrary from the point of view of physicalist concerns. Of course, such a claim doesn't exactly ensure that the evolving structure of science is relevant to physicalist concerns either. More argumentation is required of both sides of the dispute if P1 is to be clearly vindicated or rejected.

My argument for the claim that, despite the impact of socio-historical factors, the existing divisions in science can and probably do have physicalistic significance runs as follows. Although it is true that factors which are quite arbitrary from the point of view of the physicalist programme influence the conduct and structure of science, that conduct and structure do not result exclusively from such factors, epistemologically significant factors are involved as well, and sometimes those factors predominate.[29] Scientists generally seek the truth and generally they do so in intellectually responsible ways. Thus there are pressures from within science to proceed in ways that may be orthoganol to social or individual interests or to political and religious institutions. Such pressures, for example, consist of general principles of rationality and method[30] that do not necessarily line up with socio-historical factors. As a result, although factors of this latter sort are surely operative, they are not alone and they do not necessarily overpower other, more cognitively and physicalistically significant factors, although they can do so on occasion.

---

[29] See Giere (1988: ch. 8) for an engaging discussion of a clearcut case in which 'the evidence' swamped all other social and psychological factors involved in theory acceptance.

[30] E.g. general principles concerned with the importance of employing empirical evidence in the assessment of hypotheses.

Further, there are within science research questions concerning what are the best ways to individuate and relate programmes of research and branches of science. The constraints on answering these questions are largely based on considerations having to do with subject matter, methods, and patterns of success. Thus issues concerning the relations between the cognitive sciences, the neurosciences, and the various subdivisions of chemistry are usually conceived to be about how the phenomena relate to each other, how different methods of inquiry complement each other, and what strategies of research have been and are likely to be successful. They are not matters exclusively resolved by social, political, or individual psychological considerations.

Therefore the suggestion by proponents of the positive thesis that, somehow, science is structured willy-nilly by arbitrary forces and that divisions within science have little or no metaphysical or cognitive significance strikes me as flying in the face of at least some of the facts. It appears that such proponents view science as a 'first-order' enterprise with little or no epistemologically significant self-reflective activity. The point of my reply is that this is a false conception of science and that the self-reflective activity found in the institution of science concerns ontological and epistemological matters, not merely matters of funding, administration, or individual and social concern. Hence such activity is, potentially at least, a serious factor in the determination of how science is structured. One of the important goals of much scientific activity is to determine where important boundaries reside, both in nature and in knowledge about nature. Thus, despite the obvious impact of arbitrary forces on science, it is difficult to take seriously the idea that *none* of the divisions in science are ontologically or epistemologically significant. I suggest that this is the result of scientific activity itself. At a minimum, actual divisions among branches of science are likely to be imperfect reflections of significant principles and not just the reflection of arbitrary social (etc.) forces.

On the basis of these considerations, I think it is reasonable to reject the second version of the positive thesis and to endorse the first. Further, such considerations suggest that the truth of the positive thesis so conceived has no consequences regarding the truth value of P1 and that there is some reason to believe that there are significant divisions among branches of science. The strategic situation is that, although the case against P1 has not been made on the basis of the positive thesis, the objection based on that thesis has served to point out a burden that must be carried by physicalists: viz., to make good the claim that there are metaphysically and epistemologically significant divisions between branches of science, especially between physics and the rest of science. Carrying this burden is a component of the working out of the physicalist programme. And as science proceeds in accordance with the principles of physicalism, it should, more or less perfectly, resolve itself into divisions that reveal metaphysical and epistemic distinctions between physics and other branches, not merely political, administrative, and economic ones. As will now be seen, the critic of P1 doubts that such a resolution is possible; but whether or not this is true, it is clear that the positive thesis alone is not sufficient to establish such scepticism.

The negative thesis advanced by the opponent of P1 is that there are no metaphysically and epistemologically relevant principles for distinguishing physics from other branches of science. Hence there are no principled divisions of the sort required by physicalism. If such a claim is true, then P1 would be defeated and with it the entire physicalist programme. What, then, are the grounds for taking the negative thesis seriously?

Perhaps the most powerful statements of support for the negative thesis are to be found in the writings of Chomsky. The following is a typical passage:

It is an interesting question whether the functioning and evolution of human mentality can be accommodated within the framework of physical explanation, as presently conceived, or whether there are new principles now unknown, that must be invoked, perhaps principles that emerge only at higher levels of organization than can now be submitted to physical investigation. We can, however, be fairly sure that there will be a physical explanation for the phenomena in question, if they can be explained at all, for an uninteresting terminological reason, namely that the concept of 'physical explanation' will no doubt be extended to incorporate whatever is discovered in this domain, exactly as it was extended to accommodate gravitational and electromagnetic force, massless particles, and numerous other entities and processes that would have offended the common sense of earlier generations. But it seems clear that this issue need not delay the study of the topics that are now open to investigation, and it seems futile to speculate about matters so remote from present understanding. (Chomsky 1968: 83–4.)

Now, I take the point of this passage to be that there are no principles for constraining the evolution of the divisions between physics and other branches of science. Hence there are no principles for constraining the evolution of what counts as 'the physical'. At best, there are only terminological matters here that are of no substantive interest. As I have argued above, the physicalist must oppose such claims if physicalist doctrine is to be significant.

Chomsky's primary argument in support of his contentions appears to consist in an appeal to examples of how physics has been revised in the past. For example, he suggests that there was no operative principle for incorporating electromagnetic theory (EM) into fundamental physics other than a principle to the effect that the failure of attempts to reduce it to mechanics is sufficient for inclusion. The idea seems to be that, since physics is an evolving branch of science, there is no firm conception of what physics is. Hence anything can be included within physics by fiat: what is at one time thought to be a separate and reducible branch can, at a later time, be 'shifted down' into fundamental physics as a result of the repeated failure of attempts at reduction and the arbitrary decisions of scientists. The morals are supposed to be that the divisions among branches are not constrained by significant principles and that the theses of physicalism are trivial as a result.

I have two replies to this line of argument. First, a look at the example cited reveals that the attempts at reduction were quite strenuous (cf. Whittaker 1960: chapters 8–9) and that the decision to shift the boundaries to include EM within basic physics was not whimsical or arbitrary, but rather the outcome of considerable methodological debate. Further, the downward incorporation did not consist in a simple appending of electromagnetic theory to mechanics. Rather, there was substantial theoretical integration: mechanics and electromagnetic theory each make essential contributions to descriptions and explanations of many phenomena that neither can adequately account for alone. And it should be emphasized that subsequent developments reveal that this theoretical integration has become increasingly important, given the prospects for unifying electromagnetic forces with gravitational, weak nuclear, and strong nuclear forces. As I see it, the downward incorporation of electromagnetic theory into basic physics has been vindicated as physics has evolved. Thus this case of downward incorporation is best conceived of as a case in which a false version of fundamental physical theory was corrected. It surely cannot be described as a simple case of *ad hoc* revision of physics. And the claim that this shift is indicative of the arbitrary revisability of the boundaries within science is an overstatement: a patently arbitrary shift, which this is evidently not, would be needed to support the intended case.

An adequate understanding of the kind of methodological conflict exemplified by the EM case would be valuable, although to my knowledge no one has worked this out in very much detail. In the absence of such an understanding, no demonstration of the arbitrariness or *ad hoc* character of the revision exists. Therefore it is difficult to understand the appeal of the Chomskian interpretation. I do not believe that the example goes very far in showing either that the specific case of electromagnetic theory counts as an arbitrary shift or that, more generally, all divisions between physics and the rest of science are as arbitrarily malleable as Chomsky suggests.

My second line of reply is to urge the distinction between a conception of the research programme of physics and the specific theories that are included within that programme. The supporter of the negative thesis must deny the existence of such a distinction, since once it is made it is possible to see how physics can be an evolving science while the conception of what physics is remains fixed and is useful both for drawing principled boundaries between physics and other branches *and* for constraining the development of physical theory. Thus once the distinction is made it reveals how it is possible for physics to be an evolving branch of knowledge, while the doctrine of physicalism is not trivially guaranteed.

In particular, if such a distinction can be made out, as in effect I tried to do earlier in offering a characterization of physics, then it would become clear why the EM case is not an instance of the arbitrary revisability of boundaries within science, but rather an instance of the motivated revision of fundamental physical theory. The justification for the downward incorporation of EM is essentially that it is required for the successful pursuit of the aims of *physics*, and not that it is required for the *ad hoc* defence of physicalism. In the light of such a distinction, we can see why Chomsky's appeal to the EM case is gravely misguided and why no support accrues, on the basis of that case, to the negative thesis.

It might be argued that the Chomskian objection does not lose its force with the introduction of this distinction since, if *physical theory* is completely malleable and subject to any change, then a revision of that theory to incorporate any other theory could occur. Hence the 'principled' boundaries between branches would be meaningless and the physicalist doctrine would still be trivially guaranteed. The contention of the objection is that physics may well be just as the principle says, but, none the less, what can fall under the heading of 'physics' is sufficiently open-ended so as always to allow for rescuing the doctrine in the face of alleged counter-examples. More specifically, the objection is that *anything* can count as a fundamental constituent, magnitude, or process just by being included among the constructs countenanced in physics. And the requirements of the principle could always be accommodated by some rigging of the theory. For example, the requirement of universality might be handled by claiming that, say, mental properties have zero magnitude at most regions of space-time and non-zero magnitude at a very few. So mental properties are defined for all occupants of space-time, and hence satisfy the crucial constraint.[31]

The correct reply here is that the objection ignores the point of the distinction I am urging. In the presence of a clear conception of the nature of physics, the boundaries between physics and other sciences will be clear *and* revisions of physical theory will be constrained in significant ways. Thus no revision of physical theory would be allowed if it involved incorporating a body of theory that did not satisfy the operative conception of physics (i.e. did not contribute to the goals of physical theory and did not satisfy the requirements imposed upon such contributions). *Ad hoc* modifications of physical theory, such as might be exemplified by the artificial rigging of a theory, are ruled out by general principles of good method. On the other hand, however, revisions which are not *ad hoc* ought to be considered very seriously. The physicalist must allow that what counts as physics is an evolving part of human knowledge, and what are the basic magnitudes, constituents, and processes is a matter of ongoing discovery. But to allow this does not in any way commit the physicalist to the idea that any modification we like can be made if it is needed to save

[31] See Putnam (1983: 224–5) for a variant of this sort of objection.

physicalist doctrine. The problem the objection does point to is a very important one indeed: that of being able to distinguish the *ad hoc* revisions in knowledge from the revisions that are not *ad hoc*. This, fortunately, is not a special problem for the physicalist.[32] I conclude that the Chomskian appeal to the EM case is quite inconclusive and does not begin to motivate belief either in the non-existence of any significant grounds for distinguishing physics from other branches of science or in the triviality of physicalist doctrine.

The strategic situation is currently as follows. In developing the physicalist bases I appealed to the distinction between the research programme of physics and specific physical theories. Such a distinction allows that physics is an evolving branch of science without denying that physics is distinguishable from other branches of science on significant metaphysical and epistemological grounds. An abstract characterization of physics was suggested, and, given that characterization, the evolution of physical theory can be viewed neither as arbitrary from the point of view of the physicalist programme nor as simply ensuring the truth of physicalism in a trivial way. Such a principle therefore underwrites and vindicates P1. Such vindication is not, *contra* Chomsky, a matter of terminological stipulation, but rather a matter of theory construction concerning the nature of physics. And this is a matter that is part of the working out of the physicalist programme.

The burden of the physicalist is to present and defend such principles. The burden of opponents, on the other hand, is to produce arguments to the effect that no such principles are possible (i.e. to support the negative thesis), or at least to the effect that all specifically proposed principles are defective in some way. If the opponents succeed, then P1 is defeated, no principled identification of the physical bases exists, and hence the theses of physicalism would be trivial or vacuous. If physicalists succeed, then P1 is vindicated, principled identification of the bases is possible, and thus the theses will have significant content. So far, I have concluded that neither appeals to socio-historical factors in the evolution of science nor appeals to the EM case of downward shifting of theories within physics substantiates the opponents' claims.

However, there is an alternative reading of Chomsky's discussion

---

[32] See Grunbaum (1976) for an instructive discussion of *ad hoc* hypotheses.

of these matters that suggests a difficulty with P1 against which my responses to the positive and negative theses are not effective. The difficulty may be formulated thus:

> The sense in which our conception of the physical is evolving is based in part upon the idea that, at higher levels of organization and complexity, new phenomena and principles concerning them 'emerge'. And if the antecedent physical principles concerned with less complex phenomena prove not to be sufficient for accounting for the emergent phenomena, then, in order to preserve the fundamental and comprehensive character of physics, such phenomena and principles must be conceived of as physical. But if this happens, although the revision need not be seen as arbitrary or as a matter of terminology or convention, it leaves open the possibility that phenomena and principles originally conceived of as mental (i.e. paradigmatically non-physical phenomena and principles) must be re-conceived as part of the physical bases. This undermines a central motivation of the physicalist programme (viz., to understand the mental as a non-fundamental manifestation of fundamental physical phenomena).

According to this objection, even if it is allowed that revisions of the basic science need not be arbitrary, it is not ruled out that non-arbitrary revisions of the basis might lead to inclusion of 'obviously' non-physical phenomena and principles, thereby defeating a central motivation of the programme. This objection, the problem of downward incorporation, is a serious one to which the physicalist programme, as I conceive of it, appears to be vulnerable. I shall defer further discussion until Chapter 6, where I shall consider possible avenues of reply that have promise for saving the programme from the kind of self-destruction envisioned by this objection.

3.3. PRESUPPOSITION 2: DETERMINACY OF THE BASES

The second presupposition is that there are determinate physical bases to be employed for the purposes of the physicalist programme. By 'determinate bases' I mean classes of entities that are *well defined*: for any entity, there is a fact of the matter as to whether it is included in the bases of the system or not. P2 is a presupposition because the bases must exist and be determinate if the theses of physicalism are to have specific and significant content. If the bases do not exist, then the theses lack content entirely. And if there is no fact of the matter as to what objects are included in the bases, then there is no fact of the matter regarding what the theses are expressing, and hence there is no fact of the matter regarding whether the theses are true or false. Vacuous or indeterminate content, therefore, undermines the significance of physicalist doctrine and obstructs the attainment of the goals of the physicalist programme. As we shall see, critics have frequently cited such alleged difficulties in their briefs against physicalism.

As outlined earlier, the development of the physical bases proceeds in the following stages:

Stage 1: identification of a specific branch of science as physics.
Stage 2: identification of a class of theories which are formulated within that branch and which constitute the *physical* theories.
Stage 3: identification of the referential vocabulary of the language in which those theories are expressed.
Stage 4: identification of a class of all physical truths (singular and general) formulable in that language.
Stage 5: identification of an ontological structure that provides an intended interpretation of the language of physical theory.

Stages 3 to 5 provide what is required for the development of the physical bases presupposed by physicalist theses: the referential vocabulary delivered by stage 3 is essential for developing the basis for ideology, the class of truths delivered by stage 4 constitutes the basis for doctrine, and the ontological structure delivered by stage 5 is essential for developing the basis for ontology. For the purposes of the following discussion, I shall assume that my characterization of physics is defensible and that it successfully buttresses my response to the attacks on P1 lately discussed. Given this assumption and given our five-stage strategy, a number of objections designed to show that P2 is false present themselves. Adequate replies to these objections must be developed if the physicalist programme is to have even *prima facie* plausibility.

To begin with, it might be objected that the physics of the

actual world could be radically different from the physics of some other possible world. Hence to develop the physical bases in terms of our physics is either to be guilty of arbitrary parochialism or to fail to provide an adequate account of the physical bases. The issue is whether to include within the *physical* bases the objects, attributes, terms, and truths adverted to or employed in the other possible physics.

To address this objection, some clarifications are required. What, for example, is meant by 'the physics of some other possible world'? Since *physics* is a research programme that is not specified in terms of the detailed theories it produces, it is possible that the enterprise of *physics* is the same across possible worlds which can otherwise vary with respect to the physical theories true of them, and hence with respect to the objects and attributes posited by physicists in them. If there are other possible worlds, then the physical theories true of those worlds need not be the same as the physical theory true of this one.

This possibility should not be confused with some others, however. First, the physical theory true of our world concerns not only the actual states of the world, but also states which could have been instantiated given the physics of our world. Such possibilities can be represented by a 'possible worlds' semantic structure, but the 'possible worlds' represented by that structure are not those in which there is a radically different physics. Thus we shall want to keep talk of the possible states of this world distinct from talk of other possible worlds and their possible states.[33]

Second, depending upon what physical theory is true of this or some other world, many complicated structures are possible: for example, a world in which there are 'parallel universes' or a world in which there is an oscillating universe with distinct and entirely isolated epochs. Such 'universes' or 'epochs' are not distinct possible worlds but aspects of one world that satisfies one physical theory that subsumes the different structures that can or do occur. Note that, relative to different initial or boundary conditions, the

---

[33] Further, the distinction between 'real' possible worlds and set-theoretic representations of possibility as occur in possible worlds semantics must also be acknowledged. It is only the former that are involved in the discussion of the determinacy of the bases. Of course, those worlds may well be represented set-theoretically. And it is indeed controversial whether there are such worlds. See Lewis (1986) for discussion of these issues.

detailed structure of all or part of a given world may vary considerably even if there is only a single physics of that world. The aim of fundamental physical theory is the subsuming of all possible developments and elaborations of the world to which it applies. And it should be acknowledged that the laws of physics and the universal constants may evolve and vary with the particular details of the world as it unfolds. None of these scenarios should be confused with the idea that there are other possible worlds with radically different physical theories true of them.

With these clarifications in mind, the appropriate response to the objection is to claim that physicalism is indeed a doctrine that is developed in a parochial manner: it is concerned with one world at a time, and in the present case, it is our own. Thus other possible worlds with a physics different from the physics of this world do not fall within the scope of physicalist theses as framed by us. Hence the physical objects, attributes, terms, and truths of those worlds need not be included in the physical bases. Rather, such bases include only those entities that are derived from the specific physical theories true of this world.[34] On the other hand, from the point of view of some other possible world, proponents of a physicalist programme in that world would likewise develop the bases in terms of the physics of that world. The fate of their programme would hang upon the contingent facts regarding the ontological structure of that world and regarding the relevant systems of knowledge. The physicalist programmes in our world and in theirs are entirely independent, however: one could be true while the other is false. Therefore the possibility of the non-uniqueness of physical theory across possible worlds does not mean that the physicalist bases, as conceived above, are either *inappropriately* parochial or deficient. Such a possibility poses no threat to P2.

The second objection to P2 is based upon the idea that there can be more than one formulation of the physics of the actual[35] world that satisfies the characterization of what physics is and that meets

---

[34] In a different parlance (Horgan 1982, 1987; Lewis 1983), this is to say, *inter alia*, that no 'alien' physical objects or attributes fall within the physical bases: the bases include only those physical objects and attributes that are instantiated in this world or that could be instantiated given our physics.

[35] See Lewis (1986) for discussion of the meaning of 'the actual world'. Here, I use the expression to refer to *this* world.

all empirical and methodological tests. Therefore, it is concluded, there is no unique physical theory from which to derive the bases of the physicalist system, and consequently those bases suffer from deep-seated indeterminacy. To assess the force of this objection, there are a number of possible scenarios to consider. The first is one in which, although there are different formulations of physical theory, they are formulations of the same theory (i.e. there is only one physics of the actual world, but it has alternative, theoretically equivalent formulations). The second, however, is one in which there are indeed formulations of different theories that are empirically adequate and evidentially equivalent.[36] Within this second possibility, there are two further possibilities: either the different theories are compatible with each other or they are not. I shall discuss these possibilities in turn.

If the situation is one in which there are two formulations of the same theory,[37] then there is no problem with relativizing the development of the physical bases to one or the other of the formulations: there would be different characterizations of the bases, but the differences would be strictly notational (i.e. there would be one set of bases described in different terms). Such *merely* notational differences do not make any difference to the physicalist programme: the same physical objects and attributes would be included in the basis for ontology, and equivalent systems of physical truths would be included in the basis for doctrine. Alternatively, if some alleged notational difference did make a difference to the programme, then that would be grounds for thinking that the difference was not just notational in character, and therefore that the case was not one of equivalent formulations of the same theory.

But this picture is clouded somewhat, given our strategy for identifying the bases, since notational differences appear to constitute a straightforward difference in *ideology*: thus the basis for

---

[36] For theories to be empirically adequate and evidentially equivalent means that each accounts for all possible observations and they do not differ on any consideration bearing on their truth or falsity. The assumption is that each satisfies all evidential tests. Of course, these notions are problematic, as I shall discuss further below.

[37] Such a possibility requires that there should be reasonably well-defined conditions of theoretical equivalence, a requirement that is more problematic than is often supposed. See Glymour (1971), Horwich (1986), Putnam (1983: ch. 2), and Quine (1981a: ch. 2).

ideology would be different relative to the different formulations. However, if such apparent ideological differences did not have any ontological or epistemological consequences (for example, the same attributes are expressed, the same explanations are provided), then such notational differences would not count as significant ideological differences. Although there would appear to be a number of alternative physical ideologies, the differences between them would not be significant, and the choice among them would be strictly a matter of convention. As a consequence, there is no way to undermine P2 in this case, since the existence of alternative characterizations of the bases does not undermine their determinateness. Whether some object or attribute is included is unaffected by the fact that it can be picked out in different ways. In the case of alternative ideologies and equivalent systems of truths, the differences are purely conventional.

Suppose, on the other hand, that the case is not one of alternative formulations of the same theory, but one of formulations of alternative theories. Again, suppose that the different theories are empirically adequate and evidentially equivalent. In such a case we would have formulations of *different* theories, each of which satisfies the general characterization of what physics is, each of which is compatible with all possible evidence, and each of which is evidentially indistinguishable from the others.[38] Does this not pose a serious challenge to P2? For example, an objection to P2 based on this possibility might run as follows: even though physics has been pinned down, physical theory is non-unique and thus the physical bases are indeterminate, since there are, *ex hypothesi*, no possible grounds for selecting one or the other of the theories for the purpose of developing those bases.

---

[38] There are those who would deny that the imagined scenario is very plausible (e.g. because no two theories are likely to be evidentially indistinguishable), and there are those who would deny that it is coherent (e.g. because the notion of all possible evidence for a theory is not coherent), and there are those who would insist that all such cases are instances of the first possibility (i.e. the two formulations are of one theory). I do not believe that any of these claims has sufficient merit to show that the case being considered is not inevitable, but I shall not pause here to consider the issues, which are rather involved and would take us too far afield. If such cases are not real possibilities then there is no objection to P2 that can be based upon them, which will serve my purposes here quite nicely. See Boyd (1984; 1985), Lauden and Leplin (1991), Sklar (1985), Van Fraasen (1980), Wilson (1980), and the references cited in the last footnote, for discussion of the issues here.

The two variants of such a case are that the different theories are compatible and that they are incompatible, where 'incompatibility' here means that the formulations of the theories are formally inconsistent and that there is no way of reconciling them despite such formal inconsistency.[39] Now, in either case the argument to the denial of P2 is a *non sequitur*. The fact that a characterization of physics does not uniquely determine a physical theory relative to all possible evidential considerations does not entail that there are no determinate bases in terms of which the theses of physicalism can be formulated. My explanation of why this is so varies somewhat in the two cases, and I shall consider them separately.

If the two theories are compatible, then we must ask why it should be required that there is exactly one way of 'carving up' the world in order to serve the purposes of physics. A physical theory of nature is simply anything that satisfies our characterization of what physics is, and there is nothing in this characterization that precludes there being more than one true physical theory concerning the actual world. Whether there is such a theory depends upon 'the world' and not upon a philosophical bias towards there being only one set of joints to carve up.[40] Thus in the case of compatible theories[41] we can plausibly view them as providing different ways of 'cutting up the pie', each of which counts as an empirically correct and true physical theory. If this is a genuine possibility, then physicalism only requires that each theory provides grounds for developing determinate physical bases that underwrite physicalist theses. In such a case there would be more than one true physical theory, each of which provided the basis for a physicalist picture of nature and knowledge. And there would be different 'total theories of nature', each based on one of the

---

[39] E.g. within special relativity, apparently incompatible claims about the velocity of a particle can be reconciled by relativizing the claims to different frames of reference. Incompatibility, as I conceive of it here, entails that no such reconciliation strategy is viable. See Goodman (1978: ch. 7) for discussion of various reconciliation strategies.

[40] Horgan (1984: 30) appears to hold the view that there is a unique set of 'perfectly natural' joints. Of course, if Horgan is right, then there is no problem here for P2, since at most one of the theories will be the right one, although we may never be able to tell which one it is. Thus the determinacy of the bases will not be threatened by a plurality of equally acceptable candidates.

[41] Here, the reconciliation occurs, not because we have different formulations of one theory, but because the different theories can all be true of the actual world, although they are concerned with different ontological structures within that world.

alternative physical theories here hypothesized and each structured in accordance with physicalist theses. In this case P2 is not threatened, since relative to each true physical theory the bases will be determinate, pending the outcome of other objections to be considered below. What if the theories are incompatible? Either there is a fact of the matter as to which (if either) is true[42] or there is not. If there is, then, even if it is forever beyond our ken to tell which, there is no problem here regarding the determinateness of the physical bases, since, relative to whichever is the true theory, the bases are determinate (pending the outcome of other objections). Within a thoroughgoing metaphysical realist framework, the evidential equivalence of incompatible theories does not spell disaster for physicalism, since the independent world will determine which (if any) of the evidentially equivalent theories is true. Such equivalence may well mean that we are incapable of identifying the actual physical bases and that physical theorizing, though empirically successful, could be false if it were to hit upon one of the false theories.[43] In a nutshell, radical epistemic indeterminacy does not entail metaphysical indeterminacy of the sort that would undermine P2. To establish the latter sort of indeterminacy, additional argumentation is required, at least to refute various realist positions that explain why the entailment does not go through.[44]

What if there is no fact of the matter as to which of the evidentially equivalent theories is *the true physics*? (i.e. what if no realist scenario is viable?). This is not the place to explore the range of non-realist frameworks within which to comprehend this possibility,[45] and I believe that, for present purposes, it is possible to reply

---

[42] 'True' here is understood in a realist way (e.g. correspondence to an independently existing world). Non-realist conceptions of 'truth' and 'existence' fall under the second disjunct.

[43] I take Post (1987: 156–7) to be committed to this view in his discussion of 'the true physical universe'.

[44] Some realists will insist that evidential considerations may well provide grounds for choosing among any collection of theories, and thus that the scenario envisioned here is at best an unlikely possibility. See Boyd (1984), Friedman (1983), Lauden and Leplin (1991), and Sklar (1985) for expressions of this heroic realist position.

[45] E.g. frameworks involving non-realist conceptions of truth (e.g. idealized warranted assertibility) and non-realist conceptions of existence (e.g. objects and attributes are constituted by our theories). See Dummett (1978), Goodman (1978), and Putnam (1983) for versions of such views.

to the objection to P2 without making such an excursion. In the case under consideration, the alleged indeterminateness of the bases is supposed to depend upon it being possible to have incompatible physical theories that are empirically and methodologically correct relative to all possible evidence, and are such that there is no fact of the matter as to which is true in a realist sense.[46] If such a situation is possible, then it appears to be a case of indeterminateness of physical theory and hence of the physical bases. However, I think that the *physicalist*, unless a committed *metaphysical realist*, need not feel threatened by this scenario.

My view is that, in such a situation, it is possible to develop our total theory of nature and our physicalism on the basis of either or both of these apparently competing theories. Within some non-realist philosophical settings, two incompatible theories that are each evidentially correct might be viewed as, for example, being 'true in different worlds'.[47] And, in such a philosophical setting, physicalism is a doctrine concerning matters 'internal' to each system of knowledge and internal to each world. Thus 'in each world' or 'relative to each theory' the bases will be determinate and in accordance with P2, again pending the outcome of other objections. Physicalism can be comfortably accommodated within a non-realist philosophical setting in which a plurality of different, incompatible systems are acknowledged as long as its scope is restricted to the systems taken one at a time.

The main thrust of this reply is that physicalism ought to be disentangled from metaphysical debates between realists and non-realists and that physicalism ought to be conceived of as a programme that is internal to a *world* or a *system of knowledge*.[48] The meaning of 'determinate bases' in P2 must be understood in a world-relative fashion, where for the realist there is exactly one such world, while for the non-realist there can be many. But the

---

[46] This might be held, perhaps, because realist conceptions of truth are incoherent, as Goodman and Putnam contend. The idea of an independent reality that breaks epistemic ties between incompatible theories is untenable, according to their view.

[47] As Goodman (1978) might suggest.

[48] In a realist setting, this means 'internal to the system of knowledge concerned with *the* actual, independent world' (or, if some reconciliation strategy applies, 'internal to each of the systems concerned with that world'). In a non-realist setting, this means 'internal to one (or more) of the true systems' or 'internal to one (or more) of the dependently existing worlds'.

physicalist need not take sides on the deeper metaphysical issues. Rather, the bases and the system erected upon them can be conceived of within the most viable, larger philosophical framework available, whether realist or non-realist in character. Thus, contrary to many physicalists, I see no *good* reason for supposing that the fate of physicalism is inextricably linked with that of metaphysical realism.[49] And there are some very good reasons for trying to disentangle the two: for example, the enormous cognitive benefits provided by physicalism are independent of the deeper metaphysical issues.

What the physicalist must oppose is instrumentalism and fictionalism with regard to physical theory. Physicalism requires that physical theory be taken to be both true and about a world of objects and their attributes. This, however, is quite compatible, for example, with taking such a world to be constituted by our theories and with viewing physical theory as true in such a world.[50] Thus standard arguments for scientific realism and against instrumentalism (cf. Friedman 1981; Boyd 1984, 1985) can be fully endorsed by the physicalist who, none the less, wishes to stay neutral on debates over *metaphysical* realism.

The moral here is that there are many different, and more or less independent, ways of developing answers to such questions as: 'What is the nature of truth?', 'How does knowledge relate to the reality it concerns?', and 'What is the nature of existence?'. The physicalist need not take a stand on any more of them than are required for the development of the physicalist position. With respect to such matters, my contentions are that a properly developed internal realism suffices for all physicalist purposes and that such realism is independent of so-called 'metaphysical realism'. I want it to be absolutely clear, however, that I am not here endorsing non-realist conceptions of truth and reality. Such conceptions are quite problematic, as are many realist conceptions. The line I want to take is that physicalism can be developed within both realist and non-realist philosophical environments. If there are sound reasons for rejecting one or the other or both, physicalism is not affected. It is only an independent commitment to one or the

---

[49] Physicalists who appear to make such commitments include Hellman and Thompson (1975), Lewis (1983), Post (1987). See Ch. 6 for further discussion.

[50] See Goodman (1978), Putnam (1983), and Goodman and Elgin (1988) for explanation of how this might be understood.

other of these metaphysical views that could lead one to think otherwise.

My conclusion up to this point is that the first class of objections to P2 (i.e. those premised upon the possibility of alternative formulations of physical theory) do not seriously threaten P2. The next type of objection seeks to deny P2 by denying that a true physical theory can, or ever will, be available for the development of physicalist bases. Thus P2 is false, not because there are too many eligible physical theories, but because there are none.

The first version of this argument arises from consideration of the following discussion of Hempel's:

> I would add that the physicalist claim that the language of physics can serve as a unitary language of science is inherently obscure: the language of *what* physics is meant? Surely not that of, say, eighteenth-century physics; for it contains terms like 'caloric fluid', whose use is governed by theoretical assumptions now thought false. Nor can the language of contemporary physics claim the role of unitary language, since it will no doubt undergo further changes too. The thesis of physicalism would seem to require a language in which a *true* theory of all physical phenomena can be formulated. But it is quite unclear what is to be understood here by a physical phenomenon, especially in the context of a doctrine that has taken a determinedly linguistic turn. (Hempel 1980: 194–5.)

Putting aside concerns about talk of 'unitary' languages and 'a determinedly linguistic turn', both of which I think are misconstruals of physicalism, there is here a formulation of the stock objection that, since past and present physical theories are probably false and since we do not know what future or ideal physics is like, then either physicalism is premised upon a false theory and thus is itself a false doctrine *or* physicalism is premised upon an unknown theory and is thus without content. Physicalism, it is alleged, is either false or meaningless because there is no known true physical theory relative to which a viable and determinate development of the physical bases can be made.

Hellman (1985) extends and deals with this objection in a way that has also appealed to others (for example, see Post 1987). First, he considers the reply that it is some 'mild extension of current physics' that could play the crucial role in identifying the physical bases and hence in giving content to physicalist claims (cf. Lewis 1983: 361; Putnam 1983: 212). But Hellman rejects this approach on the grounds that it is plausible that even mild extensions of physics are likely to be false, and he concludes that Hempel's dilemma recurs at this higher stage. I shall take Hellman as accepting the yet stronger idea that 'any current physics' is likely to be false, and that at no point in the actual unfolding of physics will it be the case that a *true* physical theory is available for the purposes of developing the physical bases. Again, either physicalism is conceived relative to some specific physical theory, in which case it is false because it depends upon a false physical theory, or it is conceived relative to no specific theory, in which case it is without content.

Now, as a concerned physicalist Hellman suggests the following alternative approach in reply: he proposes 'to accept this objection' and to allow that 'the best we can do is work with (perhaps, mild extensions of) our best available physical theories' (Hellman 1985: 609). And he concludes: 'Therefore, I am committed to the first horn of the dilemma. That seems to me as it should be' (Hellman 1985: 610). It follows that, in Hellman's view, physicalist theses undergo changes in content each time there is a change in physical theory and hence a change in our conception of the physical bases. And he believes that physicalist theses are chronically false, because they inevitably derive their content from a false physical theory. He softens this blow with a mention of two important functions that physicalist principles can serve despite the aforementioned concessions to critics: viz., they can serve in articulating conditions of adequacy that science aims to meet, and, like other literally false theories, they can be highly instructive and very nearly or exactly true in a restricted form (for instance, when considering mind–brain relations). Thus physicalist principles, though false, can still play important methodological roles and yield significant insights.

In considering the issues here, it will be useful to identify some of the crucial general assumptions made by both Hempel and Hellman. The following are plausible candidates:

(A1) Physicalist theses have determinate content only if there is a specifically identified physical theory relative to which the bases of the system can be conceived and developed.

(A2) Physical theory is constantly evolving and is likely to be false at each stage of its evolution.

(A3) Physicalist theses are true only if the physical theory relative to which the bases are defined is true.

(A4) Thus physicalist theses are true and non-vacuous only if there is a specifically identified physical theory that is true.

Although I will grant A2[51] and I believe that A3 is correct, I think that A1, upon which the objection crucially hinges, is false. Therefore A4 is false and the objection to physicalism can be seen to rest upon a false dilemma (i.e. physicalist theses are either false or vacuous). Further, I shall argue that, despite the likely chronic falsehood of physical theory and despite our inability to discern what ideal or future formulations of physical theory are or will be like, neither the determinacy of the physical bases nor the content and truth of physicalist theses are undermined. Let us look more closely at Hellman's argument to see why this is so.

First, I agree with his dismissal of the 'mild extension of current physics' approach, but for more fundamental reasons: it is fatally vague and it is unmotivated. It is vague because proponents (for example, Feigl 1969; Lewis 1983; Putnam 1983) offer very little by way of clarifying the relevant respects of similarity that *must* be preserved by the similar future theories. Consider the following expressions of the view offered by Feigl and Putnam respectively:

This thesis [i.e. unity of science] is, of course, not only problematic, but also inevitably vague in that such a theoretical physics may have to be very different from its current stage. All that can be said at the moment is that the style of explanation might be somewhat similar to that used in the present stage of the theories of relativity, quantum mechanics, and quantum electrodynamics. (Feigl 1969: 22.)

I shall assume that the fundamental magnitudes are basically the usual ones: if no restraint at all is placed on what counts as a possible 'fundamental magnitude' in future physics, then *reference* or *soul* or *Good* could even be 'fundamental magnitudes' in future physics! I shall not allow the naturalist the escape hatch of letting 'future physics' mean we-know-not-what. Physicalism is only intelligible if 'future physics' is supposed to resemble what we call 'physics'. (Putnam 1983: 212.)

Neither Feigl's allusion to 'styles of explanation' nor Putnam's table-thumping rhetoric and his talk of the 'usual' fundamental

---

[51] For the purposes of argument, I shall grant the assumption, although I think that the arguments supporting it are, at best, controversial. See Devitt (1984: 143 ff.) and Hooker (1987: 327 ff.) for discussion.

magnitudes gives us any very clear idea of how physical theory could evolve and still be relevantly similar to current theory. What measure of similarity is to be employed? What features (if any) of current theory must be preserved by future theories, and what features could be revised? Both Feigl and Putnam do seem to presuppose that physics can be identified, and Putnam (1970) has contributed significantly to clarifying the problem of how to isolate physics from other branches of inquiry. But the above statements regarding similar future physics do not provide us with any serious alternative to the view that simply appeals to future physics: if the latter is vague, then so is the former.

The approach is unmotivated for a number of reasons. First, as Hellman argues, the likely falsity of future theories that resemble current physical theory undercuts the value of the approach for physicalists who wish to show how physicalism can both have content and be true. Second, the approach seems to violate the empirical spirit that moves both physicists and physicalists. There is real tension, for the physicalist, between maintaining an open empirical spirit towards physics and placing restrictions upon what physics can introduce for theoretical purposes. Now, I do not wish to suggest that either Feigl or Putnam or Lewis does not share this spirit, but I do think that the strategy they suggest is at odds with it, in so far as it limits the possible directions of the future evolution of physical theory relative to a specific formulation of physicalist doctrine.

Physicalists should be able to acknowledge, while asserting their position, that *physics* has changed dramatically over the past three or more centuries and that it is quite possible that it will undergo equally dramatic changes in the future, *while not ceasing to be physics*. Since physicalists accord privilege to *physics*, it should be within their ken to acknowledge this most salient feature of physical theory, without compromising their position. Thus physicalists, like physicists, should be cautious about necessarily wedding their position to any specific features of current physical theory. And there is no good reason for thinking that the theses of physicalism change their content each time a physicist makes a change of theory, dramatic or not.[52] The burden of physicalists, in my opinion,

---

[52] See Salmon (1984: 241) for ideas along these lines with regard to what he calls 'the mechanical philosophy'.

is to show how physicalism can have content while retaining its identity across changes in physical theory, not to attempt to give content to physicalism by introducing restrictions upon how physical theory can evolve consistently with physicalist theses.[53] As Lewis put it at the opening of this chapter, 'materialists' side with physics but do not take sides within physics.

Finally, I think that the 'similar future physics' approach is unmotivated because it is responsive to a false dilemma: viz., that either physicalism is relativized to current theory and is thus false *or* it is relativized to an unidentified future physics and is thus without content. Either way there is no specifiable true theory, and thus either way there are no specifiable and viable physical bases to ground the system. This dilemma assumes principle A1, which I shall discuss shortly. That is, it assumes that physicalist principles have no content if the bases are not relativized to a currently specifiable physical theory. If this assumption is false, then the dilemma is defused and there is no motivation from this quarter for requiring that future physics resemble current physics. In light of the other suspicions recently raised regarding this move, abandoning the approach should not be unwelcome.

Now, returning to Hellman's position, the problem he sees is that of how to provide a substantial conception of the physical bases in order that physicalist theses will be non-vacuous while preserving as many of the important features of those theses as possible (for example, truth, utility). I agree that this is an important problem, but I would add that there is a high premium to be placed upon being able to offer non-vacuous expressions of physicalism that are both true and compatible with the idea that, while physical theory can change dramatically over time, both physics and physicalism retain their identity. Thus, although I certainly agree with Hellman's observations regarding the methodological role and the potential for insight afforded by physicalist

---

[53] Putnam might be conflating two separate issues in the passage cited. The first is whether the physical bases can be expanded to include such evidently non-physical entities as 'Good', 'soul', and 'reference' (i.e. the problem of downward incorporation). The second is how to give content to physicalist theses. These are indeed quite different issues. And it is neither clear that the *only* way to preclude 'Good' (etc.) from the bases is to adopt the mild extensions approach to the bases, nor evident that, given its problems, the mild extensions view gains any serious support by citing the fact that it is responsive to the problem of downward incorporation. See Ch. 6 for further discussion of this problem.

principles, I shall argue that his reply not only gives aid and comfort to a much too large number of opponents, but it is also misleading in important respects. Hempel's protestations to the contrary notwithstanding, it is, in my opinion, plausible to hold that physicalism is both a true doctrine and one that is not devoid of significant content. There is, therefore, no need to hold either that physicalism is a chronically false doctrine or that it constantly changes its content each time there is a change of physical theory. My reply to Hempel is that he has posed a false dilemma. In taking the dilemma seriously Hellman has shot himself and physicalism in the foot, so to speak.

Consider first whether the claim, A2, that any 'currently' accepted physical theory is likely to be false entails that P2 and hence physicalism are false. Clearly, it does not. If there is a true physical theory that correctly describes the reality that current physical theories purport to describe, then, regardless of whether we ever hit upon such a theory, it and the reality it describes exist and constitute the physical bases required by physicalist theses.[54] Thus P2 can be true even if we never actually develop a true physical theory. And if P2 is true in this way, then the fact that every theory actually accepted by physicists turns out to be false is simply irrelevant to the question of whether physicalism is true. The theses of physicalism are true or false relative to the real physical bases, not relative to bases as conceived by our best physical theory at some given time.[55]

Hempel, in the quote above, balks at this point because he thinks that we have no clear idea of what is meant by a physical phenomenon, and hence that talk of a true theory of physical phenomena is vacuous. On the contrary, I think that we do have a non-vacuous understanding of what a physical theory is. Indeed, the question of which theories are physical is a question that *all* physicalists must address, and as I have emphasized above, the answer does not turn upon the issue of the truth or falsity of some current theory. Rather, as discussed earlier, *the research pro-*

---

[54] I am formulating the reply from the point of view of a realist framework. If non-realist frameworks are viable then it would have to be put differently. I believe that neither Hempel nor Hellman nor Post will object to the realism here.

[55] See Post (1987: 154–8) for apparent concurrence on these points in his discussion of the distinction between the 'manifest physical universe' as reflected in current theory and 'the physical universe' *tout court*.

gramme of physics seeks to answer a number of questions subject
to certain constraints. *Physical theories* are those which are put
forward to answer those questions. If, in fact, there are correct
answers to such questions, then a true physical theory is one which
expresses those answers, and such a theory delineates determinate
physical bases, whether or not we have access to it. Thus, although I shall grant for now the assumption A2, the chronic falsehood of 'current physical theory' in no way undercuts the existence
of either a true physical theory or the physical bases generable in
terms of it. What would undercut such existence claims are arguments for the non-existence of a true physical theory, but neither
Hempel nor Hellman have said anything along these lines. Therefore their arguments do not undermine P2,[56] the first horn of the
dilemma is blunted, and Hellman need not accept the falsehood of
physicalist theses.

Further, consider the idea, A1, which implies that the theses of
physicalism have no content if they are not based on a specifically
identified physical theory. Now, there is surely a sense in which it
is true that we lack knowledge of the details of what the physicalist
theses imply if we do not know a true physical theory, and hence
do not know what objects, attributes, terms, and truths are in the
real physical bases. But does this mean that physicalism lacks
content? I think not. The issue turns on being able to distinguish
between physics and physical theory and to distinguish between
what physicalists assert and what physicists assert.

Physicalism asserts the privilege of *physics* in the sense that the
objects, attributes, and truths discovered by physicists provide bases
for the dependence, supervenience, and realization of all phenomena, the determination of all truth, and the ground for vertical
explanation. Thus physicalism asserts that the phenomena in the
domain of physics and the truths concerning that domain occupy
a place of privilege in nature and in systems of knowledge about
nature. *Physics*, on the other hand, is concerned with the study of
space-time and with discovering the fundamental constituents of
all occupants of space-time and the fundamental attributes that

---

[56] It would, I think, be uncharacteristic of Hellman to deny the distinction
between our best physical theory at a given time *and* a (or the) true physical
theory. Hellman seems to be threatened with the possibility that physicalism is
really true though he can never say so. There must be a better way out of Hempel's
dilemma than this!

account for all dynamics, change, and interaction involving such occupants. *Physical theory* characterizes space-time and these constituents and attributes, and it provides the relevant account. Physical theory, therefore, also provides detailed characterizations of the bases that can occupy the roles outlined by physicalism. And if physicalism is true, then there exists a true physical theory, whether we know it or not, which characterizes the actual physical bases that in fact occupy those roles.

Now, with this distinction in hand, the following additional points about physicalism and physics will help us to see why assumption A1 is misguided. First, until pushed to the wall, as Hellman apparently thinks he is, physicalists should not give up on the idea that they are putting forth true theses about nature and knowledge. This is a matter of sound strategy for system-builders. Second, as discussed above, physicalism makes no commitments regarding what is in the domain of physics. Such commitments are to be made by physicists in the context of empirical research, and physicalists should not take sides. To claim that physicalism has a content tied specifically to current physics is to fly in the fact of this wisdom.

Third, physicalism does make claims about physics and its actual domain, *whatever it in fact contains*. Since physics is abstractly characterizable, physicalist theses have non-vacuous content, albeit relatively abstract content: they abstract away from the details of both physical theory and the associated physical bases, as they should. But such abstraction means only that the physicalist is making claims about the unknown or imperfectly known, but none the less real, physical bases which contain the objects of inquiry of the research efforts in physics (i.e. they contain the objects that physicists study and with which they actually interact). Until physics succeeds, such bases can be picked out abstractly by such phrases as 'whatever is in the domain of physics'. But, *contra* Putnam, such characterizations of a real physical domain are not vacuous since we do have independent conceptions of what *physics* and physical theory are. Such bases can also be picked out *imperfectly* by current physical theory, which will provide the best current conception of those bases. But physicalism is not making claims about the ontological commitments of false theories. Thus the theses must be construed as being about the real physical bases, whether or not current theory characterizes them correctly.

To see the plausibility of these points, consider the following possible scenario:

t1—theory T1 is false
t2—theory T2 is true
t3—theory T3 is false

Here we have the dreaded situation that physicists, at time t2, correct a formerly accepted false theory, T1, and adopt a true physical theory, T2, only to go on and revise it in favour of T3, which is false. How should we track the fate of physicalism given these developments? I submit that the only sensible way of describing what has happened is that there are in fact real physicalist bases (given by T2) which, let us suppose, underwrite true physicalist theses. At each time t1–t3, the real bases exist and physicalism is true, although only at t2 was a correct conception of those bases available. What is not sensible, it seems to me, is to say that physicalism underwent a succession of changes of content and that at t1 and t3 it is false, while at t2 it is true. Hellman seems to me to be committed to this latter interpretation. But, for the reasons outlined above, he need not be so committed.

An analogy based upon a more concrete example also suggests the untenability of Hellman's stance. Consider the empirically defeasible speculation, made by a detective, that the murderer of a certain victim is a mean person. The actual murderer is the object of inquiry in the ongoing investigation, although he or she is completely unknown to the detective, outside his or her reprehensible effects. But the content of the speculation concerns the actual murderer, and not just any particular suspect who is currently being considered. If Jones is currently under suspicion, the original speculation does not become 'Jones is a mean person'; and if Smith comes under suspicion in place of Jones, the content does not change to 'Smith is a mean person'. And the original speculation is not falsified when it is discovered that Jones is both not the murderer and a saint. The speculation clearly has significant content and a truth value, not tied to particular hypotheses about who the murderer is. The content and fate of physicalism, as opposed to specific physical theories, should be understood in a comparable fashion.

Finally, it should be observed that other doctrines, notably determinism, do appear to have stable and non-vacuous content even if knowledge of the specific theories of nature involved is not to hand.[57] Thus one can significantly entertain the thesis that, given the laws of nature, the present state of the universe is fixed once the entire history of the universe up to this point is fixed or, alternatively, once any past state of the universe is fixed, even if one does not have any knowledge of the particular true theories that give detailed content to either 'the laws of nature' or 'the present state of the universe'. And if this is true in the case of determinism, then why not in the case of physicalism? Whereas determinism derives its content from such notions as *the state of the universe, the past, the present*, and *being fixed by*, physicalism derives its content from comparable abstract concepts (for example, the various concepts involved in characterizing *physics* and *the physical*, as well as such notions as *realization, determination*, and *explanation*). What we do not know are the specific details regarding what the physical determiners and realizers are. But it is misleading to say that this entails that physicalism is without content *unless* it is buttressed by a theory that ensures its falsity.

It follows both that physicalist theses do not change their content each time there is a change of physical theory and that physicalist theses can be true, even if currently accepted physical theory is false. According to the physicalist who accepts A2, this is in fact the case. Although our knowledge of the physical bases changes with physical theory, the actual bases themselves do not. And although current theory provides the best estimate of what is in the domain of physics and thus in the bases, it neither provides the content of physicalist theses nor determines their fate. Therefore, given that some physical theory must be true if there are to be determinate physical bases and true physicalist theses, that theory need not ever be accessible to humans in order for the theses to be non-vacuous. Thus, A1 is false and the second horn of Hempel's dilemma is also blunt.

Physicalism has content in so far as the research programme of physics is well defined. And this, I have claimed, depends upon there being a clear idea of what the questions and constraints are

[57] This is not to say that the fate of determinism is not significantly affected by which physical theories are true. But that is a matter of the truth of determinism, not of its meaning. See Earman (1986) for what is surely the best treatment to date of the issues concerning determinism.

that guide the programme rather than on what specific theories provide answers to such questions within those constraints. Thus I conclude that physicalists need not and should not heroically deny their own doctrine on the grounds that its content must inevitably be derived from a false theory. In so far as physicalists assert their doctrine, they are asserting the special role of *physics* and its domain in a system of knowledge without taking a specific stand on what the details of the true physical theory are. In presupposing the existence of determinate physical bases, they only take on the further commitment that there is in fact a true physical theory whose domain and truths constitute the bases. This is as it should be.

The second and more formidable challenge to P2 premised upon the falsity of physical theory is suggested by the views of Nancy Cartwright regarding fundamental physics. I cannot here do justice to those views, especially as developed in her important book on this issue (Cartwright 1983). Rather, I must make do with some general strategic comments. To begin, consider the following passage:

In physics it is usual to give alternative theoretical treatments of the same phenomenon. We construct different models for different purposes, with different equations to describe them. Which is the right model, which is the 'true' set of equations? The question is a mistake. One model brings out some aspects of the phenomenon; a different model brings out others. Some equations give a rougher estimate for a quantity of interest, but are easier to solve. No single model serves all purposes best. (Cartwright 1983: 11.)

The core position Cartwright appears to take, on the basis of considerations such as these, is that, in *theoretical physics*, the deployment of alternative, jointly inconsistent models in the study of physical phenomena is best understood as representing the instrumental character of those models and the literal falsity of the equations that describe them. There are no true theories that address the sorts of general questions I have included under the heading of 'physics'. There are only many, more or less useful and often inconsistent, models that each have partial application with respect to some restricted domain and some specific research purpose.

In her book, Cartwright presents detailed and compelling arguments designed to show that this description captures how physicists operate in their day-to-day activity. And she presents arguments, based on detailed study of applications, for the literal falsity of fundamental theoretical statements. More specifically, she argues that approximations required in various applications of a model typically involve a move from the simplifications and literal falsehood of fundamental theory to the complexity and truth of specific claims about specific systems. On this view, the real world exists, but not as conceived or modelled in fundamental physics. Claims that are true in the models should not be taken as literally true of the world for the understanding of which those models are introduced (cf. Hacking 1983: 145, 217 ff. for discussion of this view; Cartwright 1983: 46, 165, 202 for useful elaborations).

This is not the place to discuss in detail Cartwright's claims and arguments. Rather, I shall only attempt to clarify how they bear upon P2 and to sketch out a physicalist's line of response. Taken baldly, it would appear that Cartwright's view of fundamental physics entails the falsity of P2, since the development of determinate bases for ideology, doctrine, and ontology depends upon there being a true physical theory. I should add, however, that her position is complicated by the endorsement of what has been called 'entity realism', the view that the entities studied in physics exist but, given theoretical anti-realism, they are not accurately conceived within fundamental physical theory (cf. Cartwright 1983: 92, 201; Hacking 1983: 265 ff.). Thus there *might* be room here for a determinate basis for ontology even if there are no other determinate bases.

For present purposes, I shall read Cartwright's view as implying the denial of P2 on the grounds that fundamental physics, as currently conceived, consists of literally false, jointly inconsistent theories/models that are, at best, highly useful instruments for thinking about and partially organizing a certain field of phenomena. There simply is no true fundamental physical theory that could provide the foundation for developing determinate physical bases as required by physicalism: nature and physics simply are not right for that. Hence, in the absence of determinate bases, physicalism must fail in its efforts to provide a unified picture of nature and a unified system of knowledge. This conclusion fits well with Cartwright's general vision of how things are in the world and in physics: viz., disunity reigns. That is, she has repeatedly endorsed a view of physics and of nature according to which physical reality is an aggregate of disparate and disunified facts, not a highly unified structure as imagined by proponents of the physicalist unity of nature and knowledge (Cartwright 1983: 13, 165). How can the physicalist respond to such an assault?

The first line of reply might be to argue that the current state of physics, as described by Cartwright, is only a stage and that, as physics evolves, a unified, consistent, and true fundamental theory in physics will emerge. Thus nature and physics will eventually be seen as unified in the way the physicalist imagines and it will turn out that there is a true physical theory to underwrite P2. To bolster this response it might be pointed out that arguments such as the following presented by Hacking are flawed:

Every single year since 1840, physics alone has used successfully more (incompatible) models of phenomena in its day-to-day business, than it used in the preceding year. The ideal end of science is not unity but absolute plethora. (Hacking 1983: 218.)

In addition to the questionable relevance of the study of the day-to-day business of physics to a conclusion concerning the ideal end of science, the argument is an instance of a crude, theoretically unsupported enumerative induction. Without any analysis of the sorts of cases involved or of what is included under the heading of 'physics', the significance of the argument is unclear. A simple appeal to an increase in the number of models employed tells little about why the increase occurred. And if models are introduced for radically different practical and theoretical purposes, then increases in the number of incompatible models do not necessarily signal any significant conclusions about either truth or ideals. Thus a failure to identify the different types of activities of physicists, let alone who is included under the heading 'physicist', could well mean that the induction is performed over a radically heterogeneous sample, not all members of which are relevant to the issues upon which the argument is supposed to bear (i.e. what is the ideal end of science? *or*, will a unified physical theory emerge?). As a consequence of such lack of clarity, little weight should be attached to this argument as formulated.

However, I also think that the physicalist should not place too much emphasis on the prospects for the actual emergence of a unified, true physical theory. There are perhaps many reasons why physics will never get sorted out sufficiently to produce such a theory. For example, the condition to which Cartwright and Hacking allude could well be chronic because physics just might be too hard. Or, more to the point regarding the value of the evidence they cite, there may be many different currents in physics, some moving in the direction of unity and truth and others moving in the direction of 'plethora' and falsehood. Such currents plausibly reflect very different research interests, not all of which require truth and unification. Their co-presence does not imply that physics *must* exhibit fundamental theoretical disunity and falsehood. And those physicists interested in developing abstract theories that subsume and unify all of physics are not shown to be misguided because there are other physicists interested in more concrete studies of specific phenomena for which certain types of models are useful, though false or misleading in some respects.

Further, one of the many deep issues that Cartwright raises is whether the quest for unity at a fundamental level must always falsify as it abstracts away from the diversity of concrete phenomena within its domain. I think this is, at best, an open question, and I see no definitive reason in Cartwright's discussion for physicists to abandon the ideal of unity within physics and the quest for a true, unified, fundamental physical theory. As for the physicalist, there is no case here for abandoning P2. The existence of a true physical theory behind the veil of our current knowledge and practice, which it must be acknowledged are both imperfect and reflective of many diverse pragmatic interests, has not been refuted. It has, of course, not been confirmed either. But commitment to the existence of such a theory is not entirely a matter of blind faith on the part of either theoretical physicists or physicalists. There are directions of research that are more supportive of such a commitment than those Cartwright reviews (for example, developments in the effort to unify the four forces). However, the bottom line is that the *non-existence* of a true, unifying physical theory would undermine the physicalist's programme, and that, although the case for such non-existence has not been made, the issues discussed here are indeed open. The radical disunity and falsity of fundamental physics is one way physicalism can go wrong, and physicalists must take note of that fact.

The first few objections discussed above purport to undermine P2 by showing either that there are too many or that there are too few physical theories available for use in developing the physical bases. The next objection assumes that physics is identifiable and that a formulation of a true physical theory is available. Despite such assumptions, it purports to establish that there are, nevertheless, sources of indeterminateness that undermine P2. Thus, given a theory formulation, there is no determinate way of sorting the terms in the language of the theory into those which have referential import and those which do not. And thus, given the standard strategy for identifying the bases, there are no determinate bases for ideology, truth, or ontology. This argument might be supported by the idea that theory formulations embody mathematical or other apparatus which serve functions other than those of denoting objects or expressing attributes or classes. For example, the theory may contain machinery whose job it is to facilitate deductions. Or the mathematics employed in the theory may bring with it trappings that are not relevant to the actual subject matter of the theory (for instance, solutions to equations that have no physical interpretation). Or there may be terms (for instance, numerical constants) that are simply artifacts of the mathematical formulation, having no physical significance.

However, such possibilities do not by themselves suffice to undermine P2. An additional claim is required to the effect that it is not in general possible to sort out the two classes of terms. Although there may well be no general principle that distinguishes the two cases, there is no reason to suppose that, by various means suited to particular kinds of cases, physicists cannot distinguish those elements of a theoretical formulation that have physical significance from those that do not. Of course, friends of so-called 'causal theories of reference' will wish to reply that the relevant distinctions can be made by appeal to causal relations that provide interpretations of those terms having referential import. But whether or not such an approach is sound, there is no evident reason why the physicalist must concede indeterminacy of the bases on the grounds that there is an issue regarding which symbols in a formulation of physical theory are realistically interpretable and which terms are artifacts of the formalism.

Two variants of this objection also assume that a physical theory has been identified. The first purports to show that there is indeterminacy of the referential apparatus of the language of the theory due to 'indeterminacy of translation'. The second argument allows that the referential apparatus of the language of the theory can be identified, but purports to show that there is indeterminacy of the objects and attributes that are referred to or expressed by the terms of the language due to 'inscrutability of reference'. Both conclude that one or another of the physical bases lacks the determinateness required by P2. These arguments provide a special challenge to Quine, who is both a physicalist and the primary proponent of the two key doctrines involved (Quine 1960; 1969a). In Chapter 6, I shall discuss the significance of these arguments for physicalism, although space precludes taking them on directly in this book.

The final objection to P2 that I shall now consider arises as a result of reflection upon some of the features of quantum mechanics. Consider the following comments made by Putnam:

Worse still, from the metaphysician's point of view, the most successful and most accurate physical theory of all time, quantum mechanics, has no 'realistic interpretation' that is acceptable to physicists. It is understood as a description of the world as *experienced by observers*; it does not even pretend to the kind of 'absoluteness' the metaphysician aims at . . . (Putnam 1983: 227–8.)

the materialist claims that physics is an approximation to a sketch of the one true theory, the true and complete description of the furniture of the world. (Since he often leaves out quantum mechanics, this picture differs remarkably little from Democritus': it's all atoms swerving in the void.) (Putnam 1983: 210.)

Putting aside, for the moment, the inappropriate suggestion that physicalists have monopolistic pretensions (for example, 'the one true theory'), Putnam is alluding here to the idea that quantum mechanics has features that appear to be inimical to physicalism. There are several such features, and they should be considered separately, not as a package. Putnam is also suggesting that physicalists have tended to ignore the potential for embarrassment that quantum mechanics holds for them. In so far as this is true it is condemnable and physicalists must take on such difficulties, if they are real. My aim in what follows is to chart some of these problems and to indicate some lines of reply open to the physicalist. A full-scale technical treatment of the issues is beyond the scope of this book.

What, then, are the features of quantum mechanics (QM) that are supposed to undo physicalism? The following is a partial list of features that readily suggest themselves:

- The 'observer dependence' of phenomena.
- Value indefiniteness and 'the measurement problem'.
- Indeterminism and the irreducibly statistical character of physical laws.
- Non-separability of quantum systems and states.

I shall briefly discuss each of these with an eye to mitigating any negative force they may have *vis-à-vis* P2 and physicalism generally.

'The observer dependence of phenomena' is supposed to express the idea that quantum mechanics embodies an essentially anti-realist framework: there are no facts of the matter independent of actual observations, where observation is a mind-dependent activity. So conceived, the world of QM (i.e. the physical world) is the world as experienced by observers (i.e. a mind-dependent world), and this seems to cut deeply into physicalism with respect to its concerns about objectivity and its apparent commitments to an independent world. In so far as P2 is supposed to concern bases that consist of mind-independent truths and entities, then there are no such bases if QM is the physical theory that underwrites their development. On the other hand, if quantum mechanics is treated in a purely instrumental way, having no realist commitments beyond the patterns in our experience, and thus if it is not employed in the development of the physical bases, then again no such bases are forthcoming since there is no suitable alternative theory to do the job. This consequence is surely an embarrassment for the physicalist.

There is, however, nothing in this line of objection that is highly compelling. It is at least debatable whether QM is committed to the observer dependence of all facts. For example, systems described in terms of the wave function are described in a non-observer-dependent way. And the idea that specific magnitudes take on specific values only when observed by a human mind is a misleading description of the fact that it is disturbances of a system that collapse the wave function. Mentality is not an essential ingredient in a characterization of this most perplexing phenomenon (the so-called 'measurement problem'). If the crux of the objection is that, according to quantum mechanics, physical reality is a mind-dependent reality, then it has little force if one considers the wave function as a description of reality, and it has unknown force in light of the measurement problem.

On the other hand, if the objection rests on the claim that quantum mechanics must be construed instrumentalistically since no generally acceptable realistic interpretation has been provided, then the case for this conclusion has yet to be made. However, the objection does make clear that physicalists must recognize that the physical theories underwriting their strategy for developing the physical bases must be both true and realistically interpretable. As we shall presently see, such realism does not require a classical vision of the reality which interprets the theory (for example, value definiteness, an object ontology, and satisfaction of the Bell inequalities need not be features of the interpretation). And, as discussed earlier, such realism is not to be equated with metaphysical realism.

'Value indefiniteness' refers to the idea that physical magnitudes do not have specific values until a particular disturbance of the system featuring such magnitudes takes place. For example, the position and momentum of a particle are indefinite (i.e. there is no fact of the matter regarding their value) and can only be characterized in terms of probability distributions (about which there are facts of the matter) until a measurement disturbance occurs. The results of such measurements are, of course, subject to uncertainty relations which constrain them. Thus the idea of point-valued magnitudes which characterize the exact state of a system at each point in space-time is not true to currently accepted physical theory. In so far as P2 and physicalism are committed to this classical vision of reality, QM undermines them.[58]

Freely acknowledging that if physicalism is committed to a classical view of the distribution of objects and attributes in space-time then it is undermined by QM, the physicalist can and should

---

[58] Putnam (1983: 211) appears to conflate this feature of quantum mechanics with the feature just considered. Specifically, the observer-dependence of physical phenomena is not the same as value indefiniteness. One can allow that fundamental physical magnitudes do not have definite values until 'measured' while allowing that there are observer-independent facts about real systems (e.g. that a given system can be described by the wave function in a certain way). See Earman (1986: 215–19) for a discussion of realism and quantum mechanics that bears upon this point.

eschew any such commitment. Physicalism asserts that physics has a certain place in a system of knowledge, not that physics has a certain character. It is a matter strictly internal to the conduct of *physics* whether or not physical theory provides a classical view of objects and their properties, and the fate of physicalism does not hang directly on such issues. Rather, the fate of physicalism hangs on whether, given the view of the world as presented by physics, the theses of physicalism are satisfied. If there is an objection here at all it is not that QM violates a classical view of things to which physicalism is specifically committed, but rather that, given a non-classical view, physicalist theses are false. This is a very different matter, which does not bear upon the determinateness of the physical bases, and it is a matter about which nothing can be inferred from the current objection.[59]

Now, let us consider specifically whether P2 is falsified by value indefiniteness (i.e. does value indefiniteness as understood in quantum mechanics undermine the determinacy of the bases as required by physicalism?). I think it is clear that this is not the case. First, as indicated above, the wave function gives very definite descriptions of systems in terms of probability distributions; such descriptions give a complete characterization of the system. Thus for any such description or associated probability distribution,[60] there is a fact of the matter regarding whether it is in the physical bases. What more could one ask for in characterizing such bases? Second, upon 'collapse of the wave function', systems take on more specific values for the physical magnitudes which characterize them, and such specific determinations of value are also available for inclusion within the physical bases. Again, there is no reason to suppose that determinateness of the bases is violated. As a consequence, physicalist theses will express claims about the realization, supervenience, and explanation of objects and attributes relative to bases that characterize specific physical systems in terms of quantum mechanical descriptions and attributes of either sort (i.e. probability distributions or specific determinations of a

---

[59] See Earman (1986: 226) for discussion of the claim that determinism does not presuppose value definiteness. In my view, physicalism is in exactly the same boat on this issue.

[60] The issue of the realistic intepretation of quantum mechanics is acute here. However, that is a different problem which must be dealt with, as conceded above. Whatever such distributions are or stand for (e.g. 'propensities') is what goes into the basis for ontology.

magnitude), whichever is appropriate. Neither the determinateness of the bases nor the fate of physicalism are directly impugned by such considerations.

Similar replies can be made to objections based on 'indeterminism' and 'non-separability'. As has often been pointed out by physicalists (for example, Horgan 1982: n. 17; Post 1987: 26), physicalism is *not* committed to determinism. The fact that QM is, in certain (but not all) respects, an indeterministic theory, though interesting and important, is not an immediate problem for physicalism. And the non-separability of physical systems and states, their peculiar entanglements with each other, poses no problems for the physicalist who is unfettered by classical visions. There is, for example, no reason to suppose that physicalism implies the Bell inequalities. Again, whether such facts of physical reality undermine the truth of physicalist theses is another matter. But indeterminism and non-separability do not undermine the determinateness of the physical bases or any other constraint upon them. Quantum mechanics, though constituting a dramatic revision of our view of the world, at most reminds the physicalist about what are *not* appropriate commitments of the programme.

3.4. PRESUPPOSITION 3: THE PRIVILEGED STATUS OF PHYSICS

The final presupposition that I shall consider is the claim, P3, that physics occupies a special place in science which justifies developing the physical bases in terms of its ideology, ontology, and doctrine. Specifically, P3 requires that the objects, attributes, and truths identified in physics are relevant to the metaphysical and epistemological purposes of the physicalist programme. The point of identifying P3 as a presupposition is to underscore the fact that the strategy for identifying the physical bases by looking to physics is a motivated one, not subject to serious objection. As with P1 and P2, P3 is not a logical presupposition of the physicalist doctrine. However, if the theses are to have significant content of the right sort, then P3 had better be true.

There are two kinds of issue that arise in this context. The first concerns the propriety of appealing to physics at all in developing the bases. Is there something fundamentally wrong with the approach? Examples of such issues are: does this approach trivialize the theses?, does it commit physicalists to an untenable form of conventionalism or essentialism? The second sort of issue concerns the motivations behind the programme as a whole: the primary question is why theses concerning bases developed from physics should be of more interest than theses concerning a different sort of basis, or theses concerning no unifying basis at all.

The first objection to P3 is that appealing to physics in characterizing the bases *trivializes* the theses. Since physicalists believe that they are endorsing a significant and defeasible doctrine that accords a special status to physics, such a consequence would be undesirable and would put the strategy in a bad light. However, this objection can, I think, be readily dismissed since the conditions under which physicalism would be trivialized by the appeal to physics do not obtain. For physicalism to be made trivially true by our strategy, it would have to be the case that it is somehow built into the nature of physics that the successful development of a physical theory entails the truth of physicalist theses *and* that the successful development of physical theory is beyond question. Neither of these conditions hold, however.

First, it is evident that, given the characterization of physics presented earlier in this chapter, there is no guarantee that physicalist theses concerning ontology, objectivity, and explanation are true. The successful development of a theory that accounts for all composition, dynamics, interaction, and influence involving occupants of space-time in terms of a set of fundamental objects and attributes and that is subject to the constraint of generality does not logically preclude certain forms of ontological dualism, does not guarantee the supervenience of all facts and truths upon the physical facts and truths, and does not ensure that there are physically-grounded vertical explanations of the realization of all phenomena.[61] Thus, unlike characterizations of physics framed explicitly in terms of physicalist theses (for instance, a supervenience thesis), the sort of understanding of physics I have endorsed above does not guarantee the truth of any physicalist claims.

Further, it is evident that physicalism is not simply the assertion of physics, given the very significant differences of content between

[61] Arguments for these contentions are easily generated, and I shall forgo doing so here. See the discussion in section 3.1 concerning the relation between supervenience and physics, and see Ch. 5 concerning the modal status of the theses of physicalism.

physical theory and physicalist theses. Unlike approaches that equate physics with all of natural science, thereby guaranteeing the incorporation of many problematic phenomena within the bases of a physicalist system, the approach taken above identifies physics as a proper part of the scientific study of nature. Thus there is no easy avoidance of the many difficult obstacles to the successful working out of the physicalist programme (for example, mind, meaning, and values). Finally, the suggestion by various critics that physical theory is open to unconstrained revision, thereby ensuring the truth of physicalist theses, is mistaken for the reasons discussed earlier (cf. section 3.2). The sort of comprehensiveness required of a true physical theory is not of a sort sufficient to guarantee the incorporation within physics of whatever it takes to make physicalism true. As a result, the success of physicalism is far from trivially guaranteed, regardless of whether physics succeeds in its purpose.[62]

On the other hand, the success of physics itself is, likewise, far from a sure thing. Considerations like those of Cartwright and Hacking discussed above and like Putnam's infamous meta-induction to the falsity of any current physical theory (Putnam 1978: 25), although not conclusive, do underscore the fact that the existence of a true physical theory of the sort required by physicalism is a contingent matter about which there is uncertainty. Physicalists, of course, are banking on there being such a theory, but they should acknowledge that they could be wrong about that. In no way is the truth of physicalism a trivial matter, given our approach for identifying the bases. It does not follow from the success of physics, and it is empirically dependent upon such success, which is far from guaranteed.

The second objection to P3 proceeds as follows. The strategy for developing the bases depends critically upon accepting a characterization of what physics is, and, given this characterization, many of the standard objections to physicalism have been swept aside. However, the strategy is defective since it depends upon what is either an arbitrary conventional stipulation of what physics

---

[62] I differ with Post (1987: 189), who sides with Quine and Lewis in holding that it is part of the business of physics to ensure the truth of (at least) physicalist determination principles. As argued above, I think one should distinguish 'the business of physics' from the contingent demands imposed upon physics by physicalist theses.

is, or, alternatively, an essentialist thesis concerning the nature of physics. Neither of these alternatives is acceptable.

To begin with, I allow that physicalism does rest upon the assumption of a principled division between physics and other branches of knowledge (i.e. P1), but I deny that the ground of that division depends upon an *arbitrary* conventional stipulation or an unacceptable essentialist thesis concerning the nature of physics. Rather, physical inquiry addresses the questions identified above, subject to the identified constraints. Were physicists to cease addressing questions concerned with space-time, dynamics, interaction, composition, and influence, they would cease engaging in physical inquiry so conceived. It is *this* research programme that physicalists think of as providing a unifying basis for their programme. It is a contingent matter whether anyone in fact engages in this enterprise or whether those who are currently called 'physicists' are so engaged. And physicalist theses do not concern physics conceived of in some dramatically different way. It is not a doctrine concerned with whatever happens to be called 'physics'. Nor do physicalists intend either to legislate regarding the use of a term, or to tell those who are engaged in certain research projects not falling squarely under the above conception of physics, that something is wrong with what they are doing.

My main contention is only that it is part and parcel of the physicalist programme that the bases be derivative from theories developed in the context of a certain sort of research into nature:[63] theories of *that* sort provide bases for unification, or so the physicalist claims. It is a more contingent claim of physicalists that, in fact, current science includes research of this sort, although I do not think there is much serious doubt that physics, so conceived, is a live programme, even if it is not the only research conducted under the heading 'physics'. Were either of these claims false, physicalism would, perhaps, be in trouble, but not because physicalists have made a philosophical mistake of the sort alleged by the objection. Nowhere in this portrayal of how the physicalist

---

[63] Note that, as I have emphasized, the theories developed within the programme of physics are free of formal or substantive constraints beyond those mentioned: philosophers should not legislate to physicists regarding what is a legitimate construct for use in physical theorizing. Clearly, deep revisions of physical theory are always possible while staying within the context of physics research as conceived of in the text. Thus what falls within the physical domain is a contingent matter to be resolved via the empirical interaction of researchers with the natural order.

views physics are there any illicit conventionalist or essentialist assumptions that I can see.

It might be replied that it is open to the physicalist freely to revise the presupposed conception of physics in any way required to save the doctrine from counter-example. And thus, even if the view is innocent of philosophical error, it is cleared in a way that reopens the charge of trivialization. Indeed, what does constrain the conception of physics operative within the physicalist programme? My reaction to this new charge of triviality is the same as that given earlier (section 3.2) to a similar objection based on *ad hoc* revisability of theories in physics: *ad hoc* revisions are ruled out on general methodological principles shared by all participants in the debate, while non-*ad hoc* revisions must always be considered seriously. If the conception of physics suggested earlier is defective and does not capture the kind of programme the physicalist believes will deliver a unifying basis, then it should be revised; but if revision of the conceptions is simply designed to save the programme from embarrassment, then it should be resisted. To ask more of physicalism than is asked of other responsible research programmes is unreasonable.

I conclude that the first two objections do not pose any serious threat to P3, since the strategy based on the identification of physics is neither a trivializing one nor one which has unacceptable philosophical commitments. The next few objections, on the other hand, concern the grounds for adopting a strategy that appeals to physics at all. The basic idea behind these objections is to challenge the appeal to physics, as opposed to proceeding in some other way, for the purpose of developing the *physical* bases. Earlier, I contended that there are no other viable ways of proceeding (for example, *physical₁*, *physical₂*) and indeed that our only grip on what 'the physical' consists in is to be understood in terms of the research programme of physics. The privilege accorded to physics here is based on the nature of research pursued within its confines (for instance, research concerned with identifying fundamental objects and attributes and completely general laws). There are no other approaches that effectively contribute to capturing what the physicalist values and believes.

The objector might press on by claiming that it is so much the worse for physicalism if appealing to physics is the only available strategy for identifying the physical bases. Physics, and hence the bases conceived in terms of it, simply do not warrant the privilege accorded to them by the physicalist. Some of the reasons for this challenge might include such claims as that physical theories are false, disunified, and non-objective by physicalist standards. These charges are discussed and disarmed elsewhere in this book. However, two further reasons are that no special importance attaches to a unification programme grounded in physics and that physics does not have a unique claim to providing a unifying basis. I shall consider these two lines of objection in turn.

The first line challenges the point of pursuing the physicalist programme at all. It is sometimes remarked, 'So what if psychology does not reduce to physics; physics does not reduce to psychology either.' Making liberal allowances, here, for the meaning of the term 'reduce', the point is supposed to be that the failure of general programmes that seek to relate the terms, truths, and phenomena studied in different disciplines does not constitute any serious loss. Research programmes are relatively autonomous and, although it may be interesting if there are connections among them, there is no reason to expect, require, or search for such connections.

This radical autonomy of disciplines is, I think, seriously misguided, for several reasons. The discussion in Chapter 1 revealed that there are a number of important gains that result from the establishment of certain relations between physics and other branches of knowledge. These gains are not simply gratuitous, but are central to the purposes for which systems of knowledge are pursued at all (for instance, increased understanding). Radical autonomy represents a degenerate case of system-building, and not one that can be assumed without argument. In particular, there is no general presumption on the part of practitioners within various disciplines that what they are doing is totally unrelated to what happens in other areas. Indeed, the contrary is more often presumed. And it is arguable that physicalism is a working assumption of theorists in many areas of inquiry. Further, it is quite inconceivable what the radical autonomy of the sciences would be like. An awful lot of scientific theorizing and activity in psychology, for example, would be unintelligible if it were not assumed that the mind was significantly related to physical processes in the brain (for example, the assessment of evidence concerning mental processes, the study of relations between mental processes and behaviour and of relations between mental processes and brain processes). Similar points apply to all other sciences. Glib comments like the one featured in the current objection should not be taken seriously.

The second line of objection does not challenge the value of pursuing the sort of system-building in which the physicalist is interested, but claims instead that there is no reason to prefer physicalist system-building to, say, phenomenalist system-building or system-building grounded in some other brand of psychology. Both Goodman (1978) and Putnam (1979) have claimed that the programme for 'reducing' psychology to physics is no more or less of a success than the programme for reducing physics to psychology: each, in their view, is a potentially valuable programme and each has points of strength and weakness comparable to those of the other. Allowing that physics is appropriate for the purposes of physicalist system-building, P3 is none the less challenged if it entails that physics is *unique* in being appropriate for grounding a reductive system.

There are, however, several reasons why this objection lacks force. First, it rests on the dubious assumption that the phenomenalist programme is succeeding or failing to the same extent as the physicalist programme. A proper, detailed argument against this claim is beyond the scope of the present discussion. But it would be illuminating for physicalists to develop such an argument in reply to Putnam's and Goodman's challenge.[64] I believe that such an argument can be developed, and that the objection loses force because it rests upon an unsupported and probably false assumption. However, it should be recognized that physicalists have not yet conclusively presented evidence that establishes the superiority of their programme over such competitors. This is an unpaid debt.

Second, since the ideas and values behind the two programmes are usually conceived of as quite different, their success or failure would likewise have quite different significance. For example, the

[64] In addressing the issues here it is important to consider detailed characterizations of the various types of programme since, as we saw in Ch. 2, there are many versions of physicalism and they are not equivalent with respect to content or merit. Comparable variation exists with respect to psychologically-based programmes, and blanket comments regarding physicalism and phenomenalism are simply not enough.

failure of phenomenalism would signify the demise of an epistemological programme (for instance, a certain sort of conceptual or evidential foundationalism), whereas the failure of physicalism would signify the failure to attain certain metaphysical and epistemological goals (for instance, monism of a sort, unity of knowledge with respect to supervenience and explanation). These differences mean that it is not at all appropriate to compare the two programmes. And, with respect to the attack on P3, it is therefore not appropriate to view physics and psychology as providing competing reduction bases. In so far as physicalism and phenomenalism are two programmes guided by very different ideas and values, it is only fitting that they have different bases, and thus the issue of uniqueness does not even arise in this context.

Further, there is no reason for the physicalist to oppose the appropriateness of pursuing these other sorts of programme. Conceptual, epistemological, and metaphysical unification programmes can coexist without necessarily conflicting with each other. However, the physicalist might add that, when it comes to building a system concerned with the sorts of metaphysical and epistemological purposes with which physicalism is concerned, then indeed physics is unique in being appropriate for grounding such a system.[65] It alone is relevant to such purposes, since it alone can provide the right sorts or entities for realizing, determining, and explaining all that there is in nature.

Both Goodman and Putnam would vigorously dissent from this last claim, although they would surely agree with the earlier claim about a plurality of different sorts of system-building. However, the idea that only systems grounded in physics can appropriately serve the metaphysical and epistemological purposes at which physicalism aims is, according to them, an article of faith on the part of physicalists.[66] It is, they would add, an article of faith resting upon an unacceptable form of so-called 'metaphysical realism', since only a metaphysical realist would deny the possibility of more than one successful unification programme of the sort pursued by physicalists.

Many physicalists, of course, would reply that there are simply

---

[65] See Hellman (1983), Kim (1984: 166), and Post (1987: 118, 165) for exactly such a claim.

[66] See Putnam (1981: 112). And see Ch. 6 below for discussion of some of the 'monopolistic' overtones of certain formulations of physicalism.

no serious competitors to such realist views and that the Goodman–Putnam non-realist alternative is false or incoherent. In so far as physicalism combined with realism implies that there can be only one sort of basis (i.e. a basis derived from physics) that grounds a metaphysical and epistemological system of the relevant sort, the issue here is whether there are any non-realist views that provide an alternative framework within which physicalism can be comfortably embedded and within which a plurality of other systems serving the same purposes, but with different sorts of bases, are possible. If so, then any claim to uniqueness of the physical bases with respect to metaphysical and epistemological unification rests, not upon the central ideas and values of physicalism alone, but upon their conjunction with metaphysical realism.

As I have indicated above (section 3.3) in my discussion of evidentially equivalent theories, I see no reason for the physicalist to build metaphysical realism into the programme: all its ideas and values can fit within either a metaphysical or an internal realist framework. If this is the case, then it is wise for the physicalist to retreat to neutral ground awaiting the outcome (if any is forthcoming) of the debate over these rival philosophical frameworks. Physicalism, in my view, carries no *presumption* of uniqueness. It is aimed at promoting certain ideas and values, but not necessarily to the exclusion of other programmes with similar goals. Of course, should one or another philosophical framework emerge victorious,[67] then it would be appropriate for physicalists to side with the winner. Most contemporary physicalists (for example, Boyd 1984; Field 1982; Hellman 1983; Lewis 1983; and Post 1987) tend also to be realists, and my point here is that their realism should be distinguished from their physicalism. To hold that P3, and physicalism generally, involves a claim to uniqueness on behalf of physics with respect to providing a basis for the relevant sort of unification is to incorporate realism into the position, and I claim that such realism is additional to the main thrust of the physicalism.

At this point, I tentatively conclude, pending further discussion in Chapter 6, that P3 is not threatened by the possibility of equally

---

[67] Realists and non-realists both tend to describe the views of their opponents as being 'incoherent'. In my opinion, neither side has presented non-question-begging arguments against the other. Unfortunately this topic cannot be discussed in the current project.

successful systems directed towards the same goals as is physicalism but grounded upon different bases. If realism is the framework of choice, then at most one of the competing systems will be correct and physicalists can take a stand in favour of their own. If non-realism is preferable, then a tolerant pluralism of metaphysical systems can be justifiably embraced by the physicalist, without compromising the central ideas and values of the programme.[68] Therefore, in so far as physics can deliver objects, attributes, and truths that serve the purposes of the physicalist programme, it does indeed enjoy a privileged status which justifies our strategy for developing the bases. But this privilege does not *imply* the uniqueness of physics in this regard.

[68] But a tolerant pluralism does not mean that all programmes for system-building will inevitably succeed. Although the physicalist can countenance the possibility of alternative systems, this does not mean that any such alternatives will be successful. Should physicalism prove to be the only successful unification programme of the relevant sort, then it would be appropriate to claim uniqueness without affirming metaphysical realism. In any event, such uniqueness is not implied by P3.

4

The Theses of Physicalism

All right-thinking physicists believe that macro-thermo-dynamical quantities, states and events are parasitic on the microscopic. But it turns out to be hard to characterize the parasitism in terms of identities (e.g. 'temperature is mean kinetic energy' is a glib over-simplification of a very complicated relation which may not in the end turn out to be a relation of identity).

John Earman (1986: 249–50)

A reductive explanation not only allows one to deduce the laws of the secondary theory, but also to tell us what it is about the nature of the objects of the secondary theory that brings it about that they have the properties they have and what it is about their nature that brings it about that those properties are related to each other in the way the laws of the secondary theory assert them to be related.

Berent Enc (1976: 303)

The goal of this chapter is to outline a set of theses which gives adequate expression to the core physicalist ideas and values discussed in Chapter 1. The theses must suffice for expressing the physicalist's vision of one of the important ways that things hang together (viz., physically-based unity of the universe and of knowledge that concerns it). Thus the ideas of ontological dependence, supervenience, and realization must be captured in a way that reveals, in general, how all phenomena are embedded in the physical fabric of the universe, and that circumscribes the scope and limits of influence and interaction. Principles expressing the physicalist's belief that all objective facts and truths are determined by physical facts and truths are required. And the idea that all that happens is explainable and integratable within a system of physically-based vertical explanations must be framed. In these ways, the theses should concern metaphysical and epistemological structures that will yield understanding and that can play an important role in the pursuit of research and in the development of a culture.

Two further sorts of consideration are important for pursuing this objective. First, we must heed the morals of Chapter 2 regarding the shortcomings of other formulations. The issue of how properly to express concerns about realization and non-emergence, on the one hand, and vertical explanation, on the other, must be confronted and resolved, given the demise of classical reductionism and given the essential weakness of the various supervenience relations. Second, as will become evident, there are a number of objections to the theses that must be dealt with if the programme of physicalism based upon those theses is to get off the ground. The objective is to frame theses that jointly provide an adequate expression of physicalist ideas and values, that do not fail for reasons comparable to those that undermine other formulations of physicalism, and that are not vulnerable to further objection.

4.1. ONTOLOGICAL DEPENDENCE, SUPERVENIENCE, AND REALIZATION

A formulation of the physicalist ontological position must give expression to the idea that the physical ontology is *basic* with respect to all other phenomena. I have suggested that being basic in this context has three components: ontological dependence, supervenience, and realization. My arguments in Chapter 2 were aimed at establishing that not only do the various versions of classical reductionism fail to express these ideas, but also a host of supervenience theses fail as well. Reliance upon purely linguistic formulations, purely supervenience formulations, and theses involving nomological sufficiency or equivalence is not enough for the physicalist's ontological purposes. More powerful alternatives must, therefore, be introduced.

Before proceeding with the development of such theses some preliminary comments are in order. As noted earlier, it is necessary to treat all the various ontological categories in a full development of the physicalist position. However, I shall not do this in the present project for reasons of space. Rather, the focus will be on objects and attributes. There are a number of points worth noting with respect to each of these categories.

To begin with, there are serious problems in how to treat *objects* within a physicalist system: the viability of the concept is brought under deep suspicion by developments in contemporary physics. However, I shall bypass all issues concerned with, for example, criteria of individuation, location, and boundaries of objects. Whether and how such problems, at both the level of microphysics and more macroscopic levels, are to be resolved must be left open, but they must be acknowledged as a debt of the physicalist programme. Further, I shall subsume under the concept of an object all manner of concrete individuals, whether they are natural kinds, well-defined artificial kinds, or rather unnatural objects such as the fusion of some arbitrary class of objects. Such distinctions are important for the working out of the physicalist programme, as different categories may well be treated differently, but for present purposes a more abstract approach suffices. As I shall discuss in more detail shortly, the physicalist ontological position can be adequately expressed in terms of regions of space-time and their contents, where the latter are construed in terms of the instantiations of objects[1] and attributes.

I also make a number of assumptions about attributes. My focus will be on qualitative, as opposed to non-qualitative, attributes: i.e. I shall focus on attributes that do not involve any essential reference to particular individuals. And my use of the term 'attribute' should be clearly understood to include both properties and relations. Within the former category, I include both intrinsic and extrinsic properties. As argued in Chapter 2, I see no insurmountable difficulties with the idea that an individual may possess a certain property in virtue of how things are elsewhere in the world (for example, the property of being the tallest person).

Along a different line, I take it as obvious that there are non-physical, as well as physical, attributes. Therefore I reject as false the view that every attribute is a physical attribute. This rejection is, I believe, the received wisdom resulting from the struggle with these issues over the past few decades in the philosophy of mind and elsewhere. Rather than leading to the demise of the physicalist programme, however, such wisdom should lead us to appreciate more deeply the ways in which the physical is basic. Further, I

---

[1] As an expository convenience, I shall speak of the 'instantiation' of objects in regions of space-time, although this usage is non-standard.

reject the view that, while denying the identity of all attributes with physical attributes, asserts token identity of attributes. This view founders either because it leads to the view that there are property particulars or that somehow it is possible for properties to be distinct but their instantiations to be identical: neither option is viable, in my opinion.[2]

Finally, recall that the development of the physical basis for ontology in Chapter 3 led to characterizations of (*a*) classes of physical objects and attributes, and (*b*) the class of possible physical states of the world each consisting of space-time regions and their physical contents (i.e. distributions of physical objects and attributes).[3] And similarly to the case of the physical domain, other ontological domains can be understood as being constituted by classes of objects and attributes, the members of which can be distributed over regions of space-time. Within each domain, classes of basic objects and attributes are identified, and classes of complex and higher-order objects and attributes are generated from the basic classes via appropriate generative operations.[4]

The distinction between the *actual* distribution over regions of space-time of the objects and attributes in an ontological domain and the range of *possible alternative* distributions in that domain can be understood in a way comparable to how that distinction was made for the physical domain. However, the characterization of this range of alternatives is more problematic than for the physical case since, in addition to constraints imposed by laws appropriate to that domain, there are other constraints as well. For example, 'laws' concerning non-physical attributes need not be without exceptions. As a consequence, in order to clarify the range of possible alternative distributions of objects and attributes in a given non-physical domain, it is necessary to clarify the constraints upon such exceptions. I shall return to this issue below.

Given this framework of assumptions concerning ontological matters, the formulation of the first ontological thesis of physicalism is as follows:

[2] Although I deny the above-mentioned identity theses for attributes, I allow that type or token identity theses for other ontological categories may be plausible.

[3] A region R is identical to a region S just in case R and S contain the same space-time points. Since such regions need not be connected, but can, instead, be scattered over space-time, objects too can be scattered over space-time.

[4] Identification of the objects, attributes, and operations is a matter for research within each domain.

(To) All objects and attributes that are (or can be) instantiated must be instantiated in regions of space-time.

This is the traditional physicalist demand that if any object or attribute, basic or complex, physical or non-physical, is instantiated in this world, then it must be instantiated in the 'space-time causal nexus'. Alternatively put, everything that is real in nature occurs in space-time. In addition, To constrains the *possibility* of an object's or an attribute's being instantiated: all that is possible in this world includes only those phenomena that can occur in space-time.[5]

The thesis presupposes (and I shall discuss this at length in Chapters 5 and 6) that the natural order can be distinguished from the realm of *abstracta*. And it relies upon the assumption that space-time physics is a relatively enduring feature of our conception of nature. Should that conception change dramatically, then some other way of pinning down the natural order will have to be introduced. Finally, note that To does *not* rule out ghosts and their ilk, if they can be located in regions of space-time. Although To provides an important ingredient of the physicalist's ontological view, it is clearly not the whole story.

The second physicalist ontological thesis is roughly that, associated with the actual (or possible) instantiation of any entity (i.e. object or attribute) in some region of space-time, there is a class of physical entities that are also actually (or possibly) instantiated. As stated, however, this is too rough. Focusing on attributes first, the mere co-occurrence of some class of physical attributes with a given non-physical attribute does not provide much bite to the idea that the physical ontology is basic. I have not even stipulated (nor will I) that the physical attributes and the given attribute are instantiated in the same region.[6] So, on the assumption that there is always some physical attribute or other which is instantiated, we have the consequence that, unless further conditions are added,

---

[5] Post has suggested (in correspondence) that there are apparent exceptions to this thesis that arise in the context of fundamental physical theory: e.g. (1) singularities that are not in space-time but which are boundaries of space-time ('real enough but of less than four dimensions'), and (2) more fundamental (i.e. less structured) spaces out of which space-time structures originate—such spaces are neither space-times nor in space-time. If either of (1) or (2) is real, then minor adjustments in To are required. However, neither seriously compromises the spirit of the thesis, in my opinion.

[6] In the case of causal properties, such a restriction may be required. But the physicalist need not be bound to this demand with respect to all attributes.

any attribute which is instantiated in a region of space-time will trivially satisfy this second thesis in virtue of satisfying the first thesis. What more, then, is required of the association between physical and non-physical attributes to establish the former as basic?

It is here that the physicalist must call directly for the *realization* of a given non-physical attribute, N, by the members of some class, P, of physical attributes with which it is associated. Nothing less will do, as I argued in Chapter 2. Thus the second ontological thesis is as follows:

(T1a) For each non-physical attribute, N, and for each region of space-time, R, if N is actually (or possibly) instantiated in R, then there exists a minimal class of physically-based attributes, P, such that the instantiation of the members of P does (or would) provide a realization of N on that occasion.

A *realization* of an attribute, N, on a particular occasion by a class of physical attributes, P, is a configuration of physical attributes that *constitutes* N on that occasion. How the members of P accomplish this will vary depending upon the nature of N. For example, the realization of the property of *transparency* in terms of a given physical structure that permits light rays to pass through it will differ significantly from the realization of the *content* of a mental state in terms of (say) causal relations between a symbol and the objects or attributes to which the symbol refers. By focusing on particular instantiations of the attributes involved, T1a allows that consideration of time and place can be relevant to the realization of a non-physical attribute (for instance, when an object's having a certain history is relevant to the constitution of one of its attributes). Clearly, it allows for there being different classes of physical attributes each of which provides a realization for some given N on *different* occasions: T1a is compatible with the multiple realizability of attributes.

In calling for physically-based attributes as members of P, T1a requires that all realizations of non-physical attributes are *ultimately* grounded in instantiations of physical attributes. But it accommodates the idea that there are many levels of ontological complexity and abstraction and that, at the higher end of the range, realizations of certain attributes consist of non-physical attributes. For example, the instantiation of properties of human behaviour will in many cases involve properties of bodily motion plus properties of a larger physical or social environment (as in the case of a bodily motion's being the signing of a cheque). What T1*a* requires is that all properties and relations involved in the realization of a certain non-physical attribute must be themselves either physical attributes or physically-based attributes. The ontological picture is one of a hierarchy of attributes that is grounded in the physical basis and that is structured by the relation of *realization*. Many intermediate levels of attributes may be required for the instantiation of some high-level non-physical attribute.

T1*a* also stipulates that the class P must be a minimal class of physically-based attributes. The idea is that there can be no free riders: each of the members of P must make an essential contribution to the realization of N. Thus I take it to be a burden of the form of physicalism I advocate that a distinction can be drawn between a class of relevant attributes that combine to constitute a realization of N, background-sustaining conditions that, although not constituting N, are required for its instantiation on a given occasion, and totally irrelevant attributes that neither constitute N nor provide significant background. For example, there would be no people on Earth if Earth were 20,000,000 miles from the sun. But that relational property is not constitutive of the attribute of *being a person*, although it is a background-sustaining condition of there being any persons on Earth. On the other hand, a certain very tiny rock on Mars having a mass of 2 grams is quite irrelevant to the instantiation of personhood on Earth (i.e. it is neither constitutive of personhood nor is it a significant background condition for the instantiation of personhood).

Although the class of physically-based attributes called for by T1*a* should include only those attributes that play a crucial role in constituting the non-physical attribute N on a given occasion, there is no *general* way in which the required distinctions can be drawn. What is relevant to the constitution of an attribute, what is significant background, and what is irrelevant are determined on a case-by-case basis. And it is quite compatible with T1*a* that there are holistically realized attributes (i.e. attributes which are realized by total states of the universe). These points serve to indicate that physicalism is compatible with a wide range of ways in which higher-order phenomena can be realized. However, the class, P, will in general consist of two subclasses: those attributes which are directly relevant to the constitution of N on the given occasion of instantiation, and those attributes which make up the background-sustaining conditions of N on that occasion. Both classes are essential to the instantiation of N, but in different ways. In the case of holistically realized attributes, of course, *all* physically-based attributes that are instantiated will be included in the first subclass.

It should be noted that *realization* is an essentially asymmetric relation and therefore it has at least one of the marks of a genuine dependency relation. Of course, it has more than that since *realization* is a relation which directly captures what is central to the dependency of the non-physical upon the physical: the latter constitutes the former and provides the conditions necessary for the instantiation of any attribute in this world. Thus the problem of relevance that plagued CR and the various forms of non-reductive physicalism is handled effectively by T1a. We can, for example, distinguish between a class P that realizes an attribute N and a class of nomic equivalents of the members of P that are each irrelevant to the instantiation of N: the former class has members that constitute N, whereas the latter class does not. The problem of emergent properties is also readily dealt with: realization is the direct antithesis of emergence. In requiring that the former relation holds, the latter relation is *ipso facto* precluded. Neither nomological sufficiency nor any species of supervenience relation can make that claim.

On the other hand, the realization of an attribute also provides, *inter alia*, a nomological sufficient condition for that attribute (i.e. a connective generalization which connects the joint instantiation of the members of P with the instantiation of N is true).[7] Although I argued in Chapter 2 that this is not enough for the physicalist, I want to side with those who include nomological sufficiency as among the relations required to obtain between non-physical attributes and attributes in the physical base. The realization

---

[7] Note that, consistent with the discussion in Ch. 2, I make no requirement either that the attributes in P involve at least some causal relations or that they are instantiated in the same region as the attribute N. Attributes that bear upon the realization of N can be instantiated in regions distant from the region in which N is instantiated. For example, the attribute of being the tallest mountain is realized by relation to other mountains which are both at a distance and not causally related to the tallest mountain in any significant way.

relation must have at least this much modal force if it is to avoid what is surely not in the spirit of the physicalist programme: viz., that two *relevantly similar* physical states of affairs should constitute realizations of incompatible non-physical attributes.

In T1a, the parenthetical qualification concerning 'possible' instantiations of an attribute expresses the idea that physicalist ontological theses bear upon both the actual and the possible states of *this* world. T1a asserts that any attribute that either is or can be instantiatiated in *this* world must satisfy the stated condition. In other words, as I shall argue in Chapter 5, T1a is not a thesis about other possible worlds. Rather, it is only the possible states of this world that fall within its scope. Any attribute that could be instantiated in this world and hence an aspect of some possible state of the world must be associated with a class of physical attributes the members of which are also aspects of that possible state and are jointly sufficient for realizing the target attribute.

T1a is a reductive thesis in so far as it requires that there be associated with the instantiation of any non-physical attribute, N, a specific class of physical attributes that are co-instantiated with N. Such reductionism, though modest, is indeed a form of reductionism.[8] It requires association of non-physical attributes with specific classes of physical attributes, subject to the condition that the latter provide realizations of the former. It therefore differs significantly from non-reductive supervenience claims of the sort discussed in Chapter 2. Further, since T1a requires no strict correlations or identifications of non-physical with physical attributes it is not a version of either CR or CR'. None the less, the condition constraining the association of non-physical attributes with classes of physical attributes is designed to give expression to the idea that the physical ontology is basic within nature. However, as I shall discuss below, this constraint is thought to be vulnerable to the objections that it is too weak to determine unique associations and that it is too strong to be plausible. Thus there are many who object to even this modest form of reductionism.

---

[8] A *reductive* thesis, as I understand the concept, is one which relates the members of one domain to the members of another, subject to certain constraints. Thus there can be many different sorts of reductive claims depending upon the domains involved and the kinds of constraints required. A non-reductive thesis, on the other hand, requires only that two domains be related in some global, systematic way without stipulating relations between the individual members of the domains.

Regarding the relations between T$1a$ and various supervenience theses, T$1a$ entails the global supervenience claim that if A and B are possible states of the world, alike in all physical respects, then they are alike in all non-physical respects. This is because A and B, *ex hypothesi*, will agree on all the physical attributes relevant to the realization of any non-physical attribute that is instantiated in those states of the world. Thus T$1a$ captures the idea that non-physical attributes supervene upon the physical attributes in this world. Again, although supervenience of this sort is not sufficient for expressing key physicalist ideas, it is an essential constituent of any adequate formulation of physicalist doctrine.

Further, T$1a$ implies the weak supervenience thesis that if two individuals agree on *all* their physical attributes (intrinsic *and* extrinsic), then they will also agree on their non-physical attributes. Again, this is because the individuals will agree on all relevant physical attributes bearing upon the realization of whatever non-physical attributes are instantiated. On the other hand, T$1a$ does not entail at least one version of the local supervenience of the non-physical on the physical: i.e. it does not entail that any two regions of space-time that are alike in all *intrinsic* physical respects must be alike in all non-physical respects. This is for the reason that many non-physical attributes are realized by relational properties that extend beyond the region in which the given non-physical attribute is instantiated.

It should also be kept in mind that T$1a$ associates the non-physical attribute N with a *class* of physically-based attributes, not with a single attribute. Thus, it is distinct from what Kim calls 'strong supervenience'. T$1a$ does indeed imply strong supervenience if complex physical properties can always be constructed from the attributes in P, and if I am right that there is a relation of nomological sufficiency between the instantiation of those attributes and the instantiation of N. However, strong supervenience clearly does *not* imply T$1a$ since the relation of nomological sufficiency is much weaker than realization, as I argued in Chapter 2.

Finally, whereas T$1a$ calls for the least class of instantiated physical attributes that provides a realization of each instantiated non-physical attribute, Post's local supervenience thesis, LS1, calls for the least class of physical attributes that suffices to determine a given non-physical attribute. T$1a$ implies, but is not implied by, LS1, again because, whereas the truth of T$1a$ guarantees the existence of those conditions that make LS1 true, realization is a much stronger relation than determination.

With respect to objects, I suggest the following physicalist thesis:

> (T1b) For each non-physical individual object, O, and for each region of space-time, R, if O is actually (or possibly) instantiated in R, then there exists a minimal class of physically-based objects, PO, such that the instantiation of the members of PO does (or would) provide a realization of O on that occasion.

The thesis, which focuses exclusively upon individual objects (as opposed to kinds), allows for the existence of non-physical objects. But it none the less imposes a significant constraint on what individuals can be instantiated in the world.

To say that some objects are 'non-physical' means that such individuals are instances of kinds of object that are not among those identified in physics or generable from such kinds (i.e. they are not among those in the physical basis for ontology as outlined earlier). Rather, they are instances of kinds constituted by, and individuated in terms of, various non-physical attributes. For example, tables and chairs, although realized by physical objects in our world, are non-physical in character. They are not found in the physical basis and they are constituted by various non-physical attributes (for instance, functional attributes). Further, social institutions, cultural artifacts, and works of art are all additional examples of non-physical objects that are realized by classes of physical objects in one type of organization or another.

Such non-physical objects can be natural objects (for instance, biological cells) or artifacts (for instance, pencils). They can be relatively simple or complex: for example, such biological kinds of objects as single cells and complex living systems are both non-physical objects that must be located within the physicalist framework. Furthermore, such objects can be highly unnatural fusions of objects of the same or different kinds: the fusion of any class of physical or non-physical objects is itself an individual which must satisfy the demands of T1b. Thus the conception of reality that the physicalist system must accommodate is one which countenances, *inter alia*, the many kinds of natural and artifactual objects that we encounter in daily life and that we identify in various theoretical and other cultural endeavours. The point of endorsing T$_1b$ is to identify the physical conditions that any individual must satisfy to be real. But in doing this, the physicalist is not ruling out all the many highly diverse aspects of our experience or the many kinds of objects that populate our world.

There are, of course, problems with the boundaries of objects (cf. Quine 1981b) and hence there are problems with their location in space-time regions. A resolution of such problems needs to be developed, either in terms of a stipulation that resolves the lack of clarity or in terms of tolerance of 'fuzzy' objects. I do not see that there are any serious problems for the physicalist programme with regard to either of these strategies of resolution. Given T$_1b$, the coordination of classes of lower-level physically-based objects with individual, higher-level non-physical objects is the principal problem here that must be solved as the physicalist programme is worked out. This is indeed a difficult problem, but its difficulty does not arise primarily because of problems with boundaries.

As with the condition on attributes stated in T$_1a$, T$_1b$ provides a condition for both actual and *possible* instances of non-physical objects. In no possible state of the actual world can there be an individual object that is not realized by some class of physically-based objects. From works of art, social institutions, and linguistic utterances to molecules, cells, and living creatures, each individual of these kinds of object is subject to the demands of T$_1b$. On the other hand, ghosts, gods, and demons, as generally conceived, do not and cannot exist in this world if T$_1b$ is true because none of these sorts of beings are realized by elements grounded in the physical domain.

Again, as with T$_1a$, T$_1b$ calls for associating a given non-physical object, O, with some minimal class, PO, of physically-based objects. The minimality condition signals the commitment to drawing distinctions between those objects in the world that are directly relevant to the realization of O on a given occasion, those objects that are part of the background-sustaining conditions for O on that occasion, and those objects that are irrelevant to the realization of O. Although there is no general principle or strategy for drawing such distinctions *and* although there may be a certain amount of arbitrariness in how the boundaries of O are circumscribed, there do not appear to be any insurmountable problems for physicalism posed by this commitment. Considerations of relevance and irrelevance are based upon the kind of object O is and the particular way in which O's boundaries are drawn.[9] And again, as with the class P called for by T1*a*, PO will consist of two sub-classes, the first involving objects that are directly constitutive of O and the second involving objects that are part of the sustaining background.

In calling for classes of physically-based objects, T1*b* allows that the realization of certain kinds of non-physical objects (for in-stance, social institutions) involves systems of other non-physical objects (for instance, persons). Given some non-physical object, O, it is some configuration of other non-physical objects (and their attributes) which constitutes the realization of O in this sort of case. And indeed, each of those other objects may in turn be realized by configurations of non-physical objects. The point of stipulating that it is *physically-based objects* which go into the realization of O is that all objects that realize O are themselves realized by physically-based objects and that at some point the chain of realization comes to an end in the physical basis.

As we have seen, the fundamental metaphysical relation that structures the world as understood by the physicalist is *realization* of an object O by a class of other objects PO. It is not *identity* of an object O with some physical object. There may, of course, be plausible identity statements relating an object in one theoretical domain with an object in another. But the stock philosophical examples of such identities (for example, water, genes) do not underwrite a general identity approach, even if they are plausible.[10] Physicalists who are interested in encompassing all objects that occur in nature, not just a highly restricted class within the do-mains of the various natural sciences, must recognize that identi-fication of, say, cultural objects like a work of art or an artifact is not plausibly identifiable with any physical object. *Realization* is a relation that is based upon the idea that particular objects exist in nature as a consequence of how certain types of more basic objects (and their attributes) are organized and embedded in a larger context. Again, it is the job of realization theories to clarify what constituent objects and what types of organization

---

[9] It is the job of what I shall call 'realization theories' to clarify, for each kind of object, how these uncertainties are to be managed.

[10] Although I doubt whether either of these is a plausible example of identity.

and contextual embedding are required for realization of some particular kind of individual that happens to exist in some space-time region. To sum up, T1*b* calls for an ontological structure in which all objects are embedded. The structure can have many levels of organization that are asymmetrically related by a relation of *realization* and that are all ultimately grounded in a basic level of physical objects. How objects of some given kind are realized can vary, but the critical stipulation is that, in all cases, an object exists in the world only if it is realized by a class of physically based objects. How a given object is in fact realized depends upon the details of the relevant realization theories as well as of the particular facts of the case. Thus the critical stipulation is a way of ruling out ghosts and their ilk while accommodating both the wide diversity of kinds of objects in the world and the implausibility of identifying all objects with physical objects. Along with T1*a*, T1*b* puts content and bite into the physicalist's metaphysical vision. These theses jointly entail that everything that occurs in nature is dependent upon, supervenient upon, and realized by what goes on in the physical domain.

4.2. DETERMINATION OF FACT AND TRUTH

The second class of physicalist theses are those concerned with objectivity. Based upon the discussion in Chapter 1, I make the following assumptions. Objectivity is ordinarily taken to be a property that attaches to 'truth', 'knowledge', and 'facts'. That is, it is taken to be a feature of the world as well as a feature of our knowledge of the world, and physicalists are concerned to capture both employments of the concept. Further, there are two key ideas relating to the notion of objective truth and knowledge: on the one hand, inter-subjective testability and agreement and, on the other, independence of languages, theories, and minds. The issues here run quite deep, since there is an important sense in which all knowledge depends upon subjective features (for instance, representational systems). As a minimum, however, the notion of independence alluded to is intended to rule out obvious individual features upon which certain claims may depend (for example, specific personal goals that motivate claims to knowledge), and it is intended to capture the idea that what is true is not subject to an individual's interests and desires.

For example, Quine's thesis of the indeterminacy of translation is a claim concerning the non-objective status of translation from one language to another ('there is no fact of the matter'). Even if such translations are objective in the sense of being supported by evidence available inter-subjectively, they are not, according to Quine, objective in the sense just described: different translators with different interests, sensitivities, and proclivities could, as a consequence of those differences, translate differently *but equally correctly*. Such sensitivity to individual differences is a feature of translation but not of physics, according to Quine. The reason for this divergence requires a third idea regarding the notion of objectivity: the idea of there being an objective matter of fact underlying a claim to knowledge.

According to Quine, translations do not count as objective knowledge because they are sensitive to individual differences between translators: they are not independent of the knowing subject. Physics is not sensitive to such differences: it is independent of the knowing subject. What distinguishes those claims which exhibit such independence from those which do not? Quine's, and the physicalist's, answer is that the former concern matters of fact while the latter do not. Hence physics, being concerned with matters of fact, is not sensitive to differences among investigators, while translation, being not concerned with such matters, is sensitive to such differences. This is why two translators, who differ in subjective features, can, on the basis of those differences, offer different translations which are equally correct. There is no underlying fact of the matter that makes one translation right and the other wrong.[11]

It is in this way that the physicalist's interest in objectivity focuses on the idea that the independence of certain claims from subjective factors can be understood in terms of there being matters of fact with which 'objective' knowledge is concerned. Such matters of fact offer up critical resistance to subjective variation of interest, sensibility, and the like. The world is the way it is whether we like it or not and whether we know it or not. To capture these ideas

---

[11] Strictly speaking, Quine holds that there are behavioural facts of the matter, but these behavioural facts do not suffice for fixing a uniquely correct translation. It is the residual slack that makes room for subjective variation.

within the physicalist programme, two questions must be addressed: 'Within a physicalist conception of nature, what counts as an objective matter of fact?', and 'Within a physicalist system of knowledge, what are the features of objective truth and falsehood?' (i.e. 'Within a physicalist system of knowledge, how are claims with a factual basis different from those that lack such a basis?').

The appropriate physicalist strategy for answering these questions is to require certain relations to hold between all domains of fact and truth and the physical bases. In particular, the physicalist theses concerning objectivity will characterize relations which must hold between non-physical phenomena and physical phenomena in order for the former to count as objective matters of fact, *and* between non-physical claims and physical claims in order for the former to have an objective truth value. If both these sorts of thesis are formulable and defensible, they should constitute an expression of the physicalist view that the physical ontology and doctrine provide bases for objectivity.

With respect to matters of fact, the physicalist view is that the objective matters of fact are the physical facts and any facts that are *determined by* those physical facts. The problem of formulation thus comes down to the problem of making precise the physicalist dictum that *the physical facts determine all the facts of nature*. The core idea behind this dictum is that there is an independent world of fact that has a certain structure: i.e. it is structured by relations to a physical basis. Such relations constitute the conditions of objectivity within a physicalist world. If a purported fact can be shown not to exhibit such relations, then that is a sufficient ground for the physicalist to reject it as not being objective.

In formulating the physicalist view of objectivity, I have spoken of 'facts' and invoked the distinction between 'facts' and 'non-facts'. Some readers may balk at such expressions, and a few words of clarification might be helpful. With respect to the use of 'fact talk', I endorse the view that such talk is completely dispensable in favour of other locutions. Below, I shall present a strategy for doing this. With respect to the distinction between fact and non-fact, I am sympathetic to the idea that the *conception* of any fact is dependent upon features of the conceiver. For example, the nature of representation and the differences between representational systems are quite pertinent to the nature of cognition, and hence to the differences between various cognitions of the world. But this general dependency does not undermine the possibility of drawing distinctions between fact and non-fact in what is conceived. From within any way of conceiving of knowledge and the world, distinctions between objectivity and subjectivity can be clearly delineated, even if that way of conceiving, like all others, involves parochial features of cognitive and representational systems. Thus, within the physicalist view of nature and knowledge, the conception of objectivity described above can be articulated.

To return now to that conception: if the objective facts are those that are determined by the physical facts, how is this to be expressed in a precise way? This question raises three more: 'How are we to characterize the physical facts?', 'How are we to characterize the non-physical facts?', and 'How are we to characterize the relation of determination that the former bears to the latter?'.

As outlined in Chapter 3, a possible physical state of nature is a complete[12] distribution of physical objects and attributes over the regions of the space-time continuum compatible with the laws of physics. The class of all such distributions is the class of all possible physical states of nature. I shall represent this class with a class of models, each member of which is a space-time structure in which there is one of the possible, complete distributions of objects and attributes over space-time regions. This class represents all the possible distributions of physical fact.[13]

To represent the non-physical facts, the strategy is similar. Consider the class of all non-physical objects and attributes and the space-time continuum as before. Then, a possible non-physical state of nature is a complete distribution of the non-physical objects and attributes over the regions of the space-time continuum which is compatible with the laws of nature.[14] The class of all such

---

[12] A *complete* distribution is one that is based upon an evaluation of every physical attribute and every physical kind of object at every region of space-time. I take this to be compatible with the idea that some assignments cannot be specifically made on such an evaluation (e.g. some magnitudes have indefinite values at certain regions). Instead, the evaluation may assign a probability distribution for a range of values of that magnitude.

[13] The idea behind this strategy is due to Hellman and Thompson (1975), although I depart from their approach.

[14] This does not assume that all laws of nature are exceptionless. This class will probably include states of nature that are in violation of certain 'laws', although compatible with other, more fundamental, laws. See below, in my discussion of T5, for further elaboration of this point.

distributions is the class of possible non-physical states of nature. As before, I shall represent these possibilities in terms of a class of models, each member of which is a space-time structure over which there is a complete distribution of non-physical objects and attributes. This class of models represents all the possible distributions of non-physical fact.

Given these two classes of models, it is possible to specify a third class of models, M, each member of which is a space-time structure over which there are complete distributions of both physical and non-physical objects and attributes, where the distributions are constrained by the laws of nature. This class represents all the possible pairings of the possible physical states of nature with the possible non-physical states of nature. In other words, this class represents all the possible states of nature. Given this class of models, we can characterize the relation of determination pertinent to clarifying the physicalist view that the physical facts determine all the facts of nature.

Informally, the relation of determination is expressed by the idea that once the physical facts are 'fixed', so are all the non-physical facts. The official statement of this idea is as follows:

(T2) For any possible complete distribution of the physical facts, there is exactly one possible complete distribution of the non-physical facts.[15]

In terms of the class of models M, if A and B are members of M and if A and B agree on the distribution of the physical objects and attributes over the regions of space-time, then A and B must also agree on the distribution of non-physical objects and attributes over those regions. Thus according to T2, for any possible physical state of nature, there is associated with it exactly one possible non-physical state of nature. A more formal characterization of the relation of determination which structures this class of models will not be given here, although there do not appear to be any special difficulties involved in providing one (cf. Hellman and Thompson 1975; Post 1987).

To summarize: on the physicalist view, an objective matter of fact is one which is determined by physical facts. The possible

[15] T2 is pretty nearly equivalent to the explications of truth determination found in Hellman and Thompson (1975: 558, #4), and Post (1987: 185, TT*; 1991: 118, D).

states of nature, which exhaust the possibilities regarding objective matters of fact, have been represented in terms of a class of models, M, each member of which pairs distributions of physical and non-physical objects and attributes. If M is structured in accordance with T2, then it is structured by a relation of determination (i.e. members A and B cannot agree on the physical facts while differing on the non-physical facts). Thus, for example, any alleged attribute for which it is contended that, in identical total physical states of the world, it might in one case be realized while in another not, is not an objectively real attribute of things. Such an attribute will not be among those included in the class of non-physical attributes employed in the model-theoretic construction described above. Rather, it will be rejected, on physicalist grounds, as unreal.[16]

I note that, although T2 is a consequence of T1,[17] the two theses give different sorts of insight into physicalist thought. Whereas T2 formulates a global constraint upon total states of nature and upon objective matters of fact, T1 provides (among other things) crucial insight into why T2 holds. The often sought-for understanding of why global supervenience theses are true of the world is that physical objects and attributes constitute realizations of non-physical objects and attributes. In observing this, I am not supporting those who believe that theses like T1 are *required* in order that global supervenience relations can be explained. There are more fundamental reasons for requiring T1, as I argued in Chapters 1 and 2. As Post has very ably discussed, there is no particular reason why global supervenience (or T1, for that matter) requires explanation at all. To think otherwise may be to hold on to some version of the erroneous principle of sufficient reason.[18] My claim is only that, as a matter of fact, T1 does provide insight into why T2 holds.

A second physicalist thesis concerning objectivity bears upon knowledge directly, rather than upon the objects of knowledge. The question to be addressed is: 'What are the marks of objective truth?' For the physicalist, this is to ask: 'Which knowledge claims

---

[16] This, of course, is exactly the Quinean position regarding the attributes studied in semantics, mentalistic psychology, and linguistics.

[17] Henceforth, I shall refer to the conjunction of the theses T1*a* and T1*b* simply as 'T1'.

[18] See Post (1987: ch. 2) for discussion of 'ultimate explanation'.

have a basis in physical fact for their truth value?' (i.e. 'Which claims have a truth value because of the way the world is physically?'). And the other side of the question is: 'How are such claims different from those which are not objectively true or false?' Thus the physicalist seeks a condition for objective truth and falsity. One of the virtues of classical physicalist reductionism was that it provided such a condition: viz., derivability from the physical truths was considered a sufficient condition for being an objective truth, derivability of the negation of a sentence was deemed sufficient for the sentence's being an objective falsehood. With the demise of classical reductionism the physicalist must formulate a viable alternative.

In recent years, the principle that the physical truth determines all the truth about nature has been favoured by many physicalists. It has generally been agreed that classical physicalist reductionism is too strong a thesis and that, however the principle of truth determination is to be cashed in, it must be compatible with the falsehood of such reductionism. In Chapter 2, I discussed some of the proposals that have been suggested for meeting this demand. For present purposes, I shall only offer a relatively informal version of the determination thesis; more technical developments can be found elsewhere in the physicalist literature (Friedman 1975; Hellman and Thompson 1975; and Post 1987).

Assume that a class of languages, L, is fixed and that the class of models, M, described above provides interpretations for those languages. The thesis of truth determination is as follows:

(T3) Within the class of models, M, if members m_1 and m_2 agree with respect to the physical truths, then they agree with respect to the truths, formulable in all other languages in L.

This version is similar to that offered by Hellman and Thompson (1975), although it is less technical and differs from their formulation in other ways.[19]

How does T3 serve the motivations of the physicalist programme regarding objectivity? A sentence, s, formulable in some language

---

[19] E.g. whereas their formulation relies upon the model-theoretic notion of elementary equivalence, T3 does not. Since elementarily equivalent models need not agree on higher-order truths, it is possible that their formulation does not capture the full thrust of physicalist truth determination.

L_i, has a legitimate claim to objective status only if its truth value is fixed once the physical truths are fixed: i.e. for *s* to be objective, it cannot be possible that, given a specification of the physical truths, either *s* or its denial could be true. Thus, in the terms of T3, there could not be models in M which, while agreeing on the physical truths, do not agree on the truth value assigned to *s*. As a criterion of objectivity, T3 purports to capture the idea that it is the mind independent physical facts that determine the truth value of *s*. Thus a clear distinction between objective truths, on the one hand, and judgements sensitive to idiosyncratic subjective features, on the other, can be drawn.

In this formulation, 'truths' are viewed as sentential rather than as propositional and non-linguistic. Thus it presupposes that the physical language is sufficiently powerful for expressing the physical truths that determine all other more complex or higher-order physical and non-physical truths that are expressible in the languages of L. If this condition is not satisfied, then the thesis could very well be false, for the relatively uninteresting reason that a sufficient number of the physical truths (propositionally construed) are not expressible in the physical language that happens to have been selected. If no such language exists, then T3 is false for a more interesting reason, although not in a way that undermines central physicalist ideas: in particular, T2 could still be true. Thus it is only relative to this assumption regarding the physical language that T3 is true and that it is a consequence of T2.

4.3. VERTICAL EXPLANATION AND REALIZATION THEORIES

Earlier in this chapter, I presented ontological theses requiring that there be associations between instantiations of non-physical objects and attributes and classes of instantiated physically-based objects and attributes which constitute their realizations. This *strongly suggests* that explanatory relations should exist between the physically-based phenomena and the non-physical phenomena they realize, and that the ontological theses characterize a structure which provides a framework for developing such explanations.

More fundamentally, in Chapter 1, I represented the goals of the physicalist programme as including the promotion of a number of explanatory objectives the attainment of which would increase the level of understanding that a physicalist system of knowledge yields. At the core of these explanatory aims is the provision of vertical explanations of the realization of non-physical objects and attributes. Thus all phenomena are to have explanations in terms of physically-based phenomena. A system of knowledge structured by such vertical explanations will be a physically-based explanatory system in the sense that all explanations, *of whatever sort*, will only make appeals to objects and attributes that are ultimately grounded in the physical domain, via a chain of vertical explanations. Such grounding further provides a framework for the explanation of all interaction and influence, as well as a delineation of their scope and limits. And it provides a framework for the integration of explanatory structures drawn from different domains (i.e. all such structures are embedded within the physical domain). Thus, significant constraint relations between theories developed in various branches of knowledge are implied by this aspect of physicalist thought.

In the present section, I shall formulate two theses which will serve to express the physicalist's ideas and motivations concerning explanation and which will clarify the nature of the vertical explanations suggested by T1 and sought after by the physicalist. The morals drawn in Chapter 2 must be taken seriously here. In particular, exclusive reliance upon the apparatus of classical reductionism or the various forms of non-reductive physicalism must be avoided, since explanatory structure of the sort we are after is not provided by either definitions, derivations, supervenience relations, or relations of nomological sufficiency. In addition, the implausibility of both generalized identity theses and generalized eliminative theses means that an alternative approach must be pursued. Specifically, the theses must directly call for explanations of the required sort (i.e. explanations of how non-physical phenomena are realized by physically-based phenomena).

The physicalist's basic conviction is that all phenomena occur in nature in virtue of what goes on in the physical domain. Thus the physicalist believes that a thing has the attributes it has in virtue of what it is made of, how it is put together, and how it is related to other things. In other words, we can come to understand why a thing is the way it is and why it behaves in the way it does (for example, why it exhibits the regularities it does) in terms of these three sorts of considerations. Theses T1 and T2 serve to capture this idea from an ontological point of view. The question remains how it is best expressed from the point of view of a system of knowledge structured by physicalist principles. My approach will be to frame theses that express the following idea: given that all individual phenomena, all regularities, and all instances of and exceptions to regularities which occur in nature occur in virtue of physical phenomena, there are physically-based explanations of all such phenomena.

By 'individual phenomena' I mean instantiations of objects and attributes in regions of space-time. By 'regularities' I mean nomological associations between instantiations of attributes in regions of space-time, where the associations may express causal or non-causal relations of covariation among attributes and where they may be deterministic or probabilistic in character. Mention of 'exceptions to regularities' reflects the idea that higher-level laws of nature are typically not exceptionless, although they are counter-factual supporting to some extent. This feature of certain laws need not, and ought not, be viewed as a flaw in our knowledge so much as a deeper reflection of the structure of things. Physicalism provides a natural framework for understanding why this is so.

The first explanatory thesis is as follows:

(T4) All instantiations of non-physical objects and attributes are vertically explainable[20] in terms of physical or physically-based objects and attributes.

By a 'vertical explanation' (VE) of an attribute[21] is meant a certain sort of explanatory relation between attributes at different ontological levels, where such levels are distinguished by degrees of complexity or abstraction. Such vertical explanations can be seen as providing answers to questions of the following form:

1. In virtue of what lower-level attributes did the instantiation of such-and-such higher-level attribute occur?
2. How did such-and-such lower-level attributes constitute an instantiation of such-and-such higher-level attribute?

---

[20] To say that all instantiations of non-physical objects and attributes are explainable does not imply that such explanations will inevitably be developed or even that such explanations are necessarily within our cognitive reach.

[21] For ease of discussion, I shall focus on attributes, although the points I shall make apply to objects as well.

Thus a VE is supposed to identify a class of *relevant* lower-level attributes that may be drawn from one or more lower levels, and to provide an account of how those attributes combine to constitute an instantiation of the target higher-level attribute.[22]

As T4 requires, the physicalist is concerned with there being such vertical explanations for all non-physical attributes. More specifically, the call for VEs is a call for 'token' explanation of the specific instantiations of all such attributes, in terms of lower-level attributes that are either physical or physically based (i.e. attributes that are themselves *ultimately* grounded in the physical basis by a chain of realization relations). Thus it is possible, and likely for many higher-level attributes, that none of the attributes cited in a vertical explanation are physical attributes; the requirement is only that they should be physically based. For a physicalist, then, a full answer to questions (1) and (2) with respect to the instantiation of a given non-physical attribute consists in citing certain relevant physical or physically-based attributes and in explaining that instantiation in terms of them. It should be noted that there is no requirement that different instances of the same attribute are vertically explainable in exactly the same way.

Such vertical explanation must be clearly distinguished from etiological explanations which identify the causes of some effect. *Realization, constitution*, and the like are not causal relations in any standard sense. Further, a VE is not in general a form of explanation in physics, even if physical attributes are involved. The principles that characterize how physically-based attributes combine to constitute an instantiation of an attribute are not necessarily physical laws, although such laws could, on occasion, be relevant to a VE. It is therefore a mistake to suggest that the physicalist aims at understanding all phenomena in purely physical terms, if that means that all phenomena can be comprehended within physical theory. This is a twofold mistake because VEs need only appeal to *physically-based* attributes and because such

[22] Although I am emphasizing the idea that attributes in the *explanans* of a vertical explanation are drawn from lower levels, I do not mean to rule out the possibility that a vertical explanation might appeal to a physically-based attribute at the same or higher level as the attribute whose realization is being explained. The central requirements are that all attributes in the *explanans* are physically based, thus ensuring that the target attribute is ultimately related to the physical domain, and that an explanation is provided of how that target attribute is realized on the given occasion.

explanation can involve non-physical principles that relate lower-level and higher-level attributes.[23] The fact is that there are high-level, non-physical structures, processes, etc. which are instantiated in nature, and a physicalist system of VEs aims at showing how such phenomena can be accounted for in a physically-based world. It is not, however, the job of physicists to accomplish this objective.[24]

To characterize more fully the kind of explanation called for here, I shall introduce the notion of a *realization theory* (RT). Such an idea is implicit in much of the work in the sciences and in a variety of discussions in philosophy. The idea is that of a theory, associated with a given attribute, N, that abstractly characterizes the kinds of attributes that are sufficient for the realization of N and that shows how such attributes can combine to actually constitute N in specific cases.[25] Examples of such theories include a characterization of transparency (cf. Chapter 1), a functionalist account of mental states, and a causal theory of reference. In all such cases, an appeal is made to a theory of how the instantiation of certain relevant physically-based attributes suffices for the instantiation of the target non-physical attribute.

For the physicalist, given an RT for an attribute, N, and given a distribution of physically-based attributes, vertical explanations can be generated to account for the specific instantiations of N, if they occur. Unlike the VEs that they help to construct, an RT for an attribute is a generic, abstract account of how an attribute can be constituted. Such an RT plays the crucial roles of helping to isolate and identify relevant physically-based attributes and of showing why those attributes are relevant and how they combine to constitute N. Thus RTs are essential for effectively answering questions (1) and (2) above. In this way, then, RTs are central to

---

[23] It is a red-herring issue, sometimes raised against any form of physicalism which introduces connective principles that are not part of physics, that somehow the priority and privilege of physics are compromised by appeal to such principles. Such an objection betrays a misunderstanding of what physicalism involves and of how vertical explanation works.

[24] Nor do I make any assumption that there is transitivity of vertical explanation in the sense that if there is a vertical explanation of some attribute, N, in terms of physically-based attributes, then there must also be a vertical explanation of N in terms of exclusively physical attributes. See Post (unpublished) for discussion of this issue.

[25] Similarly, a realization theory for a kind of object, O, reveals what sorts of configurations of physically-based objects and attributes constitute instances of O.

the physicalist project of explaining how non-physical phenomena are possible in a physically-based world (i.e. the project of showing how the physically-based facts about the world constitute the non-physical facts).

For example, functionalist accounts of mental states delineate the pattern of causal relations which, if instantiated in some physical *or* non-physical system, suffice for the instantiation of a given mental state, M. The RT in such a case consists of a characterization of the required pattern of causal relations without appealing to any more specific kinds of relations that might exhibit such a pattern (i.e. it is neutral with regard to what processes, if any, underlie the relevant causal relations). And the RT effectively clarifies what it is that makes it the case that the specified pattern of causal relations, rather than some other, constitutes M. Then, for a given instantiation of the mental state in question, the *physicalist* vertical explanation would consist in citing a set of instantiated physically-based causal relations and showing, by appeal to the functionalist RT, why those relations suffice for the realization of that mental state.[26]

It is important to understand that an RT for an attribute N is, in most cases, an abstract theory that, while specifying what it takes to instantiate N, does not normally make any reference to physical attributes. Rather, an RT specifies something like the essence of the attribute, although I don't wish to hang too much on the use of that term.[27] It delineates a configuration of attributes that suffices to instantiate N in any possible world, not just physically possible worlds. Thus, pursuing the example further, functionalism is compatible with dualism, although it is also clear why physicalists can appropriate a functionalist view of the mind to provide accounts of how mental attributes are realized by physical systems. According to the functionalist, the essence of a mental state, M, is a certain pattern of causal relations. Such patterns can be exhibited in a physically-based world of the sort envisioned

---

[26] No endorsement of functionalism in the philosophy of mind is intended here.

[27] I want to leave open the question of whether there could be more than one RT associated with a given attribute. Although it is clear that, relative to a single RT, multiple realization is possible (e.g. as in the case of functionalism), it is much less clear whether a given attribute can have more than one RT. See Cummins (1989) for suggestive discussion of this issue with respect to mental representation.

by the physicalist, or such patterns can be exhibited in non-physically-based worlds (for example, spirit worlds).

Further, and to switch examples, transparency can be instantiated in worlds without physical microstructures like those that underlie transparency in our world. Just so long as something is capable of being seen through, then it is transparent. Our physics provides an account of the physically-based structures that allow for the instantiation of this attribute in our world. But we should not confuse a parochial, physical account of transparency with its essence, just as we should not confuse a parochial, physically-based account of mentality (for example, in terms of neural structure and processing) with its essence. Such attributes as transparency and mentality can be instantiated in worlds that, although possibly similar to our own at macroscopic and phenomenological levels, are very different at microscopic levels. It is also evident that not every attribute is, or even can be, instantiated in every world. Thus the value of RTs lies both in revealing why it is that an attribute is or can be instantiated in a given world and in providing a way of understanding why it is that an attribute is not, and perhaps cannot be, instantiated in a given world (i.e. none of the possible configurations of attributes in that world suffice for realizing the conditions specified by the RT for that attribute).

Another feature of RTs is that there is not just one sort of RT appropriate for all attributes. Given the many different types of attribute that there are (for instance, intrinsic, extrinsic, causal, structural), there will inevitably be variation in both the kinds of attributes cited by RTs as well as the ways in which attributes combine to form an instantiation of the target attribute. In particular, since some attributes are intrinsic while others are extrinsic, it should be underscored that the realization base for an attribute need not be localized in the same region as the attribute that is realized. Whether or not a given attribute is instantiated may well depend upon what is going on elsewhere in the world. And whereas higher-level causal powers are, perhaps, constituted by the combining of lower-level causal powers, various more abstract features of things (for instance, semantic properties) will depend upon non-causal properties and relations (for instance, as in Post's Ammon example discussed earlier).

RTs provide conceptual structures for understanding how non-physical phenomena can be realized in a world. For the physicalist,

they *abstractly* characterize the kinds of physically-based structures that suffice for realizing non-physical attributes and that
provide physically-based models for higher-level theories of this
world. As a consequence of this abstractness, they allow us to see
how it is possible for there *not* to be any lower-level systematicity
in the class of realizations for an attribute: what the members of
such a class share is abstract relative to lower-level theory. On the
other hand, such RTs help to account for significant relations of
covariation between lower- and higher-level domains (for example,
relations of determination or nomological correlation). They flesh
out the fabric of the connections between domains, and in so
doing they allow us to distinguish between those connections that
are instances of realization relations and those that are only ac-
cidental or only nomologically correlated with relations of deep
dependence.

As a consequence, RTs reveal how the physical bases ground all
that takes place in our world, thereby helping to provide a framework for explanatory unity, integration, and connection between
everything that can or does occur. This includes, remarkably, the
reining in to the physicalist fold of even private or otherwise inaccessible phenomena. To account for how such phenomena can
be realized by physically-based phenomena is not to undermine
their privacy or inaccessibility. Knowing the character of the
physical basis of, say, private mental states possessing qualitative
attributes is not to know everything that is true of such states. But
this does not warrant the conclusion that physicalism is thereby
refuted. To explain why a certain attribute is instantiated is not
to possess the attribute, and thus it is not to come to know those
things that only a subject who possesses the attribute could know.
That some (for example, Jackson 1986; Nagel 1974) have tried to
parley such obvious truths into an objection is a mistake based on
a faulty understanding of physicalism, in my opinion (see P. M.
Churchland 1985 and Van Gulick 1985 for discussion).

It should be clear from the above discussion that RTs are not
explicit, physical definitions of non-physical attributes; nor are they
identity statements, statements regarding physical/non-physical
determination relations, bridge laws, or connective principles with
some weaker modal force. In fact, they need not explicitly concern
vertical, physical/non-physical relations of any sort. Rather, RTs
are, in general, more abstract than any of these other sorts of

principle. It is only when they are applied, as the physicalist requires, that they begin to clarify relations between physical and non-physical domains.[28]

It is especially important to see that RTs are not just explicit definitions in physical terms, as earlier versions of the programme had it. Even if physical definitions were forthcoming, *there would still be a need for RTs if the physicalist goals regarding explanation are to be served.* That is, there would still be pressure to understand why the instantiation of the members of a given set of physical attributes realizes a given non-physical attribute, even if the set of physical attributes can be converted into a definition of the non-physical attribute.[29] It is, in my opinion, a major shortcoming of both classical and recent versions of physicalism that the role of explicit definitions has not been very well understood in terms of the goals of the programme.

Causey (1977), for example, is at great pains to distinguish those nomological correlations which express attribute identities from those which express causal relations. His criterion for making this distinction is the existence, or lack thereof, of an explanatory account of a causal relation between the attributes in question. On my view, Causey's discussion is faulty because his dichotomy is not exhaustive. He leaves out the possibility of the realization of one attribute by another and he takes much too seriously the idea of attribute identity across domains (it is a rarity at best).[30] He has the right idea in seeking explanatory accounts, but he encourages a false picture of the inter-domain problem of how attributes in one domain relate to attributes in others. The cost is a failure to appreciate the need for inter-domain explanations in general. As a consequence, the goals of the physicalist unification programme are sacrificed by an inappropriate emphasis on identity.

[28] Although I think Putnam (1979: 609) is right in insisting that any form of physicalism must involve 'type–type' relations of some sort, he does not get the point quite right since he seems to have in mind physical/non-physical correlations.

[29] Note that this assumes that attributes are not individuated by the criterion of nomological coextensiveness which is the standard constraint upon physical definitions. If this were the individuation criterion, then definability would lead directly to identification of the attributes expressed by the defined and defining terms.

[30] The standard examples of theoretical identities, such as those involving temperature or pressure and statistical mechanical properties of an aggregate of molecules, are quite flawed, as Earman suggests in the quotation at the beginning of this chapter.

Further, as pointed out in Chapter 2, the view in Friedman (1975) that each non-physical predicate is associated with a (possibly infinite) class of physical predicates, although sufficient for his purpose of characterizing 'weak reduction', raises, but does not indicate how to answer, the question, 'Given a non-physical predicate, in virtue of what is each member of the associated class of physical predicates a member of that class?'. Simply to say that each expresses an attribute that is a realization of the attribute expressed by the given non-physical predicate begs the explanatory question. At the heart of that question is the quest for an understanding of *why* certain physical attributes realize a given non-physical attribute while other physical attributes do not. To respond only by asserting that the former attributes do while the latter do not begs the question and leaves unaddressed a large host of mysteries concerning realization. The point of T4 is to rid a physicalist system of such mysteries.

The general idea is that it does not matter for the achievement of the explanatory goals of the programme whether each non-physical attribute is associated with a single coextensive physical (or physically-based) attribute *or* with a class of such attributes each member of which is sufficient for the realization of the non-physical attribute *or* with a class of classes of such attributes where the members of each class jointly provide a realization of the non-physical attribute. Either way, the explanation of the realization of non-physical attributes by physical attributes is required if a large class of vertical mysteries are to be eliminated from a physicalist representation of nature. The role of realization theories is to provide a basis for the elimination of those mysteries. To satisfy the explanatory motivations underlying the physicalist programme, T4 requires explanations, via RTs, of the realization of all non-physical attributes by associated physical (or physically-based) attributes.

RTs are also not the same as what Post refers to as 'connective theories', which are largely empirical theories of the specific connective relations that hold between physical and non-physical attributes in this world.[31] Both sorts of theory are, in my opinion, essential to the physicalist system and they should be viewed as complementing each other. Whereas RTs provide the more abstract

---

[31] See Post (1987: ch. 5) for valuable discussion of connective theories and the roles they play in a full development of a physicalist system.

account of how attributes can be realized, an account that provides guidelines for the study of specific realizations of attributes, connective theories provide the specific answers to questions concerning how non-physical attributes are in fact realized in this world. In terms of the quest for vertical explanations of the instantiations of non-physical phenomena, RTs provide the general guidelines for how to isolate and combine relevant attributes bearing upon some specific phenomenon under investigation, whereas connective theories make the specific identifications of physically-based attributes regarding a particular instantiation, or class of instantiations, of that phenomenon.

To summarize: an RT provides an abstract account of what it is in virtue of which a non-physical attribute can be instantiated (i.e. an abstract characterization of the types of structures and processes that suffice for constitution of the attribute). The provision of RTs and the provision of vertical explanations of specific realizations of attributes drawing upon such RTs is one of the goals of research programmes concerned with understanding why higher-level attributes are realized by lower-level attributes. And the attainment of such goals relative to physically based attributes furthers the more general goals of the physicalist programme outlined in Chapter 1 by promoting the unification of knowledge via vertical explanatory connections, and thus by increasing understanding.

Comparable points can be made regarding the realization of objects in nature. T4 calls for the vertical explanation, in terms of RTs, of all particular instantiations of the various non-physical kinds of individual object. The RTs provide the general account of what it takes to realize a given kind of object, and particular vertical explanations employ the RTs to identify the relevant, physically-based objects and attributes that combine on a given occasion to constitute, and hence realize, a particular individual of the given kind. The result is an understanding of how individual objects are embedded in and realized by the physically-based fabric of nature. I shall bypass further discussion of this aspect of T4 for reasons of space.

Turning now to a discussion of the thesis concerning physicalist explanations of regularities in nature, early versions of the physicalist programme were concerned with what was called 'the unity of laws' in science (Carnap 1969). The idea was, simply, that all laws of nature are derivable from the laws of physics plus suitable bridge principles. Such derivations were thought to be sufficient for the explanation of those laws and of the regularities they expressed. Although I fully support the motivations behind this version of the programme, I am sceptical about the specific thesis proposed to serve them. That is, I fully endorse the idea that the physicalist programme is motivated by a quest for increasing the explanatory power of a system of knowledge by, *inter alia*, vertical explanation of regularities, but I reject the classical thesis that expressed this motive. The task now is to formulate an alternative.

The physicalist's conviction is that all regularities, their instances, *and their exceptions* are determined by underlying physical phenomena. This conviction was captured earlier, from an ontological point of view, by the thesis, T1, concerning the realization of objects and attributes, and the thesis, T2, concerning the physical determination of fact. The question now is: 'How is this conviction to be expressed from the point of view of a system of knowledge concerned with those facts?' To this end, I shall focus attention on the following question: 'Given a set of non-physical regularities characteristic of some domain, what are the underlying physically-based structures and regularities that realize them?' It is this and closely related questions that guide inquiry within inter-field disciplines (for example, physical chemistry, microbiology, the neurosciences). In short, vertical inquiry is concerned with the study of lower-level 'mechanisms' that underlie higher-level regularities.[32]

Although inter-field inquiry (Darden and Maull 1977) aims at the identification of mechanisms for regularities, the questions addressed by inter-field research programmes make no presumption that higher-level phenomena are *uniquely* realized by lower-level phenomena. Thus no presumption is made that exactly one lower-level 'mechanism' underlies a given higher-level regularity; rather, there may be different mechanisms that realize its various instances. As a consequence, the explanation of a regularity should be understood as the provision of members of a class of mechanisms

---

[32] The use of the term 'mechanism' here should not mislead one to think that the brand of physicalism I am discussing is a species of 19th-century 'mechanism'. The term is simply a stand-in for whatever physically-based phenomena constitute the realization of a regularity in nature. Thus by 'a mechanism for a regularity' I mean a system of physically-based objects and attributes governed by a class of physically-based regularities.

that jointly suffice to account for all the possible instances of that regularity. Given the potential open-endedness of such a class, there is no assurance that all the members will, or even can, be identified as a consequence of human research efforts.

It should be further kept in mind that, although the physicalist is ultimately concerned that there exist fundamental physical structures and processes that underlie regularities, it is allowed that there may well be many intermediate levels of organization, complexity, and abstraction.[33] For the purpose of explaining a given non-physical regularity exhibited by a given sort of system, the underlying mechanisms may themselves be non-physical in character and they may be constituted by objects and attributes drawn from several different levels of organization of the system. As emphasized in Chapter 1, the physicalist's only requirement is that any sequence of levels of realization of an instance of a regularity is ultimately grounded in the physical basis.

Alternatively put, one aim of inter-field research is to identify physically-based 'models' within which to embed higher-level phenomena. But simple embedding is not enough; an account of how lower-level, physically-based structures and processes suffice to realize instances of the various regularities exhibited at higher levels is required as well. And, as pointed out earlier in connection with T4, such explanation is *not* explanation in physics or any other lower-level discipline *per se*. It is a *sui generis* form of explanation that accounts for the vertical links between lower-level and higher-level domains. Inter-field inquiry is distinct from, although highly pertinent to, the fields that it aims to connect.

Given these preliminaries, the physicalist thesis concerning the explanation of regularities is as follows:

(T5) All instances of, and all exceptions to, natural regularities are explainable in terms of physically-based phenomena.

The idea is that the physicalist requires no more than that every possible instance and every possible exception to the regularities holding at higher levels are explainable in terms of lower-level, physical or physically-based phenomena. Such explanations consist

---

[33] The neurosciences provide an excellent illustration of how many different levels of organization and function are concurrently studied for the purpose of identifying mechanisms for the various regularities that are exhibited by the human nervous system.

in identifying the non-physical attributes involved in the regularity to be explained, identifying the relevant physically-based attributes involved in the specific instance that is to be explained, and providing an account of the relations between the physically-based attributes in question which constitute the causal or non-causal relation between the non-physical attributes.

For instances of some non-physical regularity, if an RT for each attribute involved is known, and if, as a result, the physically-based attributes that realize the non-physical attributes can be identified, and if relevant physically-based relations between those attributes are known, then an instance of a regularity is explained if the relation between the non-physical attributes is explained in terms of the relations between the physically-based attributes that realize the non-physical attributes. That is, if it is a high-level regularity that non-physical attribute A is a cause of non-physical attribute B, then an instance of this regularity is explained by citing the physical attributes C and D, which *realize* A and B respectively on that occasion,[34] and by explaining that C causes D. The physicalist thesis is that every possible instance of a non-physical regularity can be explained in roughly this way, although it is not assumed that all relations of interest are causal.

The idea that regularities studied in higher-level disciplines typically have exceptions is an important one whose significance for the concept of a law of nature has not been adequately studied. That all regularities in nature are exceptionless is untenable because, as one ascends a hierarchy of phenomena ordered in terms of complexity and the relation of realization, regularities at higher levels will typically have exceptions due to interferences from outside of a system, 'intrusions from beneath' (for example, defects in a mechanism), or failure of satisfaction of certain other *ceteris paribus* provisos (i.e. conditions required for the regularity to hold).[35] In my development of the physicalist doctrine, I assume that there can be such exceptions and that they must properly fit into a physicalist framework.[36]

[34] Among other things, C and D constitute the causal powers of A and B on this occasion.

[35] See Fodor (1975) for preliminary discussion of this neglected topic. And see Schiffer (1991) and Fodor (1991) for recent discussion of some of the issues as they bear upon psychological 'laws'.

[36] Field (unpublished) has also recently emphasized the importance of this demand on physicalism.

Specifically, the physicalist requires only that such exceptions be explainable along the same lines as the instances. For example, if an exception to a causal regularity is a case in which the antecedent, but not the consequent, attribute is instantiated, some other attribute being instantiated instead, then such an exception is explained via an account which clarifies the relations between the attribute(s) realizing the antecedent non-physical attribute and the attribute(s) realizing the non-physical attribute that replaced the expected consequent non-physical attribute *and* which explains how those relations between the physically-based attributes realize the relation between the two actually occurring non-physical attributes. In actual contexts of explanation, such vertical explanations will be embedded within a larger framework that explicitly reveals what particular intrusion, breakdown, or violation of *ceteris paribus* conditions led to the exception. Thus the explanation shows why the antecedent attribute was not accompanied by the attribute specified by the regularity, but was accompanied by the unexpected attribute instead. The physicalist view is that every possible exception to a non-physical regularity has such an explanation.

This aspect of the physicalist programme is, in my view, especially important. To acknowledge and take into account the existence of exceptions to high-level regularities, and hence the degrees of lawlikeness that high-level generalizations can possess, is to recognize how contingent, high-level structures can exhibit regularities that are *both* counter-factual supporting and yet not strict. It is a virtue of physicalism that it aims to account for these features of high-level systems. The counter-factual force of such regularities is based upon and realized by lower-level, relatively more exception-free physically-based regularities. The exceptions are explained in terms of intrusions, breakdowns, and inappropriate background conditions which are also grounded in the physical basis.[37] The embedding of a system in a larger physically-based

---

[37] A generalization will have diminished lawlikeness to the extent that it fails to take into account all the factors that *could* influence the phenomena it concerns. Such a failure is not necessarily a shortcoming, however. To the extent that there are unsystematic causes of exceptions, it would often be impractical and futile to try to incorporate all possible unsystematic influences in the statement of a law. On the other hand, idealizations away from systematic causes of exceptions can be of considerable value in gaining understanding of certain aspects of how a system works. Since the attainment of exceptionless generalizations at high levels would require the incorporation of all systematic and unsystematic causes of breakdown,

structure and the vertical explanation of the attributes involved provides an appropriate framework for such explanations. Such physically-based embedding has an additional virtue. With the study of breakdowns in the microstructure of a system and with the study of the systematic and non-systematic reactions of such microstructure to external influences, a deepened understanding of how a macro-level system works is gained. In particular, macro-level breakdown patterns can be better understood and dealt with. It is not just that the physicalist wants exceptions to be explained. The objective here is, again, increased understanding of how the system works. To ignore the non-exceptionless character of high-level regularities is to undermine such increases in understanding.

In light of this, it is easy to see that the emphasis some philosophers (for instance, P. M. Churchland 1984; P. S. Churchland 1986) place upon the methodological importance of 'smooth reductions' of one theory to another is misleading. The idea that if a theory is not smoothly reducible to lower-level, physically-based theories then it ought to be eliminated from science is flawed inasmuch as it does not take into account how exceptions to the higher-level theories arise. In my opinion, many of the arguments offered by the Churchlands, for example, in support of so-called 'eliminative materialism' are facile in this regard. Put another way, too aggressive an emphasis upon growth reductions, where the aim is to weed out the errors embodied in older theories by developing more powerful and more accurate newer theories to subsume and replace them, can lead one to mistake an important exception to a generalization for an error in theory. Structural reductions, in which the aim is to embed higher-level systems and theories within lower-level systems and theories, are geared specifically to accommodate exceptions which reveal the deeper character of what is going on. An exception to a law is not necessarily indicative of a fault. As a result, to aim singlemindedly for a system of exceptionless generalizations is very likely going to lead to missing what is going on in this regard. Although progress in science requires the elimination of error, this should not be equated with the elimination of non-exceptionless generalizations. Rather,

such generalizations are often either unattainable or undesirable. Thus, for both practical and theoretical reasons, the existence of exceptions need not undermine the importance of generalization.

the possibility of systematic or non-systematic exceptions must be accommodated and accounted for if high-level systems are to be adequately understood.[38]

As a consequence of these points about the sources of exceptions as well as points regarding the multiple realizability of high-level phenomena, T5 does not imply the derivability of all non-physical laws from physical laws. Hence it does not appear to capture the physicalist idea that the non-physical laws are explainable in terms of the physical laws in the way that the classical unity of laws thesis supposed. However, the derivability of laws is not required to capture the physicalist idea that everything that does or could happen in nature is explainable in physical terms. T4 and T5 do not leave room for something to take place without there being a physically-based explanation for it. What may go unexplained are the regularities themselves, if there is something more to explaining a regularity than explaining all of its instances. This is not a major loss for the physicalist as long as everything that either can or does take place in nature admits of a physically-based explanation. Perhaps it should have been expected that if not all attributes are physical attributes and if non-physical attributes can be multiply realized by physical attributes, then the regularities involving non-physical attributes would not in general be derivable from physical regularities.[39] And, as has recently been pointed out, the existence of genuine exceptions to higher-level regularities makes it undesirable to have derivations of such 'laws' from *true*, lower-level generalizations that are free of exceptions. Physicalists should therefore reject the classical derivational thesis as well as the thesis that all non-physical regularities are explainable in terms of physical laws, if this latter thesis requires such derivations. In their stead, T5 offers as much as the physicalist needs in order to achieve the explanatory goals of the programme. And it does so in a way that handles the problem that exceptions pose to the classical unity of laws.[40]

---

[38] I do not wish to be seen as opposing the value of growth reductions (in an appropriate sense of 'reduction'). What is required is recognition of both sorts of reduction, and attainment of a proper balance of the two in methodological arguments.

[39] See Fodor (1975) for a version of this sort of argument.

[40] See Field (unpublished) for a different approach to the explanation of high-level laws.

It should be pointed out that T5 does not preclude the possibility of finding unique underlying physical mechanisms for some non-physical regularities or of discovering exceptionless non-physical regularities. Nor does it preclude finding derivations of non-physical laws from physical laws. The point is, rather, that these are not requirements of the programme, since the explanatory goals of physicalism are served fully by T4 and T5. The aims of vertical explanation are to identify mechanisms that can individually play a role in accounting for some (or all) of the instances of a given regularity and that can jointly account for them all, and to identify the sources of whatever exceptions the regularity might exhibit. These aims are served whether there is one or many such mechanisms.

A further virtue of T5 is that it allows for the existence of 'emergent' laws in at least two senses. First, there may be laws that are not derivable, hence not predictable, on the basis of knowledge of lower-level laws. Second, there may be laws that 'kick in' at certain stages of development of the universe. RTs and vertical explanations based upon them will allow us to understand why certain phenomena occur and certain regularities hold only when certain kinds of structures have evolved. Neither of these sorts of emergent phenomena falsify physicalist theses, of course.

With respect to the integration of theory, T4 and T5 serve physicalist purposes in a number of ways. The existence of a system of explanations that is physically-based means that all explanations of phenomena can be linked, if necessary, by digging as deeply and as widely as is required to discover a physically-based ground for such linkages. Such a ground allows us to relate different ways of understanding how a system works (for example, as in cognitive science where psychology, linguistics, artificial intelligence research, anthropology, etc. are integrated in the study of cognition). And such a ground provides a framework for relating various theories in different domains (as when our understanding of economic factors is related to our understanding of social forces and events). Links between domains, especially links of a causal nature, are underwritten by physically-based relations and explanations. As I emphasized in Chapter 1, the physicalist maintains that all paths of interconnection and influence must be explainable in terms of physically-based phenomena.

It follows that the issue of the autonomy of domains of knowledge is addressed within a physicalist framework by denying that there are *any* distinct domains that are absolutely autonomous. Every domain is related to physics in ways prescribed by the theses (i.e. they are related in terms of relations of determination, physically-based realization, and physically-based vertical explanation). And every domain is related to every other domain via the common physical basis. The details of the relations between domains will vary from case to case, there being stronger constraint relations in some cases than in others. But since the collections of physically-based models and associated explanations that exist for various domains of inquiry are drawn from a unified, common, physically-based pool, it is not possible for any domain to be entirely independent of the others. In this regard, it is especially important to note that physics is not an autonomous domain in so far as physicalism requires that it should bear certain relations to all other domains.

Finally, to return to the issue of how distributions of objects and attributes in a higher-level, non-physical ontological domain is constrained by the 'laws' characteristic of that domain, we observed earlier that, since such laws are not exceptionless, the constraints are more complicated than in the case of the physical domain. Given our discussion of such exceptions and how they are to be accommodated within a physicalist system, the main constraint on such distributions is that either they are in conformity with the counter-factual supporting generalizations of the domain or they are in conformity with deeper and more comprehensive generalizations of greater nomic force. That is, any exception to a generalization must be subsumed under some more exception-free generalizations that rein in the sources of the exception.

This completes my discussion of the physicalist theses. In the next two chapters I shall present and discuss a set of metatheses concerning them and I shall frame and evaluate a number of outstanding objections against them to which physicalists ought to provide satisfactory responses.

5

The Metatheses of Physicalism

Why should not the unit of empirical inquiry be the whole of culture (including both the *Natur-* and the *Geisteswissenschaften*) rather than just the whole of physical science?

R. Rorty (1979: 201)

I believe that we need this goal [unification], not only as a guide for research, but also to serve as a standard for the assessment of the deficiencies in our understanding.

R. Causey (1981: 231)

The purpose of this chapter is to formulate a number of metatheses concerning the physicalist principles just developed. The importance of framing metatheses lies in the clarification they provide concerning the kind of theses that physicalism involves (for example, *a priori* or *a posteriori*, necessary or contingent), the kinds of considerations that are relevant to acceptance or rejection of the theses, the domains to which the theses are supposed to apply, and the ways in which the theses can be used in the conduct of inquiry. They play an important part in gaining an understanding of the significance of the physicalist programme.

During the past fifty years or so, there has not been complete agreement regarding the metatheses of the programme, although there has been something like a received or dominant view. As I shall formulate it, this 'received view' consists of four metatheses, as follows:[1]

(M1) The theses of physicalism apply to natural science.

(M2) The theses are *a posteriori* truths.

[1] Expression of such claims as (M1) to (M4) can be found in Boyd (unpublished), Causey (1977), Field (1972; unpublished), Fodor (1975), Friedman (1975), Hellman and Thompson (1975; 1977), Lewis (1983), Oppenheim and Putnam (1958), and Post (1987).

(M3) The theses both play and ought to play a regulative role in the conduct of inquiry within natural science.

(M4) The theses are contingently true.

Despite the dominance of M1 to M4, there have been voices of opposition from thinkers not entirely unsympathetic to physicalist thought. For example, Chomsky (1968; 1975; 1980) has been quite outspoken in his attacks on the alleged *a posteriori* status of physicalist theses, as well as on the ideas that such theses have a well-defined, independently specifiable scope of application and that they play the methodological roles attributable to them by physicalists. Teller (1984) has urged that the theses should be viewed as necessary truths. And Post (1987) has endorsed the idea that the theses have a universal scope and a variable modal status that depends upon the domain of application. Others, in reflecting on earlier versions of the programme, have denied that physicalist principles play much of a role in the conduct of inquiry.[2]

My plan is to review and clarify each of the component claims of the received view. It will emerge that this view is not tenable, that some revisions are in order, and that there are some serious problems to which defenders of physicalism must respond. Recognition of the defects of the received view, however, is not a disaster for physicalism. Rather, it leads to a deepened understanding of the nature of the physicalist programme and a clarification of the broader philosophical commitments that physicalists must assume. Consideration of the metatheses will also pave the way for subsequent discussion of various objections that threaten the acceptability of the theses.

5.1. PHYSICALISM AND THE NATURAL ORDER

By the 'scope' of the theses is meant the extent of their intended applicability: do they apply only to specific domains (for example, science) or do they apply to all domains of existence and knowledge? Proponents of M1 endorse the narrower construal if only because of their restricted interest in natural science, but such a restriction can also represent a form of caution as well. In addition

---

[2] E.g. with regard to biological research, a number of philosophers have opposed the idea that classical physicalist reductionism has played much of a methodological role. See Hull (1974), Schaffner (1977), and Wimsatt (1976).

to advancing relatively weak theses to avoid the problems of a predecessor, as some contemporary physicalists have done when faced with the problem of retrenchment, restricting the intended scope of applicability of the theses is another guarded way to pursue the physicalist programme. In Chapter 1, however, I underscored the physicalist's interest in a universal scope of application, and I argued that there is much to be gained from such a forward approach, although the risks are higher as well.

As a consequence, I shall endorse a very broad form of physicalism, and the first metathesis must be revised to reflect this decision:

> (M1') The theses of physicalism apply to all natural phenomena and all claims to truth and knowledge concerning the natural order.

There are, it appears, no domains that are exempt from physicalist demands. In addition to physical, chemical, and biological phenomena, physicalist theses concern the psychological, the social, the moral, the aesthetic, and the commonsense world of our everyday experience. However, one qualification has been introduced. We should understand that the theses are concerned only with the natural order, with the universe in which we live and with which we interact. Physicalism applies to *all* phenomena that are real or possible and to *all* claims that are objectively true or false in nature. This means that the theses do not apply to abstract realms, if there are any: mathematical structures, universals, propositions, and the like. This qualification will probably be met with two sorts of criticism reflecting alternative approaches to the scope of physicalism.

The first approach goes as follows. Given the vital role of mathematics in physics and given that, to construe that role properly, mathematics must be understood as consisting of truths about an abstract realm of objects (for example, sets), it is only fitting *to include* abstract mathematical structures (for example, a hierarchy of sets) within the *physical* bases of the system and to endorse an entirely unrestricted scope of applicability for the theses, a scope that includes abstract entities such as attributes, *possibilia*, and the like. The virtues of this approach are that it leads to a completely comprehensive ontology exhausting both the natural order and the realm of *abstracta*, and that it does not require any principle for distinguishing the two realms or for circumscribing the scope of the principles.[3]

However, I think this approach is faulty for at least two reasons. To begin with, it 'pollutes' the physical bases. In my opinion, 'the physical' does not mean 'the mathematical-physical'. Physicalists are (or should be) concerned with what exists in nature: i.e. with what can be spatially and temporally related to us, with that with which we can interact and by which we can be influenced, and with that of which we and the things around us are made. Physicists, of course, are in the business of studying exactly such matters. And it is for this reason that ghosts, gods, and the paranormal are genuine threats to physicalism: they appear to be *in nature* in various ways, but they are not encompassed by physics. However, sets, propositions, universals, and so on, when abstractly conceived, are not considered to be in nature at all. Nor are they within the scope of the physicist's domain of study (for example, because a physicist appeals to sets, he or she does not *ipso facto* become a set theorist). For these reasons *abstracta* do not pose serious threats to physicalism: they are not in nature (i.e. they cannot be related to us spatially, temporally or causally), nor are they properly encompassed by physics.

Put another way, the 'mathematical-physical' pollutes the physical bases because those bases are supposed to provide the constituents upon which all *natural* objects and their attributes depend and supervene and upon the basis of which all are realized. Physicalism is the view that the ontology of physics suffices for providing those constituents. But nothing in the natural order depends upon, supervenes upon, or is realized by sets or universals or propositions. Commitment to the existence of these latter sorts of entities arises out of different concerns from those of the physicalist. As a result, the claims that physics requires mathematics and that mathematics requires a realist interpretation do not (even if true) entail that the bases must include *abstracta* in order to do their job in a physicalist system.

This leads me to my second objection to the approach of the

---

[3] This is pretty much the approach taken in Hellman and Thompson (1975). Post (1987) takes a somewhat different line. He explicitly hedges commitment to abstract objects because he (correctly) holds that physicalists need not take a stand on debates between nominalists and platonists (167–8). However, he appears to hold that if platonism wins out, then the physicalist bases must include abstract objects. As a consequence of this claim, much of what I say in the text applies to Post as well as to Hellman and Thompson.

mathematical-physicalists: it saddles the physicalist with controversial and unnecessary philosophical commitments. Mathematics
and mathematical structures are tools of the physicist employed
for the purpose of modelling and representing those aspects of the
natural order with which physical inquiry is concerned. But it is
important to disentangle the issues of interest to the physicalist
from the issues of interest to the philosopher of mathematics. The
former is concerned with the place of physics in a system of
knowledge that incorporates all that occurs in nature, while the
latter is concerned with how to construe mathematical knowledge,
truth, and existence. These two sorts of issue are distinct. However the disputes in the philosophy of mathematics are resolved,
physicalism will be substantially unaffected, and vice versa. Indeed, I think that commitment to mathematical realism is clearly
a gratuitous association to the physicalist programme and that
physicalists are wise to shed it. Physicalism is formulable and
defensible independently of how one comes down on the nature of
mathematics.

I do not mean to suggest here that I differ from Hellman and
Thompson, Quine, and others who hold that sets are real, abstract
entities. The point is that such a view is not a commitment of physicalism. Were it to turn out that such realism is the *only* viable
position concerning how to understand mathematical truth and
existence, then physicalists, like everyone else, should be committed to it. But such a commitment would not be a product of their
being physicalists, and such a commitment would not require the
physicalist to include *abstracta* within the physicalist bases. The
physicalist includes in the physical bases only those entities required for making physicalist claims, as I argued above. In any
event, I know of no such consensus among philosophers of mathematics about the status of realism, and there is no reason why
physicalists must join the fray. Physicalists can, I believe, patiently
await the outcome of the debates over this issue without being at
risk and without making provisions for a possible victory by
mathematical realists.

A second sort of criticism of my restriction of the scope of
physicalist theses to the natural order is that the resulting
physicalism is a pale shadow of the exciting doctrine that would
claim that the physical does truly exhaust, determine, and explain
all that there is. 'Heroic physicalism', as I shall call it, is a programme committed to establishing that there is *nothing* that is not

encompassed by the physical bases, where such bases are not the polluted bases of the mathematical-physicalist ('the M-physicalist'). This form of physicalism is committed, perhaps, to nominalism, actualism, finitism, and other 'isms' that underwrite its radical ontological picture. Proponents of this view of the physicalist programme include both advocates and critics of physicalism. Thus the anti-physicalist will charge that both I and the MP-physicalist are cheating, each in our own special way, and that if physicalism is not illicitly spared,[4] then it is quite evidently false because it does not accommodate abstract objects. On the other hand, the advocate of heroic physicalism will pursue, in addition to the sorts of issues outlined in Chapter 1, all sorts of other philosophical problems conceived of as part of a global defence of physicalist doctrine.[5]

My reply here is that, whereas the MP-physicalist pollutes the physical bases, the proponent of heroic physicalism muddies the philosophical waters. I certainly do not deny that such a grand philosophical programme is possible to pursue and that, if it were successful it would be interesting and important. However, I deny that all the other 'isms' entailed by such an approach are necessarily commitments of a physicalist. Physicalism is concerned with a certain kind of structure that is characteristic of nature and of knowledge concerning it. But, an interest in physicalist structure requires neither commitment to *abstracta* nor commitment to their elimination. Such an interest can be pursued in either philosophical environment. Whether there are or are not *abstracta*, the study of nature will survive and physicalism will survive as a programme intimately connected to it.[6]

---

[4] It might be charged that physicalism is illicitly spared either by including *abstracta* within the bases (as the MP-physicalist does) or by viewing issues regarding *abstracta* as not within the purview of the physicalist's programme (as I do).

[5] Hartry Field's (1980) important efforts to provide a nominalist construal of physical theory come to mind. However, Field has stated in conversation that he does *not* view his work as part of a defence of 'heroic physicalism'. Indeed, if I understand him correctly, he views physicalism and nominalism as independent projects much in the same way as I am suggesting in the text.

[6] It might be asked how I can reconcile these remarks with my appeal to attributes in the development of the physicalist theses. My reply is that such an appeal is a matter of expository convenience. If it turns out that attributes, conceived of as abstract universals, do not exist, then physicalism can be expressed in ways that do not presuppose their existence without significant change of content. My own philosophical biases have exerted themselves above, but they are not essential to expression of physicalist thought. Thanks to Ed Becker for raising this question.

Now, it could turn out that platonism, or some other form of mathematical realism, is the only viable philosophical account of mathematical knowledge, truth, and existence. If so, then platonism will be an accompaniment of physicalism, although for the reasons cited above, I think it would be a mistake to include mathematical objects and truths in the *physical* bases. Or it could turn out that a nominalist construal of mathematics wins out. If so, nominalism will be an accompaniment of physicalism. In any event, prior to a definitive victory, I consider it a strategic error on the part of physicalists to side with either of these positions. Such taking of sides obscures the real concerns of physicalism *vis-à-vis* the natural order, and it means that physicalism runs the risk of going down in flames because of a gratuitous affiliation with an independent philosophical programme.

It will not have escaped notice that, in taking this stand, I have assumed the burden of defending the existence of a distinction between the natural order and the abstract realm (if there is one). And this burden concerns not only my replies to the objections just reviewed. If an independent characterization of nature cannot be framed and motivated, then a number of other issues will be adversely affected: i.e. do the theses have a well-defined content?, are there clearly defined sources of confirmatory and disconfirmatory evidence for the theses?, and is the scope of application of physicalist theses in their methodological role well defined? In short, if the natural order cannot be pinned down, then the content, empirical status, and utility of physicalist theses are threatened.

Early physicalists such as Carnap and Feigl distinguished natural science from formal science and the natural order from abstract realms. Thus the natural order was construed as a space-time manifold the extents of which put limits upon that with which we can be in causal contact: the so-called 'space-time causal nexus'. I endorse this picture for the most part. The natural order is, as currently conceived, defined by a space-time manifold, but such a realm is not conceived in terms of causal connection with humans. The natural order is far more extensive than whatever regions are defined by such local causal relations. For example, such connections fall well short of circumscribing everything that can be temporally or spatially related to us, relative to some reference frame. The extent of space-time, if discoverable at all, is a problem in theoretical physics, not a matter for *a priori* speculation. But it is reasonably safe to say that, however such research turns out, sets,

universals, and propositions are not going to be located in regions
of space-time (although their members, instances, or expressions
may be). Thus I assume the standard view that, since being located
in space-time is both a necessary and a sufficient condition for
being a part of the natural order and since *abstracta* are not so
located, they are not part of the natural order and hence not
within the scope of physicalist doctrine.

At this point, I shall consider an objection which claims that
this approach trivializes physicalism, namely that if we view the
scope of the theses as concerned with all that occurs in the natural
order and we regard the natural order as delimited by the extent
of space-time and we allow that it is the job of physics to map out
the extent of space-time, then the theses are guaranteed to be true
if physics is successful. However, this objection is readily turned
back since physicalism asserts that physics grounds all *structure* in
nature and all knowledge about nature in the ways specified by
$T1$ to $T5$, and this goes well beyond the assumptions just made
about the natural order.[7] The ideas that nature is circumscribed by
space-time and that physics maps out space-time do not have any
implications regarding whether the contents of the regions
of space-time are related to each other in the ways called for by
physicalist theses.

5.2. A POSTERIORI STATUS AND ITS PROBLEMS

The metathesis $M2$ is widely endorsed by physicalists. Consider,
for example, the following typical statements:

Broadly empirical in character, they [physicalist theses] are supported
inductively by scientific practice. (Hellman and Thompson 1977: 311.)

... physicalism is meant to be criticizable by way of observational testing.
According to contemporary physicalists, the principles of physicalism are
to be treated as high-level empirical hypotheses or generalizations ... If
phenomena turn up that resist a physicalist account even after years of
trying, the physicalist's own principles should be rejected or revised. (Post
1991: 95.)

The significance of this view is, at least, twofold. The theses are
a posteriori hypotheses subject to confirmation or refutation by

---

[7] To is guaranteed by this view of the natural order.

arguments based on empirical evidence and methodological principles of empirical assessment. Further, they purport to *truly* describe features of the natural order and our knowledge concerning it. Thus, given M1′, they apply to themselves; and, given M3, they play a role in their own empirical assessment. That is, given the universal scope of physicalist theses with respect to the natural order, and given that the theses constitute descriptions of both nature and the structure of knowledge concerning it, they must satisfy the demands that they themselves make upon all legitimate claims to knowledge of nature. In the remainder of this section, I shall discuss both these points of significance. It will emerge that there are serious problems that are fostered by M2 and that, consequently, there is considerable pressure to make revisions.

The alleged *a posteriori* status of the theses raises a number of questions: 'What kinds of evidence and arguments are pertinent to the confirmation of physicalist theses?', 'What kinds of evidence and argument are pertinent to their disconfirmation?', 'How does the assessment of the theses vary from case to case?', and 'What is the nature and status of the programme for gathering evidence and presenting arguments?'.

I shall begin with some brief comments on the third and fourth questions. Because the theses are more or less independent of each other and because they vary with regard to the types of entities and relations they concern, the kinds of evidence and argument pertinent to their evaluation varies considerably. A full treatment of the assessment of the theses must take this variability into account. In this book, however, I must avoid the full complexity of the task. Instead, I shall focus on general issues that bear upon all evidential considerations marshalled for or against the theses. With respect to the programme for gathering evidence and presenting arguments bearing upon physicalist theses, assessing its status is also essential for a full evaluation of the physicalist programme. Here too I must avoid the full complexity of the task and limit myself to a somewhat strategic discussion in Chapter 6. My focus in the following discussion will be on the first two questions identified above.

According to the received view, the primary sources of evidence for the evaluation of physicalist theses are developments in science. However, with a broadened conception of the scope of the theses, the sources of relevant evidence will also have to be suitably broadened to include developments in other areas of knowledge (for example, the study of values). Given such sources of evidence, there are at least four types of argument that might be proposed in the service of *confirmation* of the theses. First, it might be held (improbably) that only by an exhaustive study of all cases within the structure of knowledge and the natural order could one properly assess the theses and advance a convincing case for their truth. That is, all domains must be considered and, indeed, only in some ideal limit at which knowledge of nature has completely evolved can a proper assessment of physicalism be made with regard to its degree of confirmation.

Not only is this a totally impractical proposal because it defers indefinitely the empirical assessment of the theses, but it is also not a very sound one either. Many of the ontological and theoretical structures that would require assessment are highly unstable, too large, too small, otherwise inaccessible, or are too difficult to be studied by humans as separate cases (for example, large stretches of the universe, highly complicated patches of theory, singularities in space-time, highly complicated non-linear phenomena, the evolution of the universe and its 'laws', and so on). I mention this strategy of assessment to point out some of the difficulties that must be overcome by a more acceptable method:

- the temporal unfolding of the universe requiring the method to generalize over time and to take into account evolutionary considerations;
- size considerations requiring the method to reach both the very large and the very small;
- complexity of phenomena and theory that may require the method to justify reliance upon idealizations;
- inaccessible phenomena that require the method to generalize to cases that cannot be directly studied;
- possible, but not actual, cases requiring the method to underwrite counter-factuals.

Of course, it is the essence of justificatory practices in the sciences to attempt to manage these and other complications, and proponents of the received view surely have such practices in mind when they endorse M2. However, the standard responses to such difficulties and the standard scientific methods have serious problems when deployed in the context of the evaluation of physicalism, as we shall see shortly.

The next, and most popular, strategy for marshalling evidence in support of physicalist theses involves some sort of induction based on developments in science, philosophy, and elsewhere. Such an inductive strategy is clearly superior to an exhaustive case study if it underwrites solutions to the various problems listed above. The basic idea is to study the details of particular cases of relations between non-physical and physical phenomena or theories, to accumulate a number of successes for physicalism, and then to generalize over the unstudied cases. However, the nature of such inductions is rarely clarified and the problems associated with this approach are either not noticed or are ignored. For the present, I shall limit my discussion to the identification of a number of questions concerning both the cases and the form of argument vaguely appealed to by proponents of M2. At the very least, a burden lies upon their shoulders to respond to the questions and thereby to clarify their inductive strategy *vis-à-vis* physicalist theses.

One might begin by asking what principle of induction is supposed to license the inference from a sample of successes to the truth of general physicalist principles? A simple enumerative induction, surely, is not what defenders of M2 have in mind. But how, then, is the induction to proceed? The questions that follow concern the direction that such a clarification might take.

For example, what counts as a 'representative' sample of cases that is sufficient for providing the basis for an induction in support of some physicalist principle? Are there specific cases that must be included in any sample (for instance, cases of mind, meaning, and values)? Or are we permitted to sample a few core cases (for instance, cases drawn from physics, chemistry, and biology)? Are some cases more important than others? To underwrite an inductive argument in support of physicalism, it clearly will *not* do to rely upon a few obvious victories (chemistry, for example). But then a more definite map of the domain over which the theses range must be made, and some principles of proper sampling of this domain must be formulated. In particular, since the scope of physicalism extends beyond the boundaries of science, inductive samples will have to reach out beyond those boundaries as well.

How this is to be done remains unclear. I do not dispute with fellow physicalists that we have grounds for being encouraged by 'the evidence', but we must not confuse grounds for encouragement with a firm basis of support. There is a long history of programmes of research that were encouraged by narrowly circumscribed evidence only to be roundly defeated when a broader evidential base was considered (for example, behaviourist psychology). At the very least, physicalists have an obligation to present a more definite plan for developing inductive arguments for their claims.

Along the same lines, we should ask: 'What counts as a "good" case?' Not just any case from a domain will serve the purposes of a sound induction. Should it be required that only cases drawn from mature sciences and other disciplines be examined? If so, what counts as a 'mature' discipline? Are certain areas of philosophy (for example, philosophy of mind, moral theory) mature disciplines in the required sense? Must a confident evidential assessment await the maturity of psychology and the social sciences? The cost of prematurely sampling a domain is the risk of highly imperfect and contaminated data. Since the prospects of physicalism depend upon what there is and what is true in other domains beside physics, genuine counter-examples may be missed or genuine supportive cases may not be adequately developed unless advanced understanding in a domain is reached. For example, how could we ever tell if the mental is realized by the physical if we do not understand the domain of the mental? But no one I know seriously believes that psychology is a mature discipline, let alone the social sciences, aesthetics, or moral theory. Does this mean that cases drawn from these areas of inquiry do not count as relevant evidence at this time?

Not only must the domain with which the inference is concerned be well defined and mappable to some reasonable extent and not only must proper sampling procedures be clarified, but it must also be clear what are the appropriate conditions for an inductive argument and the sorts of domains with respect to which inductive methods of inference are legitimately employed. There is a significant cause for concern that certain assumptions about nature and the knowledge of nature which might be made along these lines (for instance, assumptions of uniformity) may be question-begging *vis-à-vis* physicalism. For example, the methods of inductive inference we employ might be based on a certain view of the world, a view which is itself informed by physicalist assumptions.[8] So it is imperative that any assumptions underwriting an induction should be clarified to ensure that physicalist principles are not surreptitiously supporting themselves by shaping our understanding of the population being sampled.

One serious version of this problem, to be discussed later, is suggested by metathesis M3. The normative employment of physicalist principles in the task of mapping out reality and objective knowledge may skew the evidential case in their own favour by screening out legitimate counter-examples. A proper induction must be performed upon a fair and unbiased selection of cases from the general population of interest. That physicalist theses play a role in our conception of what falls within that population, while not necessarily a problem, is a cause for concern. Those who believe that support for physicalism can be gleaned from an induction performed on past successes must, therefore, articulate the nature and presuppositions of such a procedure before it can be accepted as sound.

My own view, at this point, is that vague talk of inductions over local successes will not, by itself, suffice to support the claim that physicalist theses are *a posteriori* in character. The sequence of unanswered questions just reviewed means that physicalists have been much too casual in endorsing M2 and that reliance upon 'inductive' techniques of support is not justifiable at this time. If no other legitimate forms of empirical argument are available and if the defects just scouted cannot be eliminated, then M2 is placed in serious jeopardy.

A different form of confirmatory argument for physicalist principles runs as follows: If assuming the truth of physicalist theses leads to certain kinds of scientific or other success (for example, empirically adequate theories, novel predictions, unified theories with considerable explanatory power), then a plausible explanation of those successes is that the assumption is true. Further, if

---

[8] I cannot pause over the issues here, but the main point is that, unless assumptions are made about a population from which inductive samples are drawn, there is no reason to suppose that a given analysis of the sample provides a basis for concluding anything about all elements in the population. This issue is especially acute for consideration of inductions that attempt to cover 'tough' cases that are not specifically studied. It is with respect to such assumptions that the threat of begging the question arises.

all other available explanations are inferior, then the theses would gain in support via an inference to the best explanation. Alternatively put, the extent to which the theses reliably lead to the development of a system that exhibits various theoretical and empirical virtues determines how much confirmation accrues to them. This is because reliable principles, and the methods based on them, ought to be explained and the best explanation of such reliability is that the principles are true.[9]

The virtues of this sort of argument are similar to those of inductive arguments: it permits inference on the basis of a finite set of actual cases to the entire structure of nature and knowledge about it. However, the liabilities of inductive strategies are also shared, and there are some additional difficulties as well. Thus, analogues to the questions raised above can be posed with respect to this form of confirmatory argument. What assumptions are made about the domain to which the strategy is applied? What kinds of cases must be considered? What is a representative sample of cases? What counts as a 'good' case? In short, given a sample of successes based on physicalist assumptions, what must be assumed and what must be done in order to underwrite confidence in the full generality of the truth of the theses? Further, of course, so-called 'inference to the best explanation' and various related forms of argument[10] have been criticized by many (Cartwright 1983; Fine 1986; Hacking 1983; Van Fraasen 1980; Wagner unpublished) who believe that they are not legitimate forms of evidential argument at all. If what the critics claim is plausible, then this form of reasoning is, at best, a controversial empirical method, and its employment may mean that physicalists take on additional philosophical commitments that are not central to the physicalist programme.

Regarding this last problem, and despite my sympathies for this strategy of argumentation, I think physicalists should be wary of assuming an unnecessary epistemological commitment here. At least, it should not be assumed until there is consensus regarding the legitimacy of the method of argument. An alternative possibility,

[9] This is, of course, an adaptation of ideas and arguments found in various of Richard Boyd's writings (e.g. Boyd 1985).

[10] E.g. the view that a theory gains in degree of empirical confirmation as a result of being highly unified or as a result of possessing considerable explanatory power.

of course, is that any form of physicalism that includes M2 makes a commitment to the legitimacy of these forms of empirical argument. This, however, is not evident and requires some argumentation. With respect to the first problem with this method of argument (i.e. the unclarified assumptions regarding its employment), the difficulties and issues are no more resolved in the case of inference to the best explanation and success arguments than they are in the case of direct inductive arguments. Physicalists appear to be relying on evidential strategies without attending to the details of their employment in the specific case in hand. Until such difficulties and obscurities are resolved, this form of evidential argument is also not justifiably employed by physicalists.

The next form of argument for physicalist theses that I shall consider has been characterized by Boyd (1980), who suggests that support for physicalism can be obtained from 'possibility arguments' which undermine inappropriate scepticism and provide some plausibility for the theses. He cites three cases, drawn from contemporary science, which are alleged to demonstrate that traditional thorns in the physicalist's side are not as sharp as opponents of physicalism think. He suggests that the formulation of functionalist metatheories for psychology, the discovery of microbiological mechanisms for the genetic transmission of traits, and various developments in artificial intelligence research all provide grounds for rejecting stock sceptical arguments against physicalism by showing the *possibility* of the physical realization of such non-physical phenomena as intentionality, life, and intelligence.

I cannot pause to give full consideration to the specific merits and weaknesses of such possibility arguments. But I suggest that, as a form of *confirmatory* argument for physicalist theses, these arguments are subject to the same difficulties faced by the previous forms. Although they might achieve their purpose of undermining inappropriate scepticism regarding specific phenomena, it is not at all clear how much support they provide for the theses in their full generality, and it is not at all clear whether it is possible to specify and defend the conditions and assumptions upon which such arguments can be justifiably said to make the theses more plausible. One central issue facing all the forms of supportive argument is whether they can be justified as legitimate forms of confirmatory argument for physicalism without making question-begging assumptions. In this instance, it is not evident what must be assumed in order for the provision of possibility arguments in a handful of cases to provide support for the general theses of physicalism.

Finally, along the lines discussed in Chapter 1, it might be thought that, because of the unifying and explanatory *potentials* of physicalism, some kind of *a priori* support accrues to the theses. However, although there are considerable ontological and theoretical virtues that physicalism would promote if the theses are true and the programme is successfully worked out, I think these provide nothing more than strong *motivation* for pursuing the programme, not good grounds for believing the theses are true or that the programme will be successful. In any event, the notion of *a priori* evidential support for an empirical theory is a nebulous and suspect notion. Physicalists who support M2 are ill-advised to hang their hat on such a form of confirmatory argument.

At this point, I conclude that, although M2 is a popular metatheoretical claim regarding physicalist theses, it has not been adequately defended by proponents. In the absence of such a defence, the burden of proof lies with those who contend that the theses are indeed *a posteriori* in character. I do not, of course, deny that many of the types of evidence and arguments offered in support of physicalism can be supportive in some domains. My primary concern is whether there is an appropriate framework underwriting those arguments in the case of physicalism. I do not believe that physicalists have shown that there is, because they have not put together a coherent and defensible plan for empirically defending the theses. If no such plan is developed, the *a posteriori* status of the theses is suspect.

The problems with M2 do not end here, however. I now turn to discussion of possible disconfirmatory evidence and arguments, and I shall argue that there are severe problems with justifying the idea that the theses are empirically disconfirmable. In this context, the second question raised by the alleged *a posteriori* status of physicalist theses can be restated as follows: What are the kinds of developments in science or elsewhere that would provide an evidential basis for disconfirmatory arguments against the theses of physicalism?

The first and principal sort of developments, of course, would be the existence of explicit counter-examples to physicalist theses:

for instance, instantiations of genuine attributes that are not realized by any class of instantiated physical attributes. As natural as this sort of consideration is, it is beset with several difficulties that cast doubt upon the *a posteriori* status of the theses.

The primary difficulty is that *identifying legitimate counter-examples* is quite problematic. There are, as Field (1972: 357) pointed out a number of years ago, quite probably no 'crucial experiments' that would decide the fate of physicalism once and for all. As a consequence, there are no easy routes to the discovery of counter-examples. In addition, the mere presentation of some phenomenon that does not clearly fit into the physicalist framework ought not be taken to refute the theses unless, as a minimum, there are some assurances that the presentation is accurate, that serious attempts to incorporate have been made, and that there are no unexplored avenues worth pursuing.[11] Further, as 'overarching empirical hypotheses', the physicalist theses possess a considerable amount of 'centrality' within the total corpus of our knowledge. That is, there is much greater conservative pressure to retain physicalist theses than there is to retain any particular alleged counter-example to them. Such centrality suggests that the theses would not be easily given up in the face of putative counter-examples *unless* a viable alternative view, capable of filling the gaps left by physicalist doctrine, were available and all strenuous efforts to undermine the alleged counter-examples have failed.

Another consideration is this. As Chomsky (1975; 1980; 1988) has suggestively discussed, nature may abound with mysteries that are beyond our cognitive capacities to fathom. Such a possibility means that we may have difficulty in recognizing a legitimate counter-example to physicalism as opposed to a mystery, or merely a problem for which there exists a solution which we are capable of discovering but which we have not yet actually discovered (cf. Bromberger 1966 for seminal discussion of this sort of situation). The repeated failure in trying to establish that physicalism is satisfied in a given domain (for example, mental states involving semantic content) may provide some grounds for thinking a counter-example exists, but such grounds will be substantial only if the alternative possibilities (i.e. problems and mysteries) can be

---

[11] See below, in the discussion of M3, for further conditions that must be satisfied before rejection of physicalism is appropriate in the face of apparent counter-example.

plausibly ruled out. Deciding when 'enough is enough' is not a very clear enterprise at all, and thousands of years of conceptual and theoretical change should warn us against abandoning a problem too hastily.

So, if the main form of disconfirmatory argument is one based on the discovery of a counter-example, then the above considerations suggest that such an argument may be difficult to get off the ground, although not clearly impossible. However, there are other considerations which suggest that such a strategy may suffer from more principled difficulties. One possible further complication is that the identification of genuine counter-examples requires a satisfactory resolution of the problems with the scope of the theses. Unless we have a clear idea of what the natural order is, we will have no clear idea whether certain putative counter-examples are genuine. Something cannot be a counter-example to physicalism if it falls outside of the natural order. As the earlier discussion of *abstracta* suggested, this problem is not necessarily insurmountable.

There are two further problems with the identification of counter-examples, which call for consideration. The first, raised above in my discussion of the physical bases, is the idea that any intractable phenomenon could be downwardly incorporated into the physical bases, thereby sparing physicalism from embarrassment. The *physical*, being open-ended and evolving, could incorporate whatever is required to keep physicalism afloat. However, this line of reasoning, as stated, is defective, as I have already indicated. Among other things, it confuses keeping physicalism afloat with keeping physics afloat. That is, it is fair to say that whatever is required to do the work of *physics* is a physical entity, and what is incorporated into physics is a serious matter of theoretical and empirical demands imposed upon answers to certain research questions. But this means, among other things, that not anything at all can be arbitrarily included within the domain of the physical, and, specifically, not just anything required simply to spare physicalism from disaster.

However, a full defence of M2 does require that genuine counter-examples be identifiable and that the potential revisability of physics be understood and taken into account. And it requires that the ground rules on legitimate downward incorporation be articulated and defended. Thus the earlier points concerning the difference between physics and physicalism should not lead one to think that physicalism puts no pressure upon physics. Indeed, M3 is the claim that physicalism plays an important methodological role in the conduct of inquiry, and, as I shall discuss below, that role puts constraints upon the co-evolution of physics and all other branches of knowledge. But those constraints are not captured in the simplified way suggested by the trivial downward incorporation objection, and they certainly do not mean that physicalism is insulated from counter-example because any putative counter-example could be transformed into just another element of the physical basis. Thus there is little reason to fear for our ability to identify counter-examples because of *trivial* downward incorporation. Legitimate downward incorporation, on the other hand, needs to be more fully understood and will be discussed in Chapter 6.

A second sort of attack on the possibility of identifying genuine counter-examples emerges at this point. The methodological role of physicalist theses could mean that they can be used to screen out potential counter-examples. That is, since physicalist theses have a normative as well as a descriptive function according to the received view, in their normative function they could be used to argue that some phenomenon that does not fit into the system is not objectively real and that claims about it are not objectively true or false. Thus, such phenomena and associated claims do not pose any threat to the system, since a genuine counter-example is something that either is real or has an objective truth value *and* does not satisfy physicalist constraints.

This objection is, in fact, the flip side of the last. Whereas the earlier objection claimed that physics could be readily revised to spare the doctrine, the present objection claims that recalcitrant phenomena can always be rejected given the normative role of the theses in deciding what is real or objectively true or false. Both objections take physicalism as a fixed point which, in times of conflict, always forces a revision elsewhere in the system. As a consequence, no counter-examples to the doctrine are possible. My reply here is comparable to my reply to the last objection: both grossly oversimplify the methodological functions of physicalism, and much too hastily conclude that counter-examples to the theses are not possible. I shall defer further discussion of this point until Chapter 6, when we shall have before us a more detailed account of the normative functions and modal status of the theses. For the present, I want to suggest that this objection, when framed in a more sophisticated way, does indeed pose serious difficulties for the empirical status of the theses and must be overcome if M2 is to survive.

To summarize the discussion of the first point of significance regarding M2: both confirmatory and disconfirmatory arguments suffer from severe practical difficulties in their application as well as from some potentially serious, more principled, difficulties. As a result, the possibility that physicalist theses are not subject to empirical assessment, and hence are not *a posteriori*, must be entertained, pending adequate responses to these difficulties.

The second point of significance regarding M2 is that, given M1, the theses apply to themselves since they purport to be true claims about the natural order and knowledge concerning it. As a consequence, the idea that the physicalist theses are 'over-arching empirical hypotheses', though quite natural and widely endorsed, raises serious difficulties which cannot be easily met. In a nutshell, because the theses themselves involve semantical, intensional, and other philosophically intricate notions (for example, *explanation*, *determination*, *truth*, *realization*), viewing the theses as being *a posteriori* truths regarding the natural order presupposes that these notions, the claims employing them, and the attributes and entities they concern, are all 'physicalistically acceptable' in the ways called for by the theses (i.e. the truth values of the theses are determined by the truths of physics, there exists an RT for each of the attributes, etc.).

The issues here do not lead to definitive refutations of physicalism, but they do set the task of working out the physicalist programme at a very high level of difficulty (i.e. the success of the programme involves incorporating the theses within the very system they describe in the ways they describe). Thus these issues create a certain amount of embarrassment for someone like Quine, who has steadfastly held to both M2 and the physicalist *unacceptability* of semantic and intensional notions. For the moment, I shall only observe that M2 raises the question of how physicalist doctrine is to be located in relation to the system it describes. The specific answer it provides (viz., that the theses are an integral part of that system) has some potentially serious liabilities that I shall discuss in Chapter 6.

5.3. METHODOLOGICAL ROLES AND THE RESOLUTION OF
CONFLICT

The third metathesis, M3, expresses the idea that physicalist doctrine plays, and ought to play, a methodological role in the conduct of scientific inquiry into nature. Given the broad scope of the theses expressed by M1′, it is appropriate to revise M3 accordingly:

(M3′) The theses both play and ought to play a regulative role in the conduct of inquiry into the natural order.

This is, of course, compatible with the *a posteriori* status of the theses as well as with the denial of such status. M2 and M3′ are independent of each other. To understand M3′ further, it is important to distinguish between the normative and descriptive components of its content. That is, we must distinguish between the claim that physicalist theses do in fact play a certain methodological role and the claim that they ought to play such a role. Both claims are of importance for proper understanding and assessment of the physicalist programme. Further, it is also useful to keep in mind the distinction between *implicit* and *explicit* functioning of the theses in their methodological employment. The theses may be operative despite the fact that they are not used explicitly in deliberations (for example, they may inform a whole approach to a subject matter or they may be unrecognized premisses in a piece of reasoning). Thus in interpreting and evaluating M3′ one must not be too concrete in construing what it means or in deciding whether it is true.

My strategy in this section is to address the following questions: 'What are the roles that physicalist theses do/ought to play in the study of nature?', 'Do they in fact play such roles?', 'Ought they to play such roles?', 'Are there any difficulties with the view that the theses either do or ought to play such roles?'. In a nutshell, I shall conclude that the theses both do and ought to have a place in the study of the natural order. It will turn out, however, that the road to this conclusion is neither short nor easy.

What, then, is the methodological role of physicalist theses supposed to be? In fact, there are two quite distinct roles as follows:

(R1) Physicalist theses guide research by defining research questions and problems.

(R2) Physicalist theses provide a basis for assessing the reality of entities and the acceptability of theories and other claims to knowledge about nature.

Thus the idea behind M3′ is that, if one takes physicalist theses seriously, one will be led to ask certain questions concerning the relations between the ontologies and doctrines of the various areas of inquiry and the ontology and doctrine of physics.[12] And one will not rest satisfied unless answers to those questions are forthcoming in a way compatible with the theses. Or, in the event that no such answers emerge, the resulting dissatisfaction can be dealt with in a number of ways, as we shall see below. It is the raising of such questions and the provision of constraints upon answers which constitute the impact of physicalist theses upon the course of inquiry.

Focusing on R1, the theses pose a set of questions and problems which provide structure to so-called 'inter-field research'. Such inter-field problems (for example, the problems of micro-biology, physical chemistry, and neuropsychology) arise roughly as follows: if a theory of some class of phenomena is developed in one of the 'higher-level' domains, then physicalist doctrine entails that the objects and attributes posited by the theory are physically realizable and, in actual cases, are physically realized; that for each of the posited objects and attributes there is an RT which provides an account of how they are realized; that there are vertical explanations, grounded in physical doctrine, of the instantiations of the objects and attributes and of the instances and exceptions of the regularities identified by the theory; and that there are appropriate determination relations between the ontology and truths of physics and the ontology and truths of the theory. Such entailments provide a framework for conducting a research programme aimed at studying the physical basis of the higher-level theory and its subject matter. It does this by suggesting questions as well as the lines that pursuing answers to such questions might take. The details of such research programmes will, of course, vary widely depending upon

---

[12] Of course, because there may be a large gap between the ontology of one domain (e.g. social science) and the ontology of physics, the questions that are posed might well concern intermediate ontologies (e.g. psychology) that are assumed to be properly related to physics.

subject matter and upon how 'distant' the higher-level theory is from physics.

An example from the study of language proceeds as follows. Given the development of theories of the nature, acquisition, and use of cognitive structures underlying human language capacity, and given physicalist principles, a research programme aimed at identifying the physical basis of such structures is suggested. Such a programme will ultimately have to address such questions as: 'How are the semantic and syntactic properties of language structures realized?', 'What is the nature of the physical systems that underlie the acquisition and use of language structures?', and 'What physical processes underlie the regularities identified by the theory of language?'. Since language capacity is a relatively high-level phenomenon, research into its physical realization will typically be aimed at neurophysiological and neuroanatomical structures, states, and processes, as well as the causal embedding of language-users in the environment. The presupposition of such work is that these intermediate-level phenomena are in turn relatable to lower-level phenomena, and that ultimately they are grounded in physical phenomena (i.e. the presupposition is that such phenomena are physically based).

It is far from uncontroversial either that the theses do play, or that they ought to play, the role R1. Within the philosophy of science, there have been in recent years a number of attacks on exactly these sorts of claim with respect to the employment of the theses of classical reductionism. Whether the objections apply, *mutatis mutandis*, to the physicalist theses discussed above is an open question, given the significant differences between those theses and the theses of CR. Keeping this issue in mind, I now turn to the question of whether the theses do and ought to play such a role.

That physicalist theses do play the role, R1, is best argued for in terms of specific examples in which they implicitly or explicitly guide the thought and activity of researchers. That the history of science can deliver such examples is, I believe, beyond doubt. Thus, with regard to physicalist ontological and explanatory theses, the search for the physical bases of genetic transmission, clinical depression, chemical valence, and the phenomenological properties of classical thermodynamics and optics indicate implicit or

explicit belief that such bases exist and that they can be understood and used to explain the higher-level phenomena. A specific
reconstruction of each of these examples is, of course, required to
make good this claim, but it is beyond the scope of this project to
pursue such historical studies.

Supplementary to such positive arguments are negative arguments designed to deflect the objections of critics and to place the
burden of proof squarely on their shoulders. The first such argument is based on the claim that there are no clearly articulated
alternative construals of research into the physical bases of various
phenomena. Such research is quite widespread, and if it is denied
that physicalist theses play the role, R1, then there is a significant
gap in our understanding of current scientific and philosophical
practice. In philosophy, it is quite evident, in fact, that research
into mind, language, and values is often guided explicitly by
physicalist assumptions (Boyd 1980; Field 1975; Fodor 1975;
Railton 1986a, 1986b). And scientific research in various inter-
field disciplines (for example, biochemistry, neuropsychology) is
most plausibly viewed as being guided by an implicit or explicit
physicalist framework. The suggestion (Van Fraasen 1980; Wagner
unpublished) that general empiricist epistemological scruples suffice to account for such research is, I believe, quite implausible.
However, I allow that this issue needs to be explored more deeply.

A further negative argument focuses on the fact that a number
of sceptics regarding R1 direct their attacks against classical
reductionism. In the philosophy of biology, for example, a number
of philosophers (P. S. Churchland 1986; Darden and Maull 1977;
Hooker 1981a–c; Hull 1974; Schaffner 1977; Wimsatt 1976, 1978)
are at great pains to argue that biologists are not guided by an
interest in establishing definitional and derivational reductions,
nor are they constrained in their thinking by the very limited conception of inter-theoretic relations that is portrayed by the classical
reductionist picture. On the contrary, they argue that reductions
are seldom sought and the picture of inter-theoretic relations is
quite heterogeneous.[13]

On both points I fully agree, but I think that they are irrelevant
to the question of whether the physicalist theses of Chapter 4 play

---

[13] Comparable objections have arisen in the philosophy of psychology and the
philosophy of the social sciences. See, e.g. Fodor (1975) for an instance of the
former and Little (1991) for an example of the latter.

the role R1. In fact, biology *is* an excellent example of a discipline within which research *is* guided by such physicalist assumptions. What the critics have correctly pointed out is that reductionism is not a proper way of understanding how physicalist thought guides research. I suggest that a less restrictive reconstrual of physicalism along the lines I advocate provides a better understanding of biological research. The effort to identify lower-level processes and structures that underlie and account for some range of higher-level phenomena is guided by a physicalist view of biological systems. And the brand of physicalism outlined above leaves considerable room for variation in the types of inter-field relations. The same points can be made with respect to psychology and the social sciences as well. Thus these objections to the methodological role of classical reduction cut little ice with respect to physicalism, and instead tend to support the role of the less restrictive physicalist alternative.

Assuming that the case can be made out the physicalist theses do play the role R1, the question of whether they ought to play such a role arises. There are three strategies for supporting the claim that they ought to. Each has both merits and difficulties worth clarifying. Perhaps the most important line of argument is one that runs as follows: the employment of physicalist principles has been quite fruitful in leading to progress, while the alternatives to physicalism (for example, total autonomy of the branches of knowledge, unification programmes with different bases) are likely to be less fruitful. Hence continued use of such principles to suggest lines of research is both justified and mandated.

Admittedly, this argument requires fleshing out and support for its premises. However, the intent is clear: physicalist principles have been used with success in the past, and they should continue to be used in the absence of a better alternative. The power of this argument depends upon three things: that there has in fact been the kind of success alleged to result from employment of the principles; that it is reasonable to expect that what has been successful in the past in some areas is likely to be successful in the future in the same or other areas; and that the alternatives are indeed inferior to physicalism with respect to their potential fruitfulness in research.

With regard to the question of fact, there is disagreement. On the one hand, Goodman (1978: 5) suggests that physicalism has had only limited and partial success. Nevertheless his claim is unelaborated and unsupported, and one wonders what kind of study he has conducted to underwrite his contention. On the other hand, there are others who would claim that there is a broad spectrum of successes resulting from the methodological employment of physicalist principles. However, Oppenheim and Putnam (1958) provide perhaps the only attempt systematically to review the record of the physicalist programme in science. But this review is both dated and too restricted in its scope. As a consequence, I do not believe that a serious verdict can be passed on this issue in the absence of a detailed historical study of the pattern of successes and failures of physicalism in a sufficiently wide range of cognitive enterprises. The confidence of both critics and proponents of physicalism in the face of such a factual gap is not justified and is somewhat puzzling.

The issue of whether it is reasonable to expect that what has been successful in the past in some areas is likely to be successful in the future in other areas is a complex one. The idea that a successful principle is one that *ought* to be used generally should not be confused with the idea that principles that are successful in one area *are worth trying* in different areas until it is clear they are not appropriate. The former idea is without serious merit. The applicability of a principle in an area depends upon what is true of that area, and what is true of one area is not always true of others. The latter principle is reasonable advice, but it does not support the premiss in the argument. The issue is how to justify the assumption that it is reasonable to expect a successful principle to be successful in a new domain. It is difficult to see how one could support this claim without presupposing that physicalism is true. Anything less would at most justify giving physicalist principles a try, but not expecting that they are likely to succeed. Thus it would seem that the idea that we *ought* to employ physicalist principles cannot be justified unless the truth of the theses is granted. Needless to say, this assumption will seem question-begging in the eyes of opponents, whereas physicalists will see it as a perfectly natural reason for thinking that the principles ought to guide research.

The third critical point for the argument (i.e. that the alternatives to physicalism are less likely to be fruitful) subsumes two possible cases: the autonomy of research programmes and alternative programmes of unification grounded in different bases. In the first case, the burden of the physicalist is to establish that research conducted in a non-physicalist environment is not as likely to be successful as research that is conducted in a physicalist environment. The issue turns on whether general principles of rationality, empiricist epistemological scruples, or theoretical creativity and insight would be as effective as physicalist assumptions (combined with such principles) in, for example, turning up empirically adequate and explanatorily powerful theories. Again, the answer seems to be that, if physicalist theses are true, then it is likely that the employment of such theses in guiding research will be more successful than reliance only upon general principles of rational thought, empiricist scruples, or insight and creativity. The strength of the theses is that they would raise the right questions and point in the right directions. It would seem that, *ceteris paribus*, it is less likely, *in a physicalist world*, that research not guided by physicalist principles will also hit upon those same directions; assuming the truth of physicalism, research specifically guided by physicalist theses is more likely to succeed. On the other hand, if physicalism is false, the alternative framework is, perhaps, more likely to be fruitful, since physicalist principles would tend to lead research up various blind alleys. This is the most that the physicalist can say here, and, as I shall shortly suggest, there are qualifications even on this claim since the truth of the theses may not underwrite greater fruitfulness in some circumstances. Thus one of the physicalist's key premises (viz., that an alternative to physicalism based upon the autonomy of research programmes is likely to be less fruitful) lacks persuasive force.

The same considerations apply in the case of an alternative unification programme grounded on a non-physical basis (for instance, psychology). Employment of physicalist principles is more likely to succeed if physicalism is true. If physicalism is false, it is less clear that the alternative programme will be more successful, but that does not help the physicalist's case since the effectiveness of physicalist principles will be drawn into doubt. The moral here is that whether the methodological employment of physicalist principles is likely to be more fruitful than competing approaches in new domains depends crucially upon whether physicalist theses are true. This goes considerably beyond the original premises concerning past success and suggests that physicalists must look to a different argument to support the normative force of the claim R1.

A second line of argument for the claim that physicalist principles ought to be employed in role R1 relies directly on the truth of the theses as follows. If physicalist theses are true, then they ought to be employed in role R1. Thus, in so far as we have reason to believe that the theses are true, we have grounds for believing that they ought to be employed. As is evident, the burdens of this argument are quite heavy. First, the truth of the theses is not established, given the current situation *vis-à-vis* their assessment. Until a more systematic appraisal is conducted, physicalists cannot reasonably frame arguments that assume strong evidential support for the truth of the theses.

Second, even if the theses were known to be true, it does not follow that they ought to play methodological roles in inquiry. Truth by itself does not guarantee that a thesis will be usable or that it will be productive. These matters depend upon further facts about context, problems under investigation, researchers, and the like. Indeed, in some contexts false views are more likely to succeed for certain purposes. Thus this line of argument requires supplementation before it is to be effective. Although truth partially underwrites the argument from past utility, truth by itself does not guarantee utility in research. And we can see that even past utility and truth together may well be insufficient to guarantee fruitfulness in a new domain if the contextual or other factors pertinent to success do not obtain (for example, certain types of technology, availability of resources, level of complexity of the phenomena, development of certain kinds of background knowledge). That physicalist theses ought to be employed in inquiry seems to depend upon a number of contingent factors that the arguments so far do not take into account.

The final line of argument I shall consider is one that I have highlighted throughout this project: namely, that the potential gains of the programme provide strong motivation for pursuing it, and that in the absence of solid grounds for not pursuing it and in the absence of better alternatives, the programme ought to be pursued. Hence the theses ought to play the methodological roles attributed to them. That is, if the premises in this argument are true then there is some prescription for employing the theses in the role R1. The prescription derives from the valuable consequences of the success of the programme combined with a denial of any better alternatives and any subverting flaws.

In Chapter 1, I outlined a defence of the claim that the success of the programme would yield considerable gains, and in Chapter 6 I shall review the major objections to the programme and ultimately disarm them. If it can be argued that there are no superior alternative approaches, then the premisses of the argument are plausibly true. Of course, even if that task can be performed, there is no assurance that the employment of the theses will be successful. Promise of great rewards, like truth and past utility, does not guarantee fruitfulness of an enterprise. But this argument does not turn upon actual success; it provides a strong *prima facie* reason for holding that the theses ought to be used. If the objections can all be fended off and if it can be shown that there are only inferior alternatives, then this much normative force can be attributed to the theses. In Chapter 6, I shall address both of these undischarged assumptions.

I turn now to a discussion of the second methodological role, R2: the evaluative role of the theses. That physicalist principles play a significant role in the evaluation of theories has been endorsed by a number of philosophers (Causey 1981; Field 1972; Fodor 1975; Friedman 1975; Hellman and Thompson 1975; Oppenheim and Putnam 1958; Post 1987; Quine 1978). According to such philosophers, the theses function as a constraint upon what is to count as acceptable natural science. For example, Quine has argued that, because certain kinds of psychology and linguistics fail to conform to physicalist standards, they are not legitimate forms of scientific knowledge. He alleges to have demonstrated that, given a version of the physicalist thesis of truth determination and a broad conception of the physical bases (i.e. bases that include 'behavioural' truths), certain sorts of theories in linguistics, semantics, and mentalistic psychology fail to bear the right relations to the physical bases and hence do not concern objective matters of fact. They should therefore be banned from science.

Such reasoning gives rise to a number of questions: 'What is the structure of the methodological situation with which Quine is concerned?', 'Given such a situation, what are the possible ways it can be resolved?', 'What kinds of consideration are pertinent to resolving such situations?', 'Why and under what conditions, if at all, should such situations be taken seriously?'. I shall discuss each of these questions in what follows, focusing on the issue of the 'physicalist acceptability' of semantics as an example. This is an especially important case for physicalists since it bears directly upon the objection to the theses regarding whether they themselves are subject to indeterminacy, and hence not acceptable on physicalist grounds.

What, then, is the structure of the methodological situation to which the evaluative role of physicalist theses can give rise? In addressing this question, I shall be concerned with the thesis of truth determination, T3, both because that is the operative thesis in Quine's arguments against semantics and because it is operative in objections to physicalist theses as well. However, most of the points that are made clearly generalize to the other theses.

As I understand him, Quine advances the following informally expressed argument against the scientific legitimacy of translations · between natural languages:

1. The physical truths determine all the truths of natural science.
2. Translations between natural languages are not determined by the physical truths.
3. Thus, translations are not truths of natural science.

As I shall construe the violation charged by Quine, it is that, given the basis for physical doctrine and given a view about the nature of translation (i.e. a realization theory for translation),[14] specific translations between any two languages are not determined: i.e. there are equally acceptable alternative translations that are incompatible with each other and equally compatible with the physical truths. This is a violation of physicalist doctrine in so far as T3 requires that the physical truths 'fix' a unique translation. The key idea of 'indeterminacy' here is that, given the physics and given the realization theory for translation, there is a multiplicity of 'correct' translations that are possible. This is what indeterminacy relative to the physical basis consists in, and the argument offered

---

[14] E.g. a translation between languages, L and L′, is a meaning preserving mapping from the sentences of L into the sentences of L′, where 'meaning' means stimulus meaning in Quine's sense (see Quine 1960). In effect, such a realization theory parcels out those physically-based facts that are relevant to the realization, and hence the determination, of translational facts.

by Quine is of a form that could be applied to any domain that is alleged to fall within the scope of the theses.[15]

The general methodological situation consists of the following elements:

- There is some higher-level, non-physical theory, T, which appears to satisfy standards of epistemic adequacy (for example, compatibility with empirical evidence, simplicity, explanatory power).
- There is a class of lower-level, physically-based theories, P, which also appear to be well established.
- There is a class of realization theories, R, for the attributes posited in T.
- There is a class of physicalist theses that constrain the acceptability of claims to knowledge.
- It appears that the physicalist theses are not satisfied (for example, T is not truth-determined by P).

Note that, although I here talk of a class, P, of 'physically-based theories', this class could of course include the narrowly construed physical basis. The point is that P consists of all physical or physically-based truths that could be relevant to the physicalist acceptability of T. A crisis arises when, no matter how broadly the physical basis is construed, it does not suffice for meeting all physicalist demands on T.

Now, because physicalism, in its second role, requires that the theses be satisfied and that the relevant relations hold between P and T, something must give in the type of situation just schematically characterized. Quine opts for rejection of T, but there are other options that are possible and it is a flaw in Quine's

---

[15] Interpreters of Quine's philosophy may well object to this reconstruction of Quine's reasoning, perhaps because I have downplayed the role of his 'behaviourism' in the arguments for indeterminacy. Ed Becker has so objected in a personal communication. I grant that many of the arguments presented by Quine turn heavily upon behaviourist assumptions and that the indeterminacy that such arguments purport to establish is relative to the behavioural facts or truths. My focus in understanding Quine has been to fathom why he thinks that such behavioural indeterminacy signals there being no fact of the matter regarding translational and other semantic claims. It is here that I think Quine's physicalism enters. The reconstruction of Quine's argument in the text is an attempt to capture the more general form of argument that I think he endorses. See Quine (1969b: 303; 1978; 1979: 167) for passages suggesting this more general form of argument relying upon physicalism.

case that he does not adequately rule out the alternatives. There are, in fact, four ways to resolve the conflict:

1. Reject T.
2. Reject some member of P.
3. Claim that it is false that T and P fail to satisfy the demands of physicalism either by rejecting a member of R or by revealing a flaw in the argument purporting to establish such a failure.
4. Reject physicalism.

It is interesting that each of these options has been defended by one or another party in the debates regarding indeterminacy in linguistics and semantics, although no one has made their case stick very persuasively, in my opinion.

Quine's consistently favoured option has been that the higher-level theory (for example, translations, reference assignments, psychological theory) should be rejected in the type of situation described above. On his view, the apparent failure of translation, for instance, to be physicalistically acceptable, given current conceptions of physics and the nature of translation, is sufficient for its rejection as a legitimate form of scientific activity: it is not objective in the way that scientific enterprises ought to be. However, there are two problems with Quine's view on this matter, problems to which he has never adequately responded. First, his discussion is predicated upon a behavioural theory of meaning and translation, and, second, he has not offered reasons why it is the rejection of the higher-level theory that should be preferred, rather than one of the other options. Since my purpose here is to make explicit the nature of the methodological employment of physicalist principles, I shall not delve into these controversies. The point of importance is that a judicious choice from among the options must give balanced consideration to all of them.[16]

The second option, viz., to reject a member of P, receives its support from those (Hellman and Thompson 1975: 564; Post 1987: 194 ff.) who view physicalism as a constraint upon the adequacy of physics and of any physically-based theory. Thus if there is

---

[16] A different sort of example of this option is the decision to reject paranormal phenomena, witchcraft, UFOs, and the like. *Something* is going on in these cases, but certain conceptions of what it is are rejected because they lead to apparent violations of physicalist constraints.

some phenomenon or some theory that does not fit into a physicalist framework, this indicates an inadequacy in extant physical or physically-based theory. For example, Chomsky (1968: 83–4) can be understood as endorsing this option, since his idea that physics can always be modified to incorporate recalcitrant phenomena and theories is essentially the view that current physical theory can always be singled out as inadequate when conflict arises. But this is surely an unacceptable view of the matter; physics is not always the weak link in situations of conflict. This view grossly oversimplifies the methodological choice that must be made when physicalist theses are not satisfied in some area of research, and clarification is required of the conditions under which rejection of current physical theory is the appropriate option to choose.

The third option has several variants that must be distinguished. As Field (1975) and Friedman (1975) have both suggested, it could be (although it is a fact yet to be discovered) that T and P do relate to each other in the physicalistically required ways: showing this is an unsolved *problem*. Solving such a problem may involve establishing that the existing RTs are sufficient for relating T and P as required by physicalism; arguments to the contrary would thereby be shown to be faulty. Or solving the problem may involve developing new RTs. What was once believed to be irrelevant to the realization or determination of some phenomenon is now believed to be relevant. In the case of semantics and translation, a behavioural theory of meaning might be replaced by a theory that takes physiological, psychological, or causal factors into account. Thus if it is reasonable to think that the conflict situation is created by a problem concerning the relations between T and P, then it is appropriate to pursue efforts that will eventually issue in establishing one or the other of these two sorts of solution.

On the other hand, however, it could be that the fact that T and P are appropriately related is a *mystery* and that, although there is no violation of physicalism, it is not within our cognitive reach to know this or to show it. It is quite unclear how we could ever recognize that we were in such a situation, although it certainly seems to be a possibility. And it is difficult to see how to make this move without its appearing to be just a cheap defence of physicalism in the face of serious difficulties. However, physicalists must recognize that, although the idea that we should keep trying to solve recalcitrant problems, rather than opt for easy and premature rejections of either P or T, is an important component of the complex decision that must be made, at a certain point perseverance becomes 'perversity', as Spector (1978: 11) puts it. The judgement might have to be made that we are not faced with a solvable problem. At such a point, deciding whether one is faced with a mystery or a failure of the required connection between T and P is a difficult and perilous task, although the issues here are not peculiar to physicalism.

Finally, of course, it could be that physicalism itself is the trouble-maker and that the correct response in the conflict situation is the rejection of the operative physicalist thesis. Goodman (1978: 5) has declared the programme to be pretty much of a failure, and Putnam (1983: 221 ff.) appears to reject physicalism as incorrect, at least in the case of reference. It is nevertheless most unclear what constellation of considerations would lead one to reject physicalism rather than one of the other elements of a conflict situation. Neither Putnam nor Goodman offers carefully worked out and compelling arguments for their pronouncements. On the other hand, physicalists must remain open to the possible rejection of their view, and ought to clarify what circumstances would lead to taking such a possibility seriously.

In reviewing this range of options, my aims have been to establish that the type of methodological conflict engendered by physicalist theses is a complex one, and to suggest that recent participants in discussions of this sort of situation have not adequately explored the complexity and have tended to oversimplify the options that are available. To deepen our understanding of such situations, I shall identify some of the considerations which are pertinent to deciding whether or not such a situation has in fact arisen, and I shall tentatively suggest some considerations pertinent to deciding which option to select in a given case.

First, the methodological conflict envisioned above should be taken seriously only when the theories T and P are 'reasonably mature'. It is, of course, notoriously difficult to make this notion precise, but it is not without content. As a minimum, the theories should be empirically confirmed and well tested in a significant range of relevant circumstances, and they should fit in reasonably well with other established theories in the same and adjacent domains. The theoretical apparatus of a mature scientific theory will in many, but not all, cases be quantitative rather than qualitative, and they will have been refined to a significant degree of precision. More generally, mature theories will articulate, with precision and complexity appropriate to their subject matter, the structure of that subject matter. These considerations, though sketchy, should indicate that not just any theories which fail to comply with physicalist constraints give rise to a crisis. In many cases, and linguistics and psychology are prime examples, it is simply too early to be applying physicalist criteria of adequacy in their assessment. This is not to say that one ought not to develop such theories with an eye to compatibility with physicalism. Physicalist theses do indeed provide ideas for and constraints upon the co-development of theories in different domains. But those constraints have secondary importance during the early stages of the development of theories in most domains.

Second, no serious assessment of the physicalist connections between higher-level theories and lower-level, physically-based theories can get off the ground without extensive study of the 'inter-field' connections (i.e. study of how T and P and their respective subject matters relate to each other). Post (1987: chapter 5) has made a significant step in the direction of clarifying some of the important, general aspects of this sort of work, as have others who have focused on such specific domains as biology (Hull 1974; Kincaid 1990; Rosenberg 1985; Schaffner 1977; Wimsatt 1976), psychology (P. S. Churchland 1986; Fodor 1975; Von Eckardt 1978), and the social sciences (Kincaid 1986; Little 1991). The main point I wish to add here is that the development of RTs for the attributes studied in the various domains is an essential part of understanding how domains relate to each other, and it is crucial to an assessment of the physicalist acceptability of specific theories. As emphasized in Chapter 4, RTs are not to be conceived as definitions or nomological sufficient conditions in the style of classical reductionism or strong supervenience. Failure to appreciate this has impeded progress in understanding just how higher-level theories and domains relate to lower-level theories and domains.

Third, a clear understanding of how physicalism works (the desired result of the current enterprise) should be in hand so that premature assessments or inappropriate judgements of 'non-reducibility' or indeterminacy are blocked. Many recent discussions have been seriously marred by a failure to appreciate the nature of physicalism in contrast to classical reductionism.[17] The point of this demand is simply to require that assessments relying on physicalist principles ought to proceed with a correct understanding of what the doctrine does and does not involve.

Fourth, it is appropriate to consider physicalism as a serious constraint upon the development of knowledge only if there are good reasons for pursuing the programme, there are no compelling objections against it, and there are reasonable grounds for believing that it can be successful. Chapter 1 was aimed at responding to the first demand, whereas Chapter 6 will contribute to meeting the second and third. If physicalism is either unmotivated or not credible, then it is hardly reasonable to employ physicalist principles in the role R2. Hence it is hardly reasonable to take seriously any conflicts it engenders.

The above conditions are minimal with respect to the appropriateness of taking seriously the question of the physicalist acceptability of claims to knowledge. But it is evident that much of the debate in the past twenty or thirty years concerning the indeterminacy of semantics and translation and the unacceptability of mentalistic psychology, as well as issues concerning biology, psychology, and the social sciences, have failed to honour them. With regard to the indeterminacy of translation and related claims, the failure has been almost complete. The relevant higher-level theories are immature (i.e. linguistics and psychology), the theory of meaning as 'stimulus meaning' is unacceptable, discussions based on clear conceptions of the demands of physicalism are rare, and there has been little effort to establish either the motivations for or the credibility of the physicalist programme. For these reasons, I think that little has come of those discussions despite the fascinating nature and enormous impact of Quine's doctrine.[18]

Now, let us assume that the type of methodological situation

---

[17] E.g. the discussions of physicalism found in Nagel (1979), Rorty (1979), and Stroud (1987) are faulty in this way.

[18] E.g. Rorty's discussion of the indeterminacy thesis in Rorty (1979) strikes me as a complete failure, for the reasons just cited. And in their arguments against folk psychology, the Churchlands' reliance on the false dichotomy of ('smooth' or 'lumpy') reduction or elimination is an example of failure to argue on the basis of a proper formulation of what physicalism requires (see P. M. Churchland 1981 and P. S. Churchland 1986). On the other hand, P. S. Churchland's (1986) discussion of the co-evolution of theories in adjacent domains is an important contribution to the issues here.

described above has arisen: what kinds of considerations are pertinent to its resolution? Such considerations inevitably bear upon selecting one or another of the four possible response options. If the first option is taken (i.e. modifying or rejecting T), then principles of conservativism ought to be followed: such modification or rejection ought not leave any large gaps in our knowledge that cannot be filled or strategies for filling them are not available. That is, the balance tips towards option 1 only if substantial alternatives to T are on the horizon. In the case of mentalistic psychology, for example, the wholesale rejection of representational theories of the mind would, at the current stage of theoretical development in psychology, constitute a scrapping of the only approach having either significant successes or significant prospects. The alternatives to such theories are entirely inadequate to the subject matter involved.[19]

As I discussed earlier in Chapter 3, any modification made to physics or lower-level physically-based theories (i.e. option 2) ought not simply be an *ad hoc* revision aimed exclusively at relieving the tension created by a single problematic case. The integrity of physics, especially, ought to be maintained when revisions are made. That requires substantial integration of a proposed revision into extant bodies of physical theory and the overall programme of physical research. The electromagnetism case discussed above provides an excellent example of a revision of fundamental physics consonant with this condition. Comparable points apply to revisions of physically-based theories as well. Thus, although I allow that conditions could arise in which what was previously conceived of as not part of the physical bases is downwardly incorporated into those bases (for example, a new force), such a move requires substantial theoretical integration and motivation.

In general, the coordinated development of higher-level and lower-level theories is not captured by blithe reliance upon one or the other of these first two options. Both wholesale elimination of recalcitrant theories and hair-trigger revisions of physics are misleading with regard to the nature of the inter-theoretic problems that must be solved if an empirically, explanatorily, and physicalistically adequate system of knowledge is to evolve. The

---

[19] The untenability of eliminative materialism as either a general response to the methodological situation or a response in the case of mentalistic psychology is, I believe, based in part upon this consideration.

details of how different domains connect is the essence of physicalist system-building; and they are what makes physicalism important. It is, of course, always possible that either physics or some higher-level theory must be revised, but it is misleading to suggest that those are the primary options. Physicalism is not exclusively a constraint upon individual physical or non-physical theories (for instance, a standard of adequacy of objective truth or of the completeness of physics). It is also, and I think most significantly, a constraint upon system-building and thus on the co-development of many theories at once. There are constraint relations that hold in all directions, and the decision-making process regarding how to tinker with the system, given those constraints, is a complex one. On different occasions different parts of the evolving theoretical structure may be singled out as being in need of revision when trouble erupts.

This brings us to a further consideration in deciding upon a resolution of the conflict situation. The key to studying inter-field connections is the development of RTs for the attributes that are involved at the higher level. If the available RTs are not fully developed or otherwise inadequate, or if there are no available RTs, then the third option (i.e. to deny that physicalist requirements are not satisfied) becomes more prominent. To reject a well-confirmed physics or a well-confirmed higher-level theory on physicalist grounds when the inter-field connections are vague or poorly understood is a misapplication of physicalism. The more reasonable course is to pursue the development of RTs that may resolve the conflict by showing that none really exists, and thus to preserve the well-confirmed theories into the bargain.

In the case of the alleged indeterminacy of semantics and mental content, both Quine's behavioural theory of meaning and the popular 'projective theories of interpretation of mental states'[20] are, in my opinion, unacceptable on grounds independent of physicalism. Further, functionalist and 'physicalist' (i.e. the identity

---

[20] See Davidson (1970: 1973), Putnam (1978), Quine (1960), Stitch (1983). Such a theory has it that, unlike hypotheses in physics, biology, etc., an interpretive hypothesis regarding mental content is not an empirical hypothesis concerning the states of the individual. Rather, it is a 'rationalization' from an interpreter's perspective of the individual's behaviour in relation to the environment. Such a rationalization is, allegedly, sensitive to the *interpreter's* goals, cognitions, conceptual framework, etc. in a way that empirical hypotheses are not.

theory) realization theories concerning the nature of mental states are likewise unacceptable. The significance of these claims for the methodological conflict engendered by the indeterminacy thesis depends upon one's outlook. If it is concluded that they mean simply that we do not yet have acceptable conceptions of mind and meaning, but that such conceptions are possible, then a 'wait and see' attitude is indicated. If, on the other hand, it is concluded that a workable conception of mind and meaning is a mystery beyond our reach, then physicalist system-building may have reached an unavoidable impasse. Or, if a behavioural theory of meaning is thought to be the best we can do, then the indeterminacy arguments will in fact go through. The point is that, in order to resolve the conflict, serious attention must be given to the state of our understanding of the inter-field connections, especially the RTs. As I understand them, current debates over 'theories of content determination' are concerned with exactly this issue (cf. Cummins 1989; Fodor 1987; and Von Eckardt 1993).

Another consideration bearing on the choice of an option to pursue is that the rejection of physicalism (i.e. option 4) should not leave major gaps in our understanding of the nature and goals of cognitive inquiry in general and scientific inquiry in particular. Taking the latter first, contemporary physicalism provides a set of principles which capture a view of nature and the structure of scientific knowledge that is deeply ingrained in most scientists' understanding of science and its objects of study. Such a view constitutes a general framework for understanding how phenomena in the natural order are related to each other and of the interconnections among theories of those phenomena. As such, physicalism operates at a very deep level in shaping scientific activity. To reject this view, and it is conceivable to do so, would be to leave a large gap in both the scientist's and the philosopher's understanding of what science is about. At the least, a serious alternative conception, to which there are no major objections, must be available before the fourth option is entertained. Without such an alternative, it is likely that pursuing the physicalist programme, despite local anomalies, is superior to abandoning it. As indicated earlier and as will be argued below, the main alternatives that have been proposed (i.e. disunity of science and unity grounded upon a different basis) are clearly inadequate to the task of orchestrating scientific activity in anywhere near as rich a way as physicalism has done and continues to do. Those who flirt with rejection of physicalism cannot be cavalier with respect to this issue.

I believe that similar points apply to the programme more broadly conceived. In a wide range of cognitive and non-cognitive domains, physicalism plays a significant structuring role. For example, philosophical research in moral theory and epistemology is to a significant extent constrained by physicalist assumptions, though much of the work in those areas can proceed for a long time in a relatively autonomous fashion. This possibility indicates in part the immaturity of those research activities as well as their abstract character. However this may be, the point here is that, as currently conceived, serious theories in epistemology and moral theory which have relevance to this world must, at least when applied to this world, honour physicalist constraints. For example, epistemological theories that invoke causal relations will ultimately need to be compatible with physicalist demands upon causation. And theories of moral psychology will ultimately need to fit in with the physically-based realities of mind–brain relations and functioning.

Further, and in very different arenas, physicalist constraints operate at a deep level in the social institutions of law, medicine, and education as currently conceived and practised. For example, all three make generous use of 'scientific' knowledge that is physicalistically structured. And in medicine and education, an integrated understanding of what is going on in clinical and educational settings will incorporate physicalist assumptions about the interactions and processes involved. The abandonment of such constraints would put our understanding of those institutions, as they now exist in this culture, 'out to sea'. In all these areas, the choice to abandon physicalism should be predicated on the existence of a viable alternative metaphysical and epistemic framework.

Finally, in deciding on a course of action in the face of conflict arising from apparent violations of physicalism, the justificatory status of the programme to date should be understood. Goodman (1978: 5) declares that the status is poor, while others (Hellman and Thompson 1977; Port 1987) exude confidence. My own view is that its status is unclear for want of a serious effort at making an assessment. But this much should be agreeable: if it is true that physicalism is not justified by various evidential and pragmatic considerations, or if there are definite counter-examples, major

internal flaws, or clearly overwhelming obstacles to its being successfully worked out, then the option to reject physicalism is more
plausible. On the other hand, if physicalism is justified and does
not suffer from major defects, then that option recedes deeply into
the background. The general point is obvious: the epistemically
least valued element in the conflict situation is, other things being
equal, the first to be rejected. Of course, other things aren't always
equal, as the other considerations make clear. But physicalists,
especially, must shoulder the burden of establishing the justifiability
of their programme if meaningful choices are to be made at crucial
points of conflict.

The above sets of conditions on the appropriateness of employing physicalist theses and on the resolution of conflict engendered
by such employment should give some indication of the complexity of the issues raised by R2. That the theses have been used in
this role is undeniable, but my point here is that such use must
take place under certain circumstances only, and even then the
outcome of their use is subject to complex considerations that
presuppose considerable auxiliary knowledge. It is pretty safe to
say that many actual uses of the theses, both by supporters and by
opponents of physicalism, have been inappropriate and untutored.
Further, at the present time the situations of appropriate use may
be severely limited, given the manifest failure of many research
programmes in both science and philosophy to satisfy the conditions
of appropriateness and given the notable lack of much of the
needed auxiliary knowledge required for resolution of conflict
situations. None the less, the idea that such methodological employment ought never to occur is seriously mistaken. On the
contrary, the physicalist holds that the theses *ought* to play the
role R2 in the conduct of inquiry into nature. It is to this final
issue regarding R2 that I now briefly turn.

As discussed earlier with regard to R1, there are three kinds of
grounds for taking seriously the methodological employment of
physicalist theses: their past fruitfulness, strong support for their
truth, and their promise of yielding significant gains if their use is
successful. If real, these would indeed provide some grounds for
taking the use of the theses seriously. However, although the third
ground is not problematic (cf. Chapter 1), the first two are, as I
suggested above. Past fruitfulness of the theses is subject to debate
regarding how much they have been used and how well their

successful use generalizes to other domains. Further, both the nature of the evidence bearing on their assessment as well as their actual evidential status are unclear. Even if the truth of the theses is plausible, their actual utility depends upon contingent contextual factors that may not hold. The most that can be said at this point is that, if the programme is justifiably pursued and the evidential status of the theses can be positively clarified and the contextual factors are right, then the theses ought to be employed in the role R_2 when it is appropriate to do so. The issues concerning the justification of the pursuit of the programme and the evidential status of the theses are still open. As a consequence, whether physicalist theses ought to play the role R_2 must be left open as well, subject to a resolution of those issues. One of the purposes of this project is to defuse the objections to the programme and so to take it further along the path towards being more explicitly justified.

5.4. NECESSITY, CONTINGENCY, AND PHYSICALIST PRINCIPLES

The final metathesis of physicalism concerns the model status of the theses. Typical expressions of the received view on this matter are as follows:

> Materialism is meant to be a contingent thesis, a merit of our world that not all other worlds share. Two worlds could indeed differ without differing physically, if at least one of them is a world where Materialism is false. For instance, our Materialistic world differs from a non-materialistic world that is physically just like ours but that also contains physically epiphenomenal spirits. (Lewis 1983: 362.)

> it is worth noting that TT [T3] need not be a logically necessary truth but can be contingent (and probably is). That is, TT need not be true in every possible world, as it might have to be if it quantified over logically rather than merely physically possible worlds. (Post 1987: 187.)

On a superficial level, the significance of M4 (i.e. the claim that the theses are contingently true) is evident: the theses are not necessary truths—although they are true, they could be false. However, such claims, and the quotes above, mask considerable complexity and difficulty to which conscientious physicalists must attend. In what follows, I can only hope to make clear why there is reason for concern and, perhaps, to dispel some of it.

Some preliminary points are in order. First, it is absolutely crucial to distinguish between the issue of the modal status of the theses and the issue of their epistemic status. M2 affirms the empirical character of the theses (i.e. their *a posteriori* status). And although there are serious problems with fully working out the nature of relevant evidence and arguments bearing upon the theses, nothing I have said so far should be construed as clearly undermining M2. But this in no way guarantees that the theses are contingent. It is now widely agreed that we might come to know certain necessary truths via empirical means (cf. Kripke 1972). Thus the ways in which we come to have knowledge of physicalist theses leaves open their model status.

This point is especially important when it comes to assessing the reasons for believing that the theses are contingent. At any moment in time, we may have rather imperfect empirical knowledge of physics and the physicalist system; but such limitations do not bear upon the modal status of the theses one way or the other. Further, it should be recognized that our knowledge of what is necessary and what is contingent is likewise subject to our epistemic limitations, both empirical and conceptual, at any given time. And, because *the necessary* is not congruent with the *a priori*, we must recognize that 'conceivability' or 'inconceivability' arguments cannot, in general, be counted upon to provide conclusive grounds for believing that some claim either is or is not necessary. This all means that there is a very serious problem for proponents of M4 with respect to developing reasons for believing it to be true.

The distinction between contingent truths and necessary truths requires comment as well. There is no sharp dichotomy here; rather, there is a set of graded notions concerning ranges of possible states of affairs, drawn from some reference class, that make a statement true or false. A necessary truth is a statement that is true in all of a specified class of possibilities, whereas a contingent truth is one that, although true in some possible states of affairs within a specified class, is false in others. Thus philosophers freely talk of logical, metaphysical, physical, and nomological necessity and possibility, to name a few types. Each type is construed in terms of a certain reference class of possibilities and those possible states of affairs within the class which it subsumes: for example, logically necessary statements are true in *all* possible states of affairs, *tout court*, while physically necessary statements are those that are true in all physically possible states of affairs. Thus a physically necessary statement can be logically contingent (i.e. although it is true in all physically possible states of affairs, it is false in some other possible state of affairs).

To facilitate talk of such matters, philosophers have introduced 'possible worlds' locutions and semantics, a mathematical-logical apparatus for, *inter alia*, representing the meaning of modal terms and the truth conditions of modal statements. Within such an apparatus, the notion of a 'possible world', conceived of as an abstract relational structure or, alternatively, a maximally consistent set of propositions, plays a key role in the task of representing meanings and truth conditions. A possible world, in this sense, is a *representation* of 'an alternative possibility', one of the possible ways the world might be. Thus contingently true statements are true in some possible world while false in others, while necessarily true statements of a certain type are true in all of a specified class of possible worlds. To say that a statement is true in a possible world is to say that either it is true in the relational structure, or it is a member of the class of propositions that defines the world in question.

The point of rehearsing these ideas is to allow us to distinguish clearly between the mathematical structures that constitute the representations and the alternative possibilities they represent. For the metaphysical purposes of physicalism, the latter are critical. A possible world in the mathematical sense is a relational structure that represents a possible distribution of objects and attributes *in the world*. What are represented are genuine alternative possibilities concerning what the world is like, where the world can be many different ways, not just the way it actually happens to be. I shall henceforth distinguish between the mathematical notion of a (possible) world and the metaphysical notion of a (possible) world. The former is an element in an abstract mathematical structure, whereas the latter is an actually existing state of affairs or an alternative possibility to such a state of affairs.[21] It is the

---

[21] Since this is not a book concerned with systematically taking on the many vexed problems raised by modalities and the mathematical apparatus used to study them, I can here only try to indicate a framework that I hope is shared by most participants in discussions of physicalism. I am making basic distinctions to clarify the discussion which follows. What I cannot do here is try to provide answers to such questions as: 'What is an alternative possibility?', 'What is meant by the "actual" state of affairs?', and 'What, indeed, is meant by "a world" in the metaphysical use of that term?'. See Lewis (1986) for an important treatment of these problems.

latter notion of a possible world (i.e. the metaphysical notion) that will be operative in subsequent discussions of M4.

Further, I shall mean by a class of possible worlds, in the metaphysical sense, a class of alternative possibilities defined by a set of propositions that are true in each of the members of the class. Thus, for example, the *logically possible worlds* are defined by the logical truths, *the physically possible worlds* are defined by the laws of physics,[22] and *the nomologically possible worlds* are defined by the laws of nature.[23] If need be, one can impose additional constraints on the worlds in a given class to define more refined notions of possibility and necessity. One can stipulate, for example, that the members of a certain class of worlds are to contain only entities that meet certain conditions in addition to the stipulated truths they must satisfy. In this way, Horgan and Lewis employ the notion of 'alien' entities,[24] and stipulate that the members of the relevant class of worlds contain no such entities. The upshot is that, as mentioned earlier, given a stipulation of some class of worlds, a statement is necessary relative to that class if it is true in each of the worlds in the class, whereas a statement is contingent, if it is true in some and false in others. Given the range of variation in how classes of worlds can be specified, there is a comparable range of notions of necessity and possibility.

Now, for physicalist purposes there are at least two questions that must be addressed regarding possible worlds: 'What class of worlds do physicalist theses purport to truly describe?', and 'What is the status of physicalist theses in any other worlds beside those they purport to describe?'. Only when these two questions are addressed can the modal status of the theses be identified. My view on the first question, a view which is similar to the views of Lewis and Horgan,[25] is that physicalist theses purport to truly

[22] Since physics can vary from world to world, the idea of physical possibility is relative to a specific physics. When I speak of 'physically possible worlds' or 'physical necessity' I will generally mean physically possible or physically necessary relative to the physics of *our* world, although the parochial element here needs to be recognized.

[23] Given my discussion of the likelihood of exceptions to 'laws of nature', it is clear that the notion of nomologically possible worlds is a messy one at best. I cannot pause over the issues in the present work. Nomological possibility and necessity are also subject to the sort of parochial consideration that affects physical possibility and necessity.

[24] However, they differ in their accounts of what an 'alien' entity is. See Horgan (1982; 1987) and Lewis (1983).

[25] See also Hellman and Thompson (1975; 1977) for a different but related approach.

describe 'the actual world' (i.e. the world that you and I inhabit)
and all other possible worlds accessible from the actual world subject to (1) the laws that govern what happens in the actual world,
(2) the condition that any objects or attributes in such worlds
are located in a space-time manifold like that in the actual world,
and (3) the condition that all objects and attributes instantiated in
such worlds are either instantiated in this world *or* are realized by
objects/attributes instantiated in this world *or* are theoretically
acknowledged objects/attributes[26] *or* are realizable by theoretically
acknowledged objects/attributes. Although risking terminological confusion, I shall stipulate that the class of worlds specified
in this way is the class of 'metaphysically possible worlds'. It is
the members of this class of worlds that physicalist theses purport to truly describe. Thus the theses, if true, are metaphysically
necessary.[27]

I hope that this is not too troubling a use of the expression
'metaphysically necessary'. I have found most usages in the literature to be quite vague, and my aim here is to put some content
into that expression. Less apologetically, my aim is to emphasize
the import of physicalism: it is a doctrine which circumscribes not
only what actually happens in the world, but also what can happen in a very strong sense of 'can'. If physicalism is true, then
there is no metaphysical possibility of ghosts and gods and other
such entities, although they may be possible in some broader sense.
And if physicalism is true, then the physicalist theses are true in
each member in the class of metaphysically possible worlds. As
some have put it (for instance, Haugeland 1982: 99), physicalism
circumscribes the limits of 'genuine possibility'. I am not sure this

---

[26] That is, objects or attributes explicitly referred to or expressed by terms
occurring in laws of nature.

[27] It might be objected that, on the one hand, I have said that the metaphysically
possible worlds are those which, *inter alia*, satisfy the laws of nature and, on the
other hand, I have allowed, in thesis T5, that there are exceptions to the laws of
nature which are metaphysically possible. My reply is that an exception to a law
of nature is metaphysically possible only if it is in accord with a more fundamental
law and it satisfies the other conditions on metaphysically possible worlds (e.g. the
constraints upon realization). Thus there can be no exceptions to the most fundamental laws. The class of metaphysically possible worlds, as I have specified it, will
contain worlds involving exceptions to higher-level nomological generalizations as
long as these conditions are satisfied. It should also be observed that the notion
of metaphysical necessity involves a restriction of both physical necessity and
nomological necessity: it circumscribes a narrower class of possible worlds than
either of these other two notions.

way of putting things really helps, since there are many sorts of 'genuine' possibility. But the spirit of the comment is well taken; viz., in the world we inhabit, there is a certain sort of limit to what can happen and what can exist, and this limit is marked out by what exists and by what happens in the physical domain, and by what can happen in accord with the principles that govern that domain.

These points regarding the metaphysical necessity of the theses, however, do not defeat M4. Rather, they clarify what the contingency alluded to in that metathesis does *not* involve (i.e. it does not involve being false in a metaphysically possible world). Along these lines, it is important to understand clearly the difference between the claim that the theses are true and the claim that they are contingent. As just outlined, the theses, if true, are true in each of the metaphysically possible worlds. Therefore to show that they are *false* it is required that a counter-example be produced that involves one or more of the metaphysically possible worlds and that falsifies a physicalist thesis (for example, a pair of metaphysically possible worlds which agree on all the physical truths but differ on the non-physical truths). On the other hand, the theses are *contingently true* if, although true in the class of metaphysically possible worlds, they are false in some other possible world. Therefore to show that the theses are *contingently* true (if true at all), one must produce an example of a possible world that is *not* metaphysically possible, and which involves a failure of physicalist principles.

In summary, M4 asserts that the theses are contingently true. I have suggested that, given the intended meaning of the theses, this amounts to asserting that they are true in the class of metaphysically possible worlds (hence they are metaphysically necessary) and false in some other possible world (hence they are only contingently true relative to some larger class of worlds). It remains to be seen why proponents of M4 believe there is this larger class of worlds relative to which the theses are contingent, and why opponents of M4 believe there is no such larger class of worlds.

To begin with, it is useful to bear in mind some of the sorts of cases that most philosophers consider to be genuine alternative possibilities to the actual world: for example, a world of purely mental beings, worlds in which parallelist dualism, epiphenomenalism, or interactive dualism are true, and worlds that provide examples of inverted spectra. Each of these cases is, among other things, supposed to be a possible world in which one or another physicalist thesis is violated. If physicalism is true, of course, then none of them is a metaphysically possible world; but that is all to the good here. What is required is that each is a possibility in some genuine non-epistemic sense of 'possibility'.

Why believe these cases are real possible alternatives to the actual world, especially if, *ex hypothesi*, they are not metaphysically possible? As I have cautioned earlier, 'conceivability' arguments are to be viewed with some suspicion, since they do not provide sure-fire grounds for believing that some alleged possibility is genuine. But here one must be careful. Since there are many different sorts of non-epistemic possibility (logical, physical, metaphysical), conceivability arguments may have differential bearing with respect to the different sorts. Suspicion does not mean outright rejection. As it is not appropriate for me to make a long excursion here into these matters, I shall simply formulate the view I favour and indicate some of the caveats that require further exploration.

If what it is to be an alternative possibility of a certain sort is (1) to be represented by a maximally consistent set of propositions, or an abstract relational structure, subject to those constraints characteristic of the sort of possibility in question *and* (2) to be free of any disqualifying characteristics,[28] then there are as many genuine alternative possibilities as there are such sets or structures. There is no limit, within these constraints, on either the types of possibility or what possibilities of a given type there are. Thus if physical possibility is defined by that class of worlds *consistent* with the laws of physics, then there is a large class of physically possible worlds that includes the metaphysically possible worlds as a proper subclass. And, if nomic possibility is defined by that class of worlds consistent with the laws of nature, then there is a large class of nomically possible worlds that includes, but is not exhausted by, the class of metaphysically possible worlds.[29] This is

---

[28] A disqualifying characteristic is one which undermines the claim that a given abstract structure actually represents a real possibility. For example, although a set of propositions that includes the proposition that Hesperus is not identical to Phosphorus can be framed, it does not represent a real possibility since the proposition that Hesperus is not identical to Phosphorus cannot be true.

[29] Note that, in given cases, both physical and nomic possibility must be conceived in a parochial fashion relative to specific physical laws or laws of nature respectively.

because, in both cases, there are a number of incompatible extensions of the core class of physical or natural laws. This also assumes, of course, that not *all* such possibilities can be disqualified,
which I think is plausible.

Since each of the alternative possibilities that are usually alleged
to be in violation of physicalism (for instance, a world of purely
mental beings) falls within the purview of some legitimate class of
possible worlds other than the class of metaphysically possible
worlds (for instance, physically or logically possible worlds), the
contingent status of physicalist theses relative to these various
classes is secured. If it turns out that some of these alleged possibilities are not genuine and are mere epistemic possibilities instead, then they must be dropped from the class of genuine
alternative possibilities. But, again, I doubt there is any good reason to believe that all can be disqualified.[30] The possible worlds
that survive are members of the class of real alternative possibilities
to the actual world, which are not metaphysically possible but are
worlds in which physicalism is false. Thus physicalist theses are
contingently true relative to this class.

The first objection to M4, as I have construed it, is as follows:

> The only possible world is the actual world; there are no
> possible but unactualized worlds. Thus physicalist theses, if
> true at all, are true in the only possible world there is. Hence
> they are not contingently true. There simply is no other way
> the world could have been.

This, of course, is outright war against modalities. Although this
is no place to join the conflict, it should be observed that physicalism
is not seriously affected if such a view is correct. Indeed, whereas
the denial of any alternative possibilities means that physicalist
theses cannot be contingent, it also secures their metaphysical
necessity. Thus part of the import of M4 would be preserved, even
though it would have to be revised to reflect the lack of contingency. Physicalism is, in all other respects, quite independent of
the issues raised by this objection.

---

[30] But see below for discussion of Teller's arguments that such 'possibilities' are
not genuine. It is, indeed, a burden of the proponent of M4 to defend in detail the
claim that at least some of the alleged possibilities are genuine. My discussion of
Teller's objections is intended to carry this burden to some extent, although a full
discussion should cover a wider range of such possibilities than is considered
below.

However, I do want to register my disapproval of this radical actualism. General uneasiness about modalities, counter-factuals, and the like is not sufficient for us to abandon them. Much utility in the philosophy of language, the philosophy of science, and metaphysics results from talk of possibility, necessity, and possible worlds. The burden of proof lies squarely on the shoulders of those who want to scrap a seemingly useful set of concepts and logical apparatus, or to interpret them in a non-literal way. I doubt if this burden can be carried.

Next, I want to pause over Haugeland's views concerning the modal status of physicalist principles (Haugeland 1982: 99). Essentially he despairs of ever being able to pin down the class of worlds that physicalist theses are supposed to concern, and he proposes instead that such theses be viewed as an important contribution to our understanding of what is genuinely possible. Thus if we begin with the class of physically possible worlds, then that subclass in which physicalist theses are true is the class of genuinely possible alternatives to the actual world. The theses 'cull' the original set in a way that identifies more specifically what are the real alternatives to the actual way the world is.

Although there are significant differences of detail and emphasis between Haugeland's approach and the one I favour, there is substantial agreement as well. As is clear from Chapter 2, I certainly differ with him on what the theses of physicalism are. And, as my discussion above indicates, I see no good reason for giving up the task of trying independently to isolate the class of possible worlds that the theses purport to be about (i.e. the metaphysically possible worlds). On the other hand, if he can tolerate my calling the restricted class of worlds culled by physicalist principles 'the metaphysically possible worlds', then the bottom line is not substantially different in our two cases. Physicalist theses are true of the metaphysically possible worlds, but only contingently true of those worlds since there are other possible worlds in which they are false.

Finally, I want to discuss the views of Paul Teller concerning the modal status of the theses. He presents perhaps the only serious challenge to M4 that has been advanced to date. Teller (1984) has heroically argued that physicalist theses, if true at all, are necessarily true. He is, as far as I know, alone in this stance, although it should be clear that I have some sympathy for his claim to the extent that it can be understood along the lines I have suggested above. However, he appears to go well beyond claiming only metaphysical necessity for the theses. I cannot, in a short space, do full justice to Teller's arguments, and a few comments will have to suffice.

In his important paper dealing with these and other matters, Teller presents both negative and positive arguments against the claim that physicalist principles are contingent. On the negative side, he points out that arguments from logical consistency do not guarantee the existence of a possible world because of 'Hesperus–Phosphorus' type cases. And he disarms a pair of arguments offered by Lewis and Van Inwagen purporting to establish the plausibility of the existence of 'alien' possible worlds that falsify physicalism. Neither argument, he claims, suffices to establish the existence of the required sort of world. I have no quarrel with his assessment of these arguments. I want to add, regarding the relevance of consistency, that despite its not being sufficient to guarantee the existence of a possible world, demonstrating the consistency of a certain set of propositions does provide a reasonably strong *prima facie* case for such an existence claim. The fact that certain identity statements are necessary, although they have negations that are not logically inconsistent, means that additional criteria are relevant in those cases (i.e. disqualifying characteristics must be demonstrated). It does not mean that consistency fails to provide a strong reason for believing that a certain alleged possibility is real. Thus, if Teller is to refute claims for the existence of certain possible worlds that establish the contingency of physicalism, he will need to present specific arguments to show that those worlds, appearances to the contrary, are not really possible. He purports to do just this with his positive arguments.

One such argument focuses on the point of 'possible worlds' talk which, according to Teller, is roughly that it 'merely facilitates consideration of what might be, of counter-factual cases' (Teller 1984: 153). The import of counter-factual cases is that they 'are a way of saying true things about actual objects' (Teller 1984: 153). *What* we are saying about such objects when we employ counter-factuals is as follows:

The actual world displays a certain configuration of properties. We can envision alternative configurations. These alternative configurations may be limited by various restrictions, corresponding to actual limitations of regularities that characterize things and the way they go together. Corresponding to various kinds of restrictions we speak of associated kinds of possibilities: technical, personal, nomic, physical. We can also think of loosening the restrictions altogether, asking what configurations of properties go together in the absence of the limitations imposed by the state of our technology, our personal, physical, mental, and emotional strengths, the composition of matches and organisms, and (though this last might be questioned) in the absence of the laws of physics. *But in doing so we are still indirectly saying something, though less and less specific, about the real world and the nature of the things and properties in it. We are saying something about what it is to be an object in the real world, what it is for an object to have a property, and what is involved when objects and properties display one or another configuration.* (Teller 1984: 153, emphasis added.)

The argument then proceeds with the claim that 'there is no room left for possible worlds with things so alien that they do not bear, however indirectly, on the furniture of the real world' (Teller 1984: 154). But, according to Teller, only worlds with alien properties or objects would falsify physicalism. Thus physicalism is not false in any possible world, and it is therefore a necessary truth about the actual world.

Regarding this argument, I disagree with Teller on at least three points. First, it is doubtful that *the* point of 'possible worlds' talk is aptly construed in the narrow way that Teller suggests. Why is it not also a useful purpose of such talk to conceive of worlds that are radically different from our own? It seems to be an undue limitation to insist that it is idle to postulate possibilities which are 'not genuine' alternatives to the actual world because they do not concern what is actual. Thinking about inaccessible possibilities is something we apparently can do, and it is something that does have theoretical value in, for example, the analysis of attributes, the semantics of natural language, and an understanding of human conceptual ability. I, of course, do not dispute Teller's lucid characterization of *a* purpose of 'possible worlds' talk. What I dispute is that his is a characterization of the only purpose of such talk. He has at least begged the question here, and since the argument depends crucially on this point, it loses its force.

Second, I disagree with the claim that talk of inaccessible possibilities never says anything about the actual world. The notion of 'aboutness' is, of course, notoriously problematic. However, a possible configuration of objects and attributes that we 'cannot get to from here' is still an alternative possibility to the actual world. It is simply one that does not satisfy the constraints upon metaphysical possibility (in my sense) and so it might not concern objects or attributes that could (metaphysically) be realized in this world. But this does not entail that such a possibility cannot be about the actual world. Why, for example, should talking of a possible world in which there are only spirits not be 'about' the actual world, since it is, for example, another way in which actual mental properties can be instantiated? Below, I shall consider Teller's specific argument against such a possibility; pending that discussion, such a possibility (if it is real) does, in a serious sense, concern the actual world, since it concerns properties that are instantiated in the actual world, albeit in a different way (i.e. they are properties of alien objects in the hypothesized possible world, whereas they are properties of physically-based objects in the actual world). Such a possibility is one in which physicalism would be false. And yet it is one which involves alien objects and hence is not metaphysically possible. Such a possibility is, therefore, sufficient for establishing the contingency of physicalist theses.

Third, I think Teller is mistaken in his assumption that *only* worlds with alien objects and properties would falsify physicalism. On the contrary, physicalism is falsified if, in a world containing only *non-alien* objects and attributes, any one of T1 to T5 does not hold. T1 would fail, for example, if there really are ghosts and if, although they occur in the natural order and hence are not alien, they are not constituted by physically-based objects and their attributes. And T2 or T3 would fail in a world of non-alien objects and attributes just in case the appropriate determination relations fail to hold between physical attributes or truths and non-physical attributes or truths. However, for the purposes of the argument against the contingency of physicalist theses, the issue is precisely whether physicalism is falsified in worlds populated by alien objects and attributes. Those who believe that the theses are contingent contend that they are false in at least some such worlds, whereas Teller contends that there are no such worlds to falsify them. This is the claim he must ultimately defend.

Turning to Teller's more specific arguments against the standard examples of possible worlds in which physicalism is supposed to be false, I note that he reiterates the point that logical consistency is no guarantee of possibility, a point with which there is no serious disagreement. But he buttresses the point with a diagnosis of the power of our imaginings concerning such possibilities. One reason we are allegedly so impressed by such examples is that physicalist theses are *a posteriori* in character. Empirical evidence bears upon whether we should accept them as true. If one tends to equate the *a posteriori* with the contingent, then one will not be inclined to dismiss consistent thoughts about spirits and gods as not genuinely possible. Perhaps so, but I and other defenders of M4 are not the victims of such a mistaken equation.

Another suggested reason why our ability to conceive of spiritual worlds and the like has force is that there is a tendency to make a different epistemological error: the error of simply not recognizing those logically consistent thoughts or descriptions that mask necessary falsehoods. Just as one might be taken in regarding the claim that Hesperus is not identical to Phosphorus, one might be taken in, for example, by the claim that non-physical but mental beings are possible. This is the error of confusing a 'possible thought' with a 'possible world'. Although the thought is real, the world is not. However, again, neither I nor other defenders of M4 believe that this is what is going on when we affirm the existence of such a possible world. I acknowledge that there is a problem with identifying the genuinely possible worlds; but I believe the burden is squarely on those who think that certain consistent sets of propositions do not describe real possibilities to show this by revealing specific disqualifying characteristics *in each case*. As in the Hesperus–Phosphorus case, so it is in the case of the spiritual world: both require specific explanations for why the imagined possibility is not genuine. Teller's case turns entirely on his specific arguments against the possibility of spiritual worlds and the like. None of the arguments he offers provides a *general* solution to his problem.

Teller's positive argument against the existence of a logically possible world with 'mentally discriminable non-physical angels' is formulated as follows:

Mental states necessarily interact with the physical world. Necessarily, that which interacts with the physical is physical. So, necessarily, mental states are physical. (By 'is physical' I understand here being embodied by, constituted by, or realized by physical states, or the like). (Teller 1984: 156.)

As Teller observes, one can question both premises in the argument. Indeed, I think the first is blatantly question-begging while the second is totally unsubstantiated. Needless to say, Teller doesn't see things this way.

With respect to the idea that mental states necessarily interact with the physical world, Teller offers two supportive considerations. First, he asserts that nothing would be a belief, desire, or other mental state unless it had direct or indirect potential for guiding interaction with the physical stuff of the world; but as a view about the nature of the mental, this is entirely question-begging. Perhaps to be a mental state a state must have potential for guiding interaction with an environment, but what grounds are there for requiring that such an environment should be physical? Second, Teller's answer to this question seems to be that the alternative to his contention, 'Berkelianism', is an unsound metaphysical position against which every physicalist has persuasive objections. In addition to rejecting Teller's rather glib rendering of Berkeley's views (viz., all appearance 'would be an idle appearance only'), we should ask why such a view is *necessarily* false. Surely Teller begs the question here, as well by assuming that, because he and other physicalists reject crude idealism, such a view *could not be true*. Such a possibility is precisely what is at issue.

With respect to the idea that, necessarily, that which interacts with the physical is physical, Teller argues as follows:

> But if we did have a 'medium' which related inputs to outputs with systematic causal connections, what reasons would we have for saying that this medium was *not* physical, governed, perhaps, by different laws from those which govern matter in the real world? (Teller 1984: 157.)

Teller's general point is that patterns of causal relations between mental and physical phenomena within some medium would, *a fortiori*, be physically-based causal relations. That is, in any world in which mental states are causally related to the physical via some medium, the medium would be governed by 'the physics' of that world and hence would be physical. Here Teller moves too quickly, however.

First, if the physics of the world in question is different from the physics of the actual world, then it seems we would have a case of a possible world in which physicalism relative to the physical bases of the actual world would be false. Here, the parochial aspect of physicalism with respect to the bases, as discussed in Chapter 3, must be properly taken into account. Even if it is correct to say that 'physics' in that world would incorporate the relevant causal structure, that only means that physicalism relative to the bases developed in terms of that physics would not be threatened by the case in hand. However, our physicalism concerns the physics of our world and the bases developed in terms of it. Relative to *those* bases, physicalist theses would be false in the world under discussion, a world which is not metaphysically possible because it involves a different physics from our own. Hence, rather than undermining the contingency of the theses, Teller may well have established it.

Second, however, even if we consider the issue from the point of view of physicalism as developed in the world in question, Teller still needs to say more to make his case. At issue is the complicated problem of the downward incorporation of mental phenomena into the physical bases (i.e. if some phenomenon must be appealed to so that physics can do its job in a given world, then that phenomenon is an element of the physical bases in that world). As I alluded to in Chapter 3 and will discuss in Chapter 6, in order for such downward incorporation to occur legitimately, the relevant phenomena (for example, causal transactions and the medium in which they transpire) must be integratable within the physics of the world in question. Is it a foregone conclusion that they will be so integrated? I, but not many other physicalists, am open to the possibility that the mental could be so integrated. But I do not see any grounds for thinking that it must be integratable. Perhaps the physics of the given world is inadequate for its job. Teller needs to say more about why any possible world exhibiting mental–physical causal transactions must also be a world in which the enterprise of physics succeeds and physicalist theses are true. Thus I conclude that he has not shown that it is necessary that anything which interacts with the physical is itself physical. And, even if we allow for the sake of argument that the mental must interact with the physical, he has not shown that nonphysical beings with mental states are not genuinely possible.

Further, he considers the case of mental states that are embodied by a medium that does not interact with the physical at all. In this case, he falls back on his general argument against 'utterly alien worlds': they are ontologically idle (i.e. they do not do the work that talk of such worlds must do to be legitimate). In this case, the imagined idleness results from the failure of the mental states to interact with the physical realm. Hence they are utterly alien to our world, and there is no reason to take them to be genuinely possible. But here, I think, the argument lacks force, for two reasons. It depends upon the general argument against the postulation of alien worlds which, as I have already suggested, is question-begging. And, it depends upon the idea that the purely mental is utterly alien because it does not interact with the physical. But this idea is also open to suspicion for the reason cited earlier; viz., what non-question-begging reason is there for thinking that mental states must interact with a *physical* environment, as opposed to an environment which might or might not be physical? If the latter is a correct view of the mental, then a purely mental world would not be *utterly* alien, since it would be a possibility concerned with mental properties of the actual world. And, even if it is utterly alien, Teller is still in need of an argument to establish that it is not possible. As a consequence, I see no *viable* argument here against the possibility of a world in which the mental does not interact with the physical, but does exhibit some causal relations. Hence Teller has failed to show that a world of mental beings that do not interact with a physical environment is impossible.

I conclude that Teller has not formulated any conclusive arguments against the logical possibility of worlds in which physicalism is false. I agree with him that, assuming the truth of physicalism, no such world is accessible from the actual world in the sense of being metaphysically possible. But, as I have argued, metaphysical possibility is not the only sort of possibility. Although I surely have not proved beyond doubt that there are logically possible worlds that falsify physicalism, I have shown, I think, that the burden is squarely upon those who believe that the alleged cases of such worlds are not genuine possibilities. If the necessity, *tout court*, of physicalism is to be established, persuasive arguments of the required sort are still needed. In my opinion, M4, as I have interpreted it, stands firm.

5.5. SUMMARY AND CONCLUSIONS

In this chapter, I have discussed the main tenets of the received view concerning the metatheses of physicalism. The scope of the theses was revised from a relatively restricted one concerning natural science (M1) to a universal scope covering all aspects of the natural order (M1'). The main issues arising in this context were, first, that the natural order needs to be pinned down in some relatively precise way, and second, that the existence of *abstracta* (for example, universals, propositions, mathematical entities, *possibilia*) needs to be understood in relation to physicalism. Is a physicalism that countenances *abstracta* a physicalism worth having? Is physicalism committed to nominalism?

Regarding their epistemological status, the theses are usually conceived of as being empirical or *a posteriori* (M2). But it is evident that there are serious problems with the identification of legitimate forms of confirmatory and disconfirmatory evidence and arguments. A prime reason for these difficulties is that physicalism plays both a normative and a descriptive role in inquiry, and the normative role threatens to undermine the *a posteriori* status of the theses. In addition, the discussion revealed that, since the theses purport to describe the natural order and knowledge that concerns it, they must be self-satisfying. Given their semantic and other intensional content, it is, at the very least, controversial whether they can in fact meet the very demands they make.

The methodological employment of the theses in guiding inquiry as called for by M3 is a paramount component of the received view. With the expanded scope of the version of physicalism I am entertaining, M3 was revised to M3'. In both versions of the metathesis, there are at least two distinct roles the theses can play. They raise questions and suggest lines of research (R1) and they constitute constraints upon the answers to such questions and upon the development of an integrated system of knowledge concerning the natural order (R2). The actual employment of the theses, especially in role R2, turns out to be an intricate matter of weighing a number of considerations relevant to whether it is appropriate to employ the theses in a particular context. And if the theses are appropriate for use in a given context, the decision regarding how to resolve conflicts that they might precipitate is equally complicated: there are no immediate consequences of the apparent failure of a given theory or phenomenon to satisfy physicalist demands. The principal problem with regard to M3′ is that it is not clear that the theses are justifiably employed in the roles R1 and R2, given the many difficulties afflicting physicalist doctrine and given that the various justificatory arguments are problematic. Further, in the light of M2, there is a pressing need to identify a proper balance between the descriptive and normative functions of the theses so that the theses are not trivially self-supporting.

Finally, with respect to their modal status, the theses are generally believed to be contingent. This, I argued, is an ambiguous label at best and that metathesis (M4) requires interpretation in order to remove the ambiguity. I concluded that the theses are metaphysically necessary though logically contingent. The most pressing issues here are, first, that of characterizing and justifying the existence of logically (but not metaphysically) possible worlds in which physicalist theses are false, and second, that of circumscribing the class of metaphysically possible worlds.

If we combine these complications and difficulties raised by the metatheses with those raised by the presuppositions (for example, the possibility of downward incorporation of the mental, the indeterminacy of the bases, and the monopolistic overtones of the privilege accorded to the physical bases), we have a rather substantial catalogue of problems that must be dealt with if physicalism is to be even remotely acceptable or justifiably employed in scientific and philosophical research. The function of Chapter 6 will be to take on these and other difficulties in an effort to establish that it is plausible both that they can be dealt with and that physicalism emerges from this rather rough treatment relatively unscathed.

6

Assessment of the Physicalist Programme

> The reason I am going to focus my attack on materialism is
> that materialism is the only *metaphysical* picture that has
> contemporary 'clout'.
>
> Hilary Putnam (1983: 208)
>
> The physical world consists entirely of physical facts. What is
> not a physical fact is not part of the physical world. And
> physicalism is the thesis that the physical world is the only
> world there is, or the only world that is real.
>
> Barry Stroud (1987: 264)

My aim in this chapter is to conduct a partial, interim assessment
of the physicalist programme. Although several problems have accu-
mulated over the past three chapters, I am optimistic that they can
be solved. Thus this chapter should be viewed as contributing to
a defence of the acceptability of the programme. If nothing else is
accomplished, my hope is that a number of pressing difficulties for
physicalism will have been clearly articulated, and hence made
more amenable to analysis and evaluation.

In the language of Chapter 1, there are three dimensions of
assessment that need to be scrutinized: adequacy, acceptability,
and success. The first concerns how well a formulation of physicalist
principles expresses the motivating ideas and values of the programme, while the second concerns how well the formulation stands
up to critical assessment in the form of objections and empirical
evidence. The final dimension bears upon the prospects for actual
success in 'working out the programme': just how likely is it that
a system of knowledge can be constructed which conforms to
physicalist theses? I shall discuss each of these dimensions of assessment in turn.

6.1. ADEQUATE EXPRESSION OF THE CORE IDEAS AND VALUES

With respect to the adequacy of a formulation of physicalism, three principal areas were highlighted in Chapter 1: ontology, objectivity, and explanation. The theses formulated in Chapter 4 satisfy the criteria of adequacy in each of these areas. The ontological concerns of physicalism can be expressed in terms of the ontological dependence of all phenomena upon physical phenomena, the supervenience of all phenomena upon the physical domain, and the realization of all phenomena by physical phenomena. Of special importance is not only that everything that exists in nature is physically grounded, but also that all connections and influences within the natural order have a physical basis as well. Theses T0 and T1 guarantee the dependence and the realization requirements, whereas T2 explicitly calls for satisfaction of the supervenience requirement. There is nothing fancy in the way in which these theses satisfy the demands. I believe that directness is a virtue here, given the dismal history of formulations advanced to express physicalist ideas and values while relying on relations (for example, definitions, derivations, supervenience) which are only indirectly related to those ideas and values.

Physicalist concerns about objectivity revolve around the core idea that the physical facts and truths suffice as a ground for all objective facts and truths in nature, in the sense of 'objective' highlighted in Chapter 1. *Within a physicalist system*, once the physical facts and truths are fixed, then so are all the facts and truths. Theses T2 and T3 explicitly express this idea. Of course, the objective status of the physical bases of the system has been challenged, and below I shall take a stand, on behalf of the physicalist programme, to meet this attack. But, subject to this caveat, the adequacy of the theses to express the key ideas in this area is straightforward.

Explanatory concerns require that, within a physicalist system, it should be explainable how each non-physical phenomenon is realized by physical or physically-based phenomena. All objects and attributes either are physical or are explainable in terms of physically-based objects and attributes (i.e. objects and attributes that are grounded in the physical domain via a chain of vertical explanations). Theses T4 and T5 explicitly call for such explanations, and hence satisfy the requirement. Again, directly calling for explanations is the most reliable and, arguably, the only way to ensure that a physicalist system, satisfying the theses of physicalism, will exhibit the requisite explanatory structure. Other, seemingly more economical, formulations in terms of derivations or supervenience simply do not do the necessary work, as I argued in Chapter 2.

I conclude that, relative to the criteria of adequacy that I believe properly captures what the physicalist programme is about, the formulation suggested in Chapter 4 is an adequate formulation, whereas other extant formulations, such as those reviewed in Chapter 2, are not. The question of whether the theses suffice to provide a 'minimally adequate' formulation is another matter, however. Since T1 implies T2 and T2 implies T3, given modest assumptions, the latter two theses can be dropped. None the less, and regardless of whether other entailments hold between the theses given certain assumptions, I think it is more perspicuous and useful to include all of T0 to T5 in an official statement of the physicalist doctrine. Such a comprehensive formulation means that fundamental ideas and values maintain a high profile.

6.2. CHALLENGES TO THE ACCEPTABILITY OF PHYSICALISM

The *acceptability* of a formulation of physicalism concerns how well it stands up under critical assessment: i.e. how well it fares in the face of objections and how well it is supported by evidential considerations. In what follows, I shall focus primarily on various objections to physicalism, some of which I have already identified and discussed in a preliminary way. As noted above in Chapter 5, the evidential status of physicalist principles has never been systematically and thoroughly assessed. It is a shortcoming of my own assessment that it will not fill this gap, but the scope of the current project must be restricted in some ways.

The following are what I take to be the principal outstanding objections with which physicalists must come to terms:

- The theses are inappropriately monopolistic.
- The theses are 'watered down' in that they are compatible with the existence of abstract objects, non-physical attributes, and irreducible relations between physical and non-physical domains.

- The theses are not 'self-applicable', as they must be if they are to be in accordance with the metatheses.
- The physicalist conception of objectivity is defective.
- The programme, as characterized, is compatible with the 'downward incorporation' of mental and other 'patently non-physical' phenomena into the physical bases of the system.
- The theses lack the *a posteriori* epistemological status attributed to them by M2; instead, they are trivially true.
- It is unreasonable to be a physicalist *either* because the theses are 'utopian' *or* because there are alternatives that are equally justified and without the liabilities of physicalism *or* because the theses are obviously false.
- Physicalism is incomplete; it will inevitably leave some aspect of existence out of account (for example, mind, meaning, values).

There are no doubt other objections that can be advanced, some of which I have already discussed in earlier chapters. However, if the physicalist can effectively deal with the objections on the above list, the programme will be in a more secure strategic position *vis-à-vis* critics.

The Monopolistic Overtones of Physicalism

Physicalism has often been vehemently charged with being inappropriately monopolistic in one respect or another. The objections are frequently framed in polemical terms, and often they depend upon inaccurate and distorted formulations of the physicalist doctrine. Consider the following quite typical contentions:

The uncompromising physicalist who would exclude certain phenomena [e.g. meaning, values, mind] from the physical world and the more accommodating physicalist who would include them by analysing or identifying them in purely physical terms both share an idea of the physical world and its exhaustiveness. It is meant to be the only world there is. (Stroud 1987: 265.)

So any physicalism with teeth must insist that there are no processes, anywhere . . . except those brought about by the action of physical forces. But this places a powerful constraint on explanation outside physics: no explanation of non-physical processes can be accepted unless the causes and laws whose operation it invokes are special cases of physical causes and laws . . . To the extent that in insisting on the determination of all truth by physical truth, we are insisting that nothing can happen unless something physical happens, and that nothing physical can happen except through the actions of physical forces, we compromise the explanatory autonomy of all enquiries. So it seems to me that physicalism in more than name only has an inevitable tendency towards physicalist *monopolism*, the philistine hegemonism to which any monistic metaphysic is tempted . . . (Campbell 1988: 360.)

. . . the materialist claims that physics is an approximation to a sketch of the one true theory, the true and complete description of the furniture of the world . . . The appeal of materialism lies precisely . . . in its claim to be *natural* metaphysics, metaphysics within the bounds of science. . . . metaphysical materialism . . . [is] the dominant contemporary form of scientism . . . one of the most dangerous contemporary intellectual tendencies. (Putnam 1983: 210–11.)

His [the pluralist's] typical adversary is the monopolistic materialist or physicalist who maintains that one system, physics, is pre-eminent and all-inclusive, such that every other version must eventually be reduced to it or rejected as false or meaningless. (Goodman 1978: 4.)

As is evident, there are many different forms that this sort of objection to the programme can take. There are, indeed, many ways that a programme can be monopolistic. Making the relevant distinctions is important both for advancing sharp criticisms of physicalism as well as for presenting effective replies.

From the quotes above, one can extract, at least, the following distinct forms of monopoly that are allegedly characteristic of physicalism:

- The only facts are those expressed in physical language.
- Every fact is a physical fact; the physical world is the only world there is.
- Everything has a physical explanation.
- Physics is the only autonomous explanatory domain; physicalism compromises the explanatory autonomy of all domains other than physics.
- Any putative entity or property that is not reducible to physics is unreal and eliminable.
- Physics offers a unique and complete theory of the world.
- Physicalism is a form of scientism.
- Physicalism is a form of metaphysical realism.

According to the critics, such claims constitute serious offences on the part of physicalists and are grounds for rejection of the programme. I shall argue that, when suitably clarified, each claim is either not applicable to the physicalist programme as I have presented it or is not a problem with the programme, even if it does apply.

My task here is, to a considerable extent, easier than it might have been since John Post has extensively discussed several of the possible monopolistic overtones of inappropriate formulations of physicalism (Post 1987: 203–6). I concur with much of Post's discussion, but, as will become evident, I have some disagreements with his view as well. Post identifies three significant forms of monopoly to which physicalists ought *not* pretend: universal descriptive power of the physical language, absolute priority of the physical domain, and universal reducibility to the physical domain of all properties and terms. On the other hand, he affirms the comprehensiveness of the physical domain with respect to ontology (i.e. everything is identical with something in the physical inventory of entities) and with respect to determination relations (i.e. all facts and truths are determined by physical facts and truths). He affirms the conditional priority of physics with respect to a certain purpose (i.e. identification of objective truth). And he affirms a form of metaphysical realism[1] which entails the independent existence of the world (i.e. the sum total of everything there is). These views constitute a basis for response to many of the charges of monopoly expressed above.

The charge that physicalism is committed to the monopolistic idea that all facts are *expressible* in a physical language attacks a form of physicalism that is not even true in the domain of physics, as I argued in Chapter 2. Assuming that the facts about nature are constituted by distributions of objects and attributes over regions of space-time, the descriptive resources of the language of physics do not suffice for describing all physical objects or for expressing all physical attributes. Hence they do not suffice for expressing all physical facts. And if non-physical facts are similarly constituted by distributions of non-physical objects and attributes, then the descriptive resources of the physical language will also fail to express all the non-physical facts of nature that are, according to the

---

[1] As I use the term, 'metaphysical realism' does *not* entail that there is one true theory of the world, *contra* Putnam's usage. Rather, metaphysical realism is simply the affirmation of the mind, language, and theory independence of reality. See below for further discussion.

physicalist, determined by and realized by the physical facts. Physical terms do not, in general, suffice for picking out non-physical kinds of objects and non-physical attributes.

Even if it is granted that (*per impossibile*) all non-physical objects, attributes, and facts are identical to physical objects, attributes, and facts, there is no guarantee that there would be a physical term for describing or expressing each one. On the other hand, it is conceivable, under such an assumption, that all the non-physical facts could be identical to physical facts that were expressible in the physical language. But this possibility should not provide much comfort to the heroic physicalist who endorses the universal descriptive adequacy of the physical language, since the assumption on which it is predicated is untenable and, in any event, the problems with describing the physical bases are devastating to the approach. Thus the universal descriptive power of the physical language should be dropped from characterizations of physicalism by both proponents and opponents of the programme.

Further, the *identification* of every fact with a physical fact is no part of physicalist doctrine. This follows directly from the admissibility, within a physicalist system, of non-physical attributes and kinds of objects, so long as they comply with the physicalist theses. Whether this 'waters down' physicalism or makes it 'too easy' are questions I shall address shortly. But no matter how those questions are answered, I suggest that both proponents and opponents of physicalism should henceforth refrain from including such generalized identity theses in the physicalist doctrine.

What of the charge that physicalism is committed to the monopolistic claim that *everything has a physical explanation*? If the claim means that everything has an explanation *in physics*, then it is a mistaken attribution and no problem for the physicalist. On the physicalist view, physics does enjoy a place of privilege, but that privilege does not involve the idea that physics is a universal explanatory science (i.e. the idea that it explains everything). Physical theory does have full coverage in the sense that it concerns all regions of space-time, and it does concern all dynamics, interaction, and composition of the physical phenomena that occur in those regions. As a consequence, with the exception of fundamental phenomena, physics does purport to explain all that goes on in the physical basis of the physicalist system. However, there are non-physical phenomena (for instance, social and economic phenomena) which, although realized by the phenomena in the basis, are not part of the subject matter of physics. Explanations of what goes on in the basis are not, *ipso facto*, explanations of what is realized by the phenomena in the basis. Alternatively put, higher-level phenomena, although ultimately realized by physical phenomena, do not necessarily have explanations which are framed in purely physical terms or which appeal to physical laws.

Rather, explanation of these realized phenomena can be of many types, none of which need be types of explanation offered by physicists. Thus one type involves appeal to other non-physical (for example, social, economic) phenomena that causally explain a given phenomenon. Another is the vertical explanation discussed in Chapter 4, which involves appeals to physically-based objects and attributes that combine to realize the given non-physical phenomenon. Within a physicalist system, the types of explanation of non-physical phenomena do not stop here. Subject to satisfaction of the theses, there is no limit to the patterns of explanation that are compatible with physicalism. So not only is it *not* the case that everything must have an explanation in physics, but there is wide-ranging diversity in the types of explanation that non-physical phenomena can have.

If what the objector means to attribute to the physicalist is the claim that everything has a certain sort of explanation that appeals to, or is ultimately grounded in, the physical basis (i.e. vertical explanation), then, with the exception of basic physical phenomena, physicalism does indeed require this. But such explanations are not in general explanations in physics. They employ realization theories that show how non-physical phenomena are realized by configurations of physical, or physically-based, objects and attributes. Is this sort of physicalism unduly monopolistic? I think not, since as just observed this requirement does not rule out other forms of explanation. As I argued in Chapters 1 and 4, it is a form of explanation that has considerable value. Thus, although physicalism does impose a constraint upon all other explanations (i.e. that they only appeal to physically-grounded attributes, objects, processes, etc.), the objector needs to make a case for why such constraints are inappropriate. Vague charges of monopoly won't do, since explanations are always constrained in one way or another. What non-question-begging reason is there for rejecting physicalism simply because it imposes constraints on all the other, various sorts of explanation? If physicalism is true, then such constraints are indeed quite appropriate.

Let us consider further the objection that physicalism unduly constrains explanation in non-physical domains, and hence that physics is *the only autonomous explanatory domain*. Perhaps it does not suffice to allay the concerns of the objector by making quick comments about the truth of the doctrine, its compatibility with many forms of explanation, and the general presence of constraints on explanation. It might, none the less, be objected that, although there are general demands that explanations must meet (for example, logical consistency), the physicalist demand is a special one which inappropriately constrains the conduct of inquiry in programmes of research concerned with non-physical domains. As I understand him, this is just what Campbell is complaining of in the passage quoted above.

In my estimation, this is not an apt criticism of physicalism, for at least two reasons. First, physics, like all other domains, is subject to the constraints imposed by physicalist principles, which are architectural principles that relate the base to the higher levels. It is the structure as a whole that is of prime importance. The base must suffice to support the edifice just as much as the higher levels must satisfy the constraints imposed by relations to the base. Thus, if anything, the objector must claim that physicalism is unacceptable because it requires that there are *no* autonomous explanatory domains: physicalist constraint relations work in both directions.

Second, why is it objectionable that physicalism imposes *certain* constraints upon inquiry? The critic owes a more developed argument for the inappropriateness of actual physicalist constraints, given that no explanatory domain is autonomous if 'autonomy' means freedom from all constraint. Surely, self-consistency and consistency with other truths are minimal constraints imposed upon all explanation. And the idea of isolated domains of inquiry, free of all more substantive constraint relations to other domains, is not true to actual inquiry in science and elsewhere. Indeed, constraints are often discovered in the course of inquiry, even if they are not immediately presumed to exist (for example, relations between psychology and neuroscience in the study of perception). If the critic is to make the case here, it must be shown why physicalist constraints upon inquiry are inappropriate.

The burden of the critic here is not insignificant, as there are several reasons why physicalism, although constraining, does not put a strait-jacket on the various non-physical areas of research. Thus the constraints, although ultimately relevant, do not seriously come into play in the early stages of a research programme. As discussed in Chapter 5, physicalist principles actively play their normative methodological role only under certain conditions. Until such conditions are met, a direction of inquiry can be pursued in relatively autonomous fashion until it gets seriously on track. Even then, the relevance of physicalist considerations to ongoing research will vary with the domain (for example, chemistry versus economics), although all must satisfy the basic requirements of the programme.

Physicalism does *not* require that all inquiry be concerned with the physical bases of phenomena or that all explanation be of only one sort. The constraints imposed by the theses leave considerable room for research into high-level, non-physical phenomena and for the development of a variety of explanatory forms (for example, rational explanation, teleological explanation, functional explanation). It is a myth promoted by critics of physicalism that the programme requires that all explanation be just like that in physics, or, as Campbell put it in the quote above, that all explanation involves causes and laws that are 'special cases' of physical causes or laws. If this latter charge means that all such causes and laws are physically based, then more needs to be said to show that this is a bad and overly constraining requirement. In particular, exactly which patterns of explanation are inappropriately excluded? If the charge means something else, then more needs to be said to clarify what it means and to show that it is indeed an inappropriate constraint actually imposed by physicalism.

The requirement that all explanation be physically grounded does require that all attributes and objects posited for the purposes of explanation be realized by physically-based objects and attributes, that all truth be determined by physical truth, and so on. But such physicalism is compatible with considerable variation in the sorts of theoretical structures that can be erected upon the physical basis. Wide variation in languages and 'conceptual schemes' introduced for describing and thinking about the natural order is permitted so long as physicalist theses are satisfied. Physicalism simply does not require that there is exactly one way of conceiving of and describing the natural order. Thus physicalist structure does not preclude wide-ranging variation among theoretical systems, and such 'theoretical pluralism', along with the possibility of diverse patterns of explanation, makes the charge of monopoly highly overstated.

It may deflate the charge of monopoly a little more to recall that it is an empirical matter whether or not physicalist system-building, subject to the constraints imposed by the theses, will be successful. According a place of privilege to the physical bases ought not be construed as implying an *a priori* claim to privilege. That is, until further notice the physicalist's theses are subject to the test of evidence, and at most they purport to be *part* of the truth about the natural order. Indeed, if the theses are true of nature, then any constraints they impose are justified by general methodological principles concerning the compatibility of truths about a domain. Given the logically contingent and *a posteriori* status of the theses, this can hardly be called 'inappropriately monopolistic'.

Finally, a comment on the alternatives to the physicalist view is in order here. Such alternatives fall into at least three camps. The first includes those alternatives which advocate an alternative unifying basis (for example, phenomenalism), while the second includes those which advocate complete autonomy of domains. Both will be discussed below in the course of considering other objections. However, I think that if there is a burden of proof lying anywhere it is with the advocates of these sorts of alternatives. This is not to say that physicalism does not have a burden to carry: that is what this book is about. But it is to say that the burden of alternative unification schemes is, in my opinion, clearly greater, and that the view of research domains as enjoying complete relative autonomy is neither accurate nor desirable. Whether either of these types of alternatives can overcome their liabilities is, I think, seriously in doubt. Surely an 'anything goes in research' mentality should be rejected; what is left over, then, is the serious question of which constraints are appropriate. Physicalism is one proposal regarding such constraints, and there is reason for thinking that it competes favourably with these other proposals, as I shall argue below.

There is, of course, a third alternative to physicalist unification: viz., that domains are more or less autonomous depending on the case and that the types of constraints (if there are any at all) that one domain imposes on another likewise vary from case to case. A sensible physicalist will, I think, agree with the spirit of this position up to the point that it implies there are no universal constraints upon inquiry into the natural order. Thus the physicalist can and should allow that there are various ways in which, and extents to which, domains are constrained by each other. What distinguishes the physicalist from the staunch advocate of this third alternative view is the assertion, subject to *a posteriori* assessment, that physicalism provides a class of universal constraints upon our understanding of the natural order. This is not the dogmatic and *a prioristic* assertion of a monopolistic metaphysics; it is the expression of a hypothesis about one way of unifying our understanding that offers substantial benefit to us. The moral here is that, when considering objections to physicalism, it is important both to reflect deeply upon the nature and prospects of the alternatives *and* to keep in mind the nature of physicalist theses and the many degrees of freedom physicalism allows in the conduct of inquiry.

I want to turn, now, to the dichotomy often attributed to physicalism that *either a phenomenon/theory is reducible to physics or it is unreal/untrue and eliminable.* As is evident in the quoted passages above, this is sometimes cited as an aspect of the monopolistic pretensions of the physicalist programme, and so is a possible ground for its rejection. I should note here that some self-avowed physicalists do recite this dichotomy, often shortly before pronouncing a death sentence upon some putatively real phenomenon or some putatively true theory. See P. M. Churchland (1985: 45) for an example of the wielding of this sword against folk psychology.

My main complaint with the assertions of both critics and proponents of physicalism concerning this dichotomy is that either they speak of 'reducibility' without any clarification or they employ an inappropriate conception of what physicalist 'reduction' involves. If any of the variants of classical definitional and derivational reduction are intended, then the dichotomy is a false one. There are other options for how high-level phenomena and theories relate to lower-level, physically-based phenomena and theories in addition to either classical reduction or elimination. On the other hand, I believe that physicalism does involve some reductive relations: specifically, each instantiation of any non-physical object or attribute must be associated with the instantiation of the members of a class of physical objects or attributes that are jointly sufficient to realize it. This is indeed a reductive relation between higher-level domains and the physical basis, although it is clearly not an instance of CR or CR'. In light of the methodological roles of the theses of physicalism, a failure of reducibility in this sense would mean that a certain putatively real object or attribute should be dismissed as not objectively real.

There are several reasons why this is hardly the monopolistic tyranny suggested by the critic. As observed above, the *a posteriori* and logically contingent status of the theses mean that this sort of judgement to reject some phenomenon is not done on the basis of *a priori* requirements dogmatically handed down. It is a matter of empirical discovery that such failure to be reduced has deep significance bearing upon our judgements about what is real, true, and possible in nature. And, as emphasized in Chapter 5, methodological judgement made in the context of apparent failures of reducibility are more complicated than the simplistic dichotomy, 'reduce or eliminate', suggests. Sometimes conditions are not appropriate to make any sort of judgement, while in others they may lead one to explore the other *three* options available in such cases (i.e. reject the physical base, reject the claim of failure of reduction, or reject physicalism itself). Those opponents and proponents of physicalism who rely upon the simple dichotomy, even if they do supply appropriate understanding of 'reducibility', typically fail to take into account the complexity of the methodological situation raised by apparent failures to satisfy reducibility requirements.

Finally, it should be remembered that 'reduction', after all, does not itself mean the same thing as 'elimination'. Rather, it is a form of relation between one domain of entities and another, subject to certain constraints. Different domains and different constraints will lead to different reductive relations with different significance. But such relatedness between domains is not evidently a bad thing; on the contrary, such relations are an important form of knowledge (i.e. knowledge that consists in an appreciation of how things hang together). If physicalism is true, then such reductive relations as it requires should be grasped, appreciated, and taken into account in appropriate contexts. If it is properly concluded that some apparently real phenomenon fails to satisfy the requirements and ought to be eliminated as a consequence, then that too is an important piece of knowledge that no amount of anti-physicalist rhetoric can mask. Those critics who speak in a derogatory fashion about 'reduction' and 'elimination' need to move to a more sophisticated understanding of those terms, of the judgements concerning them, and of the physicalist programme, if they are to mobilize compelling objections.

Some critics, especially Putnam (1983: 210–11), have attacked the idea that, according to the physicalist, *physics purports to provide the one true and complete theory of the world*: i.e. it is alleged that physics aims at discovering a *unique* and *complete* theory of the natural order, the one theory that suffices for saying all there is to say. However, this objection fails because the idea that *physics* aims at a unique and complete theory of nature is not appropriately attributable to the physicalist. Hence there are no grounds here for a charge of undue monopoly.

With regard to the *uniqueness* of physical theory, for example, there is no reason why the physicalist cannot allow, at least, that it is *possible* for there to be alternative formulations of physical theory which are each empirically adequate, which satisfy other relevant principles of evaluation equally well, and which do the theoretical work of physics. Such alternative theoretical structures can be of at least two sorts that are not damaging to physicalism: either they are alternative formulations of one theory (Quine 1975; 1981a) or they are formulations of different theories which are not incompatible, although they do 'cut up the world differently' (Field 1982). 'Reconcilable' alternative formulations of physical theory that are evidentially equivalent are not a problem for the physicalist. The consequence of such permissiveness for the physicalist programme is that different physicalist ontological and theoretical structures are possible, each of which satisfies the demands of physicalist principles and each of which is grounded in its own physical bases. As a result, the physicalist ought not to be viewed as holding that there is a unique physical theory of nature.[2]

With regard to the alleged *completeness* of physical theory, the issue is less clear, if only because the term 'completeness' is ambiguous. If the completeness of physics means that physics can express all there is to say about the natural order, then such a claim of

---

[2] The possibility of irreconcilable, evidentially equivalent physical theories raises more serious issues for the physicalist (e.g. the question of the objectivity of the physical bases). See Ch. 3 and below for discussion of such issues.

completeness is no part of physicalism as I see it. Other descriptive languages and theories are *required* to express many of the truths concerning the natural order. Indeed, the facts and truths expressed in the language of physics are often of little interest or relevance with respect to many human concerns (for example, aesthetic experience, moral questions, human relationships). In any event, not everything can be expressed linguistically (cf. Post 1987: 247 ff.); the language of physics cannot express all truths about the world if some things (for instance, aspects of subjective experience) are not expressible at all.

On the other hand, if the completeness of physics means that physical theory provides 'full coverage'[3] in the sense of providing bases that suffice to ground a physicalist system in the ways I have discussed in earlier chapters, then in that sense physical theory is complete; but this is hardly to say that physics 'says it all'. Rather, physics provides bases for the realization, determination, and explanation of all that there is in nature. But much of what there is in nature and much of what can be said about nature do not fall within the domain of physics. Therefore just as uniqueness of physical theory is not a requirement of physicalism, neither is completeness (in the sense intended by the critics), although the universal scope of the theses with regard to the natural order is a requirement of the programme.

The charge of 'scientism', like that of 'philistine hegemonism', is a highly polemical accusation lacking in serious content or force, since none of the standard meanings of that term are applicable to the physicalist programme: for example, scientific knowledge provides a unique and total understanding of the natural order, scientific truths have unconditional priority over any other sort of claim, scientific method is the only way to gain knowledge. The first idea is as inappropriate as is the more restricted claim discussed above regarding the uniqueness and completeness of physics, and for essentially the same reasons. That is, there need not be one system of scientific knowledge, and it is not the case that all there is to say about nature is expressible in terms drawn from some class of scientific languages. Further, the idea that the truths of science have unconditional priority in all contexts has been

---

[3] Quine (1978) uses the expression 'full coverage' in a somewhat different, but related, sense.

soundly dealt with by Post (1987: 203 ff.), who brings out many of the reasons why such a view is misguided. As in the case of the truths of physics, there are many contexts in which scientific truths are of little interest or relevance and can be downright harmful (for example, in many clinical or other circumstances of human suffering).

Finally, the notion that 'scientific method' is the only method for acquiring knowledge is lame for at least two reasons: it is doubtful that there is any such thing as 'the scientific method' and, even if there is, there are kinds of knowledge for which scientific method is an inappropriate means of acquisition (for example, knowledge concerning the human condition and aspects of human experience, knowledge of aesthetic and moral facts). Physicalists need not take a strong stand on this issue one way or the other since the programme is compatible with either view. But if physicalists must be aligned with one side or the other, surely it is with the side of pluralism of method. It seems to me to be a definite burden of those who would charge physicalism with 'scientism' to clarify what they mean and to show how it is that physicalism, properly conceived, is guilty of the charge.

The final objection, alleging monopoly, is that physicalism implies a form of 'metaphysical realism' which is a monopolistic metaphysical stance that calls for the mind, language, and theory independent existence of objects and their attributes. There is, on this view, a ready-made world awaiting our discoveries, our descriptions, and our conceptualizations. Such an independent world has a certain character, whether we know it or not (cf. Putnam 1983: chapter 12).[4]

This version of metaphysical realism is alleged to be monopolistic in so far as it improperly excludes legitimate alternative forms of unification programme (Putnam 1979). Physicalism, when combined with metaphysical realism, purports to be concerned with

---

[4] Putnam has often attributed to metaphysical realists, in addition to the thesis of independence, both a correspondence theory of truth and the idea that there is one true theory of the world. However, the characterization in the text is the most upon which contemporary physicalists are likely to concur, and it is all that is required to mobilize the charge of monopoly. See Field (1982) for important distinctions among the senses of 'metaphysical realism' as that term is employed by Putnam. I concur with Field that clarity is served only if the semantic, epistemological, and metaphysical theses that Putnam has combined under a single term are separated out and dealt with individually.

the only objective reality there is, and it imputes a unity within the diverse aspects of that reality. As discussed in Chapter 3, physicalism of this sort precludes alternative unification schemes grounded in, for example, psychology, unless such schemes are of a different sort (for example, evidential or conceptual as opposed to metaphysical and explanatory). If physicalism is true, then, according to the metaphysical realist, the world exhibits a physicalist structure and unity that is not compatible with alternative metaphysical views such as phenomenalism or other forms of idealism. If every attribute is *realized* by physical attributes, it cannot also be the case that every attribute is *realized* by mental attributes.

Further, in the case of alternative, *incompatible*, empirically adequate, but evidentially equivalent, physical theories (if there are any), the metaphysical realist contends that at most one is true, even if we could never tell which it is: the world breaks the tie. This is subject to the caveat, discussed in Chapter 3, regarding alternative formulations of a physical theory and the possibility of alternative, but *compatible*, physical theories of nature. These possibilities are compatible with physicalism and metaphysical realism and could lead to a plurality of physical bases. But this does not affect the main thrust of the metaphysical realist understanding of physicalism: subject to the caveat, there is a unique physical basis.

Many physicalists (for example, Boyd 1980; Devitt 1984; Hellman 1983; Lewis 1983; Post 1987) do, I believe, concur in this appraisal of physics and physicalism. For example, consider how Post characterizes the priority of physics within a physicalist system:

But the priority physics enjoys is highly conditional. If we are interested in the kind of objectivity, comprehensiveness, and explanatory power that physics pursues, then the truths of physics take on a corresponding priority, in light of which it seems significant that all other truths are determined by them. (Post 1987: 165.)

And consider his general approach to *truth*:

The concept is a realist one, according to which truth depends on how things are in the world, not on human or other consciousness or understanding . . . According to a realist concept, truth may be said to be 'invariant', or objective . . . Realist truth is not perspectival but truth period. (Post 1987: 19.)

And, further, his comments on the likely limits of our physical knowledge:

Physicalists, then, by virtue of their prior commitment to objective truth, are committed to a true physical universe beyond the manifest universe as conceived by today's physics. Physics may or may not someday reach a stage at which the manifest universe, as conceived at that stage, is the Universe. Even if it reaches it, there may then and forever be no way of knowing for certain that it has done so. (Post 1987: 158.)

Objective truths for Post consist of truths that are invariant across perspectives and that depend upon 'how things are in the world'. In the case of physics, objective truths correctly describe the Universe as it is, apart from any conception we may have of it.

Post is explicitly *not* committed to a correspondence theory of truth, or the idea that physics describes *the* way things are, or the idea that there is one true theory of the world. And he certainly does not hold that there is only one way that the objective truths can be unified. He cannot be painted with any of those brushes. But he can, I think, be fairly viewed as being committed to metaphysical realism in the narrow sense I have characterized above. Indeed, he explicitly incorporates such realism into his formulation of minimal physicalism (Post 1987: 189). Thus, as I understand him, Post affirms the view that there is an independent reality of objects, attributes, and kinds that awaits our discovery and that is the object of our inquiry. Assuming physicalist principles are true, therefore, he can be viewed as rejecting as false any *competing* unification scheme that is *incompatible* with physicalism.[5] And he can be viewed as taking a realist stand on the problem of empirically adequate, evidentially indistinguishable, but irreconcilable, physical theories: viz., at most one is true. When there are incompatible alternatives of either sort (i.e. unification systems, physics), metaphysical realism does not tolerate more than one victor.

There are, then, monopolistic overtones to even Post's version of physicalism. He endorses a form of realism that asserts the existence of a ready-made world[6] of objects and attributes. Physical

---

[5] See Post (1987: 213, 329, 332) for discussion that suggests this view. Of course, non-literal renderings of alternative systems and differences in the goals of physicalist and other systems allow for reconciliation. The issue is whether two incompatible systems, literally interpreted and having the same objectives, can both be accepted.

[6] The phrase is borrowed from Putnam (1983).

phenomena and truths enjoy a type of privilege not enjoyed by any other domain or body of doctrine: if we are interested in identifying the basis for a certain sort of unification,[7] then physics is the only place to look.[8] And if there are irreconcilable, but equally acceptable, physical theories, the independent and ready-made world determines which, if any, is true. But these sorts of monopoly are distinct from those crude forms considered and rejected earlier (for example, the uniqueness and completeness of physics, scientism). None the less, for Post, physicalist unification is the only one *of its kind*.[9] Such 'monopoly' is a consequence of the realism held to be a component of physicalist doctrine.

Now, the thrust of Putnam's objection is that metaphysical realism is not acceptable, and hence any form of physicalism which incorporates or implies it has false (or incoherent) monopolistic pretensions.[10] The unintelligible view of the metaphysical realist is to be replaced, according to Putnam, by the view that all objects, attributes, and truths are 'theory dependent'. That is, all objects and attributes are constituted by our concepts and theories, all truth is system-relative truth, and reality is always system-relative reality (cf. Putnam 1987: 20; 1983: 224–5). Thus the world is not 'ready-made' and waiting for us to discover it, describe it, or conceive of it. Rather, worlds are created by us! Both Putnam and

---

[7] The ideas and values outlined in Ch. 1 above capture my version of the relevant sort of unification. See Post (1991: ch. 5) for an alternative discussion.

[8] See Post (1987: 118). Of course, if we are interested in something else (e.g. aesthetic excellence, healing the sick), then we might be better advised to reflect upon different aspects of existence and knowledge. Post (1987; 1991) has frequently emphasized the variety of types of unification and the importance of context and goals in determining which type is most appropriate on a given occasion. And he certainly endorses the *a posteriori* and contingent status of physicalist theses. Therefore, in suggesting that Post's physicalism is monopolistic, I do so with only a very specific feature in mind (viz., the sort of realism characterized in the text).

[9] Post does allow that there are very likely other sorts of unity which do not compete with physicalism and which are therefore legitimate even if physicalism is true. And he does consider various competitors to physicalism and rejects them all on the grounds that they fail to satisfy criteria of truth. However, I believe that the passage in Post (1987: 332) strongly suggests that *if* both physicalism and a genuine competitor were to satisfy the criteria of truth, then a crisis would be precipitated which would require either rejection of one of the competitors or a successful reconciliation. This I take as strong evidence of a metaphysical realist framework.

[10] See Putnam (1983: ch. 2 and 12; 1981: ch. 3) and Goodman (1978: ch. 7) for statements of their attacks upon metaphysical realism.

Goodman are careful to point out that this is not an 'anything goes' form of relativism (Putnam 1981; Goodman 1978), but it is none the less a radical departure from the spirit of traditional metaphysical system-building.

On such a view, even if the physicalist programme is successful in building a system structured by physicalist principles, such success does not preclude construction of an alternative, incompatible metaphysical system (i.e. a system serving the same unifying purposes as the physicalist system) that is equally successful (Putnam 1979: 603; 1981: 112; Goodman 1972: chapter 1). Such a system could be of either sort to which I have alluded: viz., either a system grounded in a radically different type of basis (for example, phenomenalism) or a system grounded in a physical basis that is not reconcilable with the basis of some given physicalist system. On this non-realist view, both physics and physicalism have only a world-dependent grasp of what is real, possible, and true. There may well be other *actual* worlds in which some other metaphysical or explanatory unity holds. If this view can be maintained, then at least the conditional priority that Post endorses must be further conditionalized: only relative to an interest in objective truth and reality within a particular physicalistically structured world do physics and physicalism enjoy privilege, priority, and a claim to uniqueness. The priority of physics and physicalist unity has a somewhat different character in a non-realist philosophical environment. It is conceived of as world-relative priority and it allows for the possibility of alternative worlds with radically different structures and bases.

A number of issues emerge at this point:

1. Does physicalism imply metaphysical realism?
2. Is metaphysical realism a cogent view?
3. Are there cogent alternatives to metaphysical realism?
4. What is the significance for physicalism of the answers to questions (1) to (3)?

It is not within the scope of the current project to take on all these issues in great detail. Rather, my intention in what follows is simply to say enough to forestall any objections to the physicalist programme without incurring too many significant philosophical debts.

To begin with, unless one explicitly incorporates a realist thesis into physicalist doctrine there is no reason to think that physicalist theses themselves *imply* metaphysical realism. Metaphysical realists, for example, have never demonstrated that realism is a necessary truth, and hence that all forms of non-realism are untenable.[11] Assuming some non-realist positions are coherent, it is plausible that the theses of physicalism are compatible with some such non-realist framework: for example, physicalism is a doctrine that can be true in some of the worlds that are constituted, while being false in others. Within such an approach, its significance is world-relative: that is, the key ideas and values of the physicalist programme are promoted in some 'actual' worlds, but not in others.

Metaphysical realism is, of course, a standard accompaniment to most historical and contemporary formulations of physicalism. Indeed, much of traditional metaphysics has generally been meta-physical realist in spirit even when taking on an idealist cast: minds and ideas in the mind, though mind dependent in an obvi-ous and trivial sense, exist whether we think so or not. So it is not surprising that physicalism (or materialism) has generally taken a realist form. But I am suggesting that metaphysical realism is a gratuitous association to the formulation of the doctrine, since it is not evidently implied by the theses and since it is not otherwise required for expression of key physicalist ideas and values. Fur-ther, unless one has in hand knock-down arguments against all non-realist positions, or at least against the compatibility of physicalism with any viable form of non-realism, metaphysical realism cannot be directly incorporated into physicalist doctrine in a non-question-begging way. Physicalists need not take on such a philosophical liability when there is so little to be gained.[12]

Unlike Putnam, Goodman, and other critics of metaphysical realism, however, I am not persuaded that the view is unintelligible. There are no compelling and non-question-begging arguments *against* metaphysical realism of which I am aware. On the other

---

[11] Of course, the issue of whether to adopt a realist or a non-realist framework may not be appropriately conceived as one concerned with truth; such a conceptualization may be inherently question-begging. However, I cannot pause here to consider the issues in greater depth. The main issue under consideration in the text is whether physicalism can be cogently embraced within either of these sorts of framework. How best to evaluate such frameworks is an issue that must be left open.

[12] These comments do not apply to scientific realism, which is a commitment of physicalism. But scientific realism, like physicalism, can be construed as an internal doctrine within a larger non-realist framework.

hand, I am sympathetic to the possibility of developing a cogent non-realist[13] position, perhaps along the lines Goodman has sketched out.[14] And I suggest here that, pending more effective arguments against non-realism, realist opposition to such philosophical programmes is also quite question-begging. The significance of such a philosophical impasse has yet to be adequately understood, although I cannot pause here to discuss it further.

However, the significance of such an impasse for the physicalist programme is evident: back off from the debate. *If* it is true that the conflict concerning metaphysical realism and its alternatives is at an impasse, and *if* it is true that physicalism is compatible with both metaphysical realism and some significant non-realist alternative, then the debate over metaphysical realism has no deep significance for the development of physicalist systems. As a consequence, physicalists are well advised not to take on a commitment to either side of the debate, at least when they are formulating, defending, and pursuing the physicalist programme. Put another way, if metaphysical realism and some non-realist alternative are both viable, then physicalism should be understood as a programme that can be embedded in either philosophical environment. *Reality, truth, explanation, existence, objectivity,* and so on are all comprehensible within both sorts of framework. Likewise, physicalist structure and doctrine can have a place in both as well. As a consequence, physicalists who are also metaphysical realists ought, in my opinion, to keep the doctrines distinct from each other.

So far, I have contended only that, assuming the coherence of realist and non-realist philosophical frameworks, physicalism ought to be neutral between metaphysical realism and some of its alternatives, since it is compatible with both sorts of approach. Of course, if one or other of the approaches is incoherent or otherwise untenable, then physicalism will inevitably become aligned with the survivor. If *all* such philosophical frameworks are untenable, then physicalism may well have to 'stand alone'.[15]

---

[13] I use 'non-realist' as the term of choice for alternatives to metaphysical realism; Goodman (1978) uses the term 'irrealism' for his version. I prefer such terms to the more aggressive 'anti-realism' because it is within the purview of non-realism/irrealism to view things realistically from within a system. So-called 'internal realism' is acceptable on such accounts.

[14] It is not possible here to do justice to Goodman's views. See Goodman (1978; 1984) and Goodman and Elgin (1988).

[15] See Fine (1986) for one way of understanding this latter option.

There is a further set of issues that are pending: namely, whether alternative unification programmes can be successfully developed and whether alternative, irreconcilable, but evidentially equivalent, physical theories do exist. I cannot pause here to address these issues, but I think it evident that their significance will be construed differently in the different philosophical contexts compatible with physicalism. Thus, assuming that the answer to both questions is 'Yes', the physicalist who is also a metaphysical realist will contend that at most one of the alternative physical theories is true and that at most one of a class of incompatible unification programmes is viable. The independent world makes both of these claims true, even if we could never determine which, if any, of the alternatives was the correct one in either sort of case. The non-realist, on the other hand, will acknowledge the value of both physicalist and non-physicalist unification programmes and will allow that any such programme that is successful is correct, its doctrine being true in the world that is constructed. For example, both physicalist and phenomenalist[16] unification programmes could be correct, if they are both capable of being successfully completed. Physicalism would be true of the physicalist world, whereas phenomenalism would be true of the phenomenalist world. Similar points apply in the case of alternative physical theories: each is true in its world and is a potential ground for physicalist system-building in that world.

Returning, now, to the charge of inappropriate monopoly made by opponents of the conjunction of physicalism and metaphysical realism. The charge can be made to stick if, indeed, metaphysical realism is unintelligible or if, as I have been suggesting, metaphysical realism is in gratuitous association with the physicalist programme. Since I do not believe that the case for unintelligibility has been made out, the question of inappropriateness turns on whether there are strong reasons for the physicalist to incorporate metaphysical realism within the physicalist doctrine. And, for the reasons discussed above,[17] I conclude that there are no such reasons and that metaphysical realism is indeed a gratuitous association.

---

[16] To compete with physicalism, phenomenalism must not be construed as a purely epistemological or conceptual unification programme, however.

[17] Viz., that there are no non-question-begging arguments against the intelligibility or acceptability of non-realism and that there is no reason to suppose that physicalist theses imply metaphysical realism.

Physicalists need not *presume* the uniqueness and the independence of 'the world', and thus the uniqueness of their unification programme *vis-à-vis* the purposes it is designed to serve. The successful completion of a physicalist system would be an enormous and significant accomplishment whether or not there are other systems that do the same work equally well.

To summarize the discussion of charges of monopoly against physicalism, none of the various objections contending that it does have inappropriate monopolistic pretensions succeed. Physicalism has universal scope in the sense that it concerns all aspects of the natural order. It is reductive in the sense that objects and attributes are realized by specific physically-based objects and attributes. And it is a programme aimed at developing a system that exhibits a deep unity in nature and knowledge, a unity that is expressed by the principles To to T5. However, physicalism is compatible with a wide-ranging pluralism of objects and attributes, modes of description and explanation, and philosophical environment. It is compatible with limitations upon the descriptive adequacy of physical language, the explanatory power of physical theory, and the importance of physical ontology, description, and explanation in many contexts. Physicalism, unadulterated by metaphysical realism and its alternatives, makes no claim to uniqueness and does not require the mind/language/theory independence of 'the world' that it is about; *nor does it require nonuniqueness and dependence*. The realism/non-realism debate lies deep in the philosophical background of the physicalist programme. Thus physicalism, properly construed, need not be saddled with any of the burdens to be carried by proponents of either side, any more than it must carry the burdens of the other forms of monopoly just reviewed. Individual physicalists may well develop their version of physicalism in one philosophical environment or another, but the programme can be profitably pursued in environments of either sort. The choice from among such environments represents an independent philosophical commitment.

'Watered-down' Physicalism

The next class of objections to physicalism concerns its content: the form of physicalism being suggested is so *watered down* as to make the programme rather uninteresting and much too easy. Consider such remarks as the following:

The first disappointment concerns number (and sets and mathematical objects in general). An ontology which announces that it recognizes no entities but physical ones, and promises an uncompromising naturalism, takes on, one would have thought, some obligation to find a naturalistic interpretation for mathematics. At least Platonism seems to be ruled out. But when he needs sets or numbers, Post just helps himself to them—even allowing Platonism, if that's your preference. The physical suddenly becomes the mathematical-physical. (Campbell 1988: 359.)

Properties, being abstract, could not be accepted by a physicalism that says that every thing that exists is a physical thing. The need for abstract objects is a more serious difficulty than many physicalists seem to suppose. (Stroud 1987: 266 n. 2.)

Any psycho-physical science would need laws or general statements connecting the physical and the psychological. They could not be expressed in purely physical or purely psychological terms. Both sorts of terms, and therefore both sorts of facts, would be essential. The truth of any such laws would therefore conflict with an exclusive physicalism. (Stroud 1987: 276.)

True physicalism, or so it seems to some, requires that one reject abstract objects of any sort, that one deny the existence of non-physical facts, objects, and attributes, and that one deny the existence of 'irreducible' physical/non-physical relations. To relax physicalism in any of these ways is, allegedly, to so dilute the thrust of the programme as to make it unrecognizable as physicalism and to strip it of its metaphysical significance. Although this sort of concern impresses certain critics of the programme, and perhaps even some proponents, there are no serious problems here. The objections rest upon either false attributions to the programme or confusion about what the alleged permissiveness involves.

With regard to *abstracta*, I refer the reader back to my discussion in Chapter 3 concerning the physical bases and my discussion in Chapter 5 concerning the scope of the theses. Although I reject the Hellman and Thompson (1975) expedient of incorporating abstract mathematical structures into the physical bases, I do not believe that the physicalist must specifically reject such structures or any other *abstracta* that may exist (for example, universals, propositions, etc.). My stance is that the physicalist programme is concerned with the natural order exclusively, and I take it as common ground that if there are abstract objects then they are not part of the natural order, even if some of their members are. Thus they fall outside the scope of the theses. Given this delimitation of scope, the physicalist programme can tolerate, without losing its significance, the acknowledgement of whatever abstract objects are required for understanding either the natural order or any other domain of knowledge (for instance, mathematics). If it is a part of such a framework to include (or exclude) abstract entities, then physicalism will be developed in the manner of that framework. But, physicalism, given its specific concerns, is compatible with Platonism, nominalism, etc. The programme is not watered down as a result *because* abstract entities are not part of the natural order,[18] and *because* physicalism is not best viewed as a completely general philosophical programme. As a result, *abstracta*, unlike ghosts, do not pose a threat to physicalism.

The idea that any version of physicalism that acknowledges *non-physical* objects, attributes, or facts is too heavily diluted has also occurred to critics. As Stroud expressed it at the beginning of the chapter, this seems to result from the false claim that, since physicalism requires that the only world is the physical world, the only facts are physical facts. Or it may result from the belief that physicalism is simply the view that all objects and attributes are identical to physical objects and attributes (i.e. a completely general identity thesis). Either way, the charge results from a misconstrual of the motives of the programme (i.e. its key ideas and values). Certainly physicalists want to say that the actual world is a physical world, but not in a way that denies the existence of non-physical objects and attributes, and not in a way that identifies such entities with physical entities. The version of physicalism outlined above shows how this can be accomplished in a way that hardly makes the programme 'too easy'. Given the adequacy of that version, the idea that such a view is watered down has little force when one considers what such adequacy means and what must be accomplished if the programme is to be successful (for example, development of realization theories for all objects and attributes that are instantiated in nature). Requiring that all phenomena be physically based does not, *ipso facto*, lead to a highly limited view of what

---

[18] This is so, I maintain, even if there are reasons why one *must* be a realist about mathematical objects in order properly to construe the role of mathematics in physical theorizing.

there is, and physicalists ought not be viewed as denying all that is beautiful, important, and meaningful in the world. Indeed, one of the primary goals of physicalism is to develop an understanding of how aesthetic, moral, and semantic phenomena are possible in a physically-based world. The programme may be too hard, but it is surely not watered down and insignificant.

That there are 'irreducible' physical/non-physical relations within a physicalist system has also provided grounds for some to think that this form of physicalism is of limited force (see, Gillespie 1984: 114 ff.; Horgan 1984: 21, 22–3; Levinson 1984: 99–101 for discussion). It is suggested that full-blooded physicalism requires that the physical bases *alone* must realize, determine, and explain all other aspects of the system, and that realization, determination, and explanation by the physical bases plus 'bridging principles' of some sort are contrary to the true meaning of the programme.

A number of points are worth making in reply. First, it should not be thought that all relevant bridge principles *must* express physical/non-physical relations. This is a holdover from classical reductionistic versions of physicalism and should be shed forthwith. A realization theory, as I conceive of it, provides a 'blueprint' for how configurations of certain objects and attributes, physical or non-physical, suffice to realize some given object or attribute. Such RTs might, for example, only appeal to causal or other relations making no specific reference to *physical* objects or attributes. That is, RTs can abstract away from a physical or other sort of lower-level base. Indeed, within a physicalist system, the base provides all the specific configurations that could realize a given object or attribute in the actual world and the RT gives us theoretical insight into why this is so. But the RT is not, in general, tied to the physical basis. Rather, it provides insight into fundamental facts concerning the constitution of objects and attributes. Such facts provide a metaphysical backdrop against which the physicalist, *or any other*, programme concerned with the realization of objects and attributes is formulated. Such facts do not water down physicalism any more than do logical or mathematical facts.

Second, however, as discussed in Chapters 2 and 4, the physicalist grants that there are connective generalizations that relate specific configurations of lower-level, and ultimately physical, objects and attributes to higher-level objects and attributes. Such connective generalizations provide (at least) metaphysically necessary, sufficient conditions for the realization of higher-level entities. And such connective generalizations represent metaphysical facts into which realization theories give insight. It is, however, quite unclear how the presence of such bridging relations is supposed to reduce the determining force of the physical basis. Rather, such principles clarify and make more explicit the character of that force.

Thus an RT (when applied to a physical domain), or a classical bridge law, or an identity statement for that matter, each describe what is at least a metaphysically necessary relation between physical (or physically-based) entities and non-physical entities. In any metaphysically possible world in which a certain physical configuration occurs, a certain non-physical object or attribute will be instantiated. What makes this so in any given case are the physical objects and attributes; and a different configuration would probably yield different consequences regarding what is realized. More generally, the particular distribution of physical objects and attributes throughout our universe realizes the sort of universe in which we live, and an account of why this is so which depends either upon RTs or upon necessary connections between physical (or physically-based) and non-physical objects and attributes does not weaken this. It clarifies it in terms of fundamental metaphysical truths, at least some of which are common ground for both physicalists and non-physicalists alike (namely, RTs).

Comparable points can be made with respect to the determination and explanation of non-physical phenomena by physical (or physically-based) phenomena. It waters down neither the determination of non-physical facts and truths by the physical bases nor the explanation of non-physical phenomena in terms of physically-based phenomena to allow that there are metaphysical facts and truths (for example, RTs and the connective generalizations they license) that are aspects of the full metaphysical and theoretical framework. It is, in my opinion, quite misleading to suggest that the presence of such facts and principles means that the physical bases 'need help' in their realizing, determining, or explanatory roles. Rather, such facts and principles reveal why the bases have the realizing, determining, and explanatory significance they do.

Finally, it should be noted that, given their *necessary* modal status, realization theories and connective generalizations of the form CG2 are trivially *determined* by the physical basis. This is so for any necessary truth, whether metaphysical, nomological, physical, or logical necessity is involved. The current objection is deeply misguided as a result, since it is clear that the physical truths determine all such 'bridge' principles, in addition to the more specific facts and truths to which such principles apply.

I conclude that, if my discussion of the neutrality of physicalism *vis-à-vis* abstract objects is correct, and if I am right that pluralism about objects and attributes is a strength rather than a weakness of the programme, and if the existence of RTs and associated connective generalizations does not compromise the force of the physical bases in their determining, realizing, and explanatory roles, then physicalism as I have presented it is not inappropriately watered down. And, given the above discussion regarding the alleged monopolistic pretensions of physicalism, I further conclude that physicalism does not have to take the form of either an overly strong and inappropriately exclusive programme or a weak-kneed and uninspired one: the alternative I have outlined is neither unduly monopolistic nor too weak to be worth considering. However, there are many more objections to surmount before we are out of the woods.

Physicalism is not Self-Satisfying

The next objection is that physicalist principles fail to satisfy the very demands that they themselves place upon all legitimate claims to knowledge about the natural order. According to M1′, the theses apply to all that is real, true, and possible in nature. According to M2, the theses are empirical truths that concern the natural order and knowledge about it. And according to M3′, the theses provide evaluative criteria for assessing the legitimacy of all claims about nature. It follows that the theses fall within their own scope of application and should be used to assess their own legitimacy. That is, the theses must be self-satisfying if they are to be legitimate claims to knowledge. This means that there are at least three requirements that the theses must meet: (1) for each of the attributes to which they appeal (i.e. *truth, realization, explanation,* and *determination*) there must exist appropriate RTs, physically-based realizations, and physically-based explanations of the right sort; (2) realization, determination, and explanatory relations between physical and non-physical entities must be determined by the physical bases; and (3) the truth of the theses must be determined by the physical truths.

The strategic situation is therefore as follows. The metatheses M1′ to M3′ imply that the theses must be self-satisfying, whereas the objector contends either that there are certain attributes appealed to by the theses that cannot be fitted into a physicalist system, that the truth of the theses is not determined by the physical truths, or that physical/non-physical relations are not determined by the bases. There are at least three possible lines of reply to this scenario: to reject one or another of the metatheses, thereby removing the requirement that the theses must satisfy themselves, to reject the claim that the theses fail to satisfy themselves, or to reject the theses of physicalism on the grounds that they fail the very test they set for any claim to objective existence and truth. Opponents of physicalism will incline to the conclusion that the theses cannot meet their own demands and are self-refuting. In what follows I shall argue for the second line of response (viz., to reject the objector's claim that the theses are not self-satisfying) and against the other two.

I shall begin by considering the first idea of rejecting one of the metatheses and hence the claim that they should be used in their own assessment. Either the theses do not fall within their own scope of application or they do. If they do not, then of course there is no question of their satisfying themselves. However, M1′ and M2 imply that they do fall within their own scope, and M3′ therefore requires that they be self-satisfying. For this to be otherwise, a restriction of the scope of the theses is required, *or* some reason for thinking that the theses are not *a posteriori* truths about the natural order must be found, *or* the methodological role of the theses in assessing claims to knowledge must be rejected. Since *ad hoc* restrictions of scope are, of course, unacceptable, and since there are good reasons for endorsing M1′ (as discussed above in Chapter 5), it should not be qualified.

As a consequence, to pursue this line further, some grounds for rejecting either M2 or M3′ are needed. Although at one time I considered rejection of the empirical status of the theses to be the best strategy (Poland 1983: chapter 5), especially given the problems with the evidence for physicalism that I mentioned above and will discuss further below, I now do not believe that such a drastic revision of our conception of the theses is necessary. As I shall argue below, the problems with the evidence present no special difficulties for physicalism. Thus I see no good reason for revising M2. On the other hand, rejection of M3′ is also too drastic a measure in the current circumstances. Because of the significance of the methodological functions of the theses, both historically and with respect to the point of the programme, elimination of those functions would be to subvert the programme. Therefore rejection of one of the metatheses, and thus rejection of the requirement that the theses satisfy themselves, is not a viable path for avoiding the current objection.

This means that the physicalist must meet the challenge of the objector head on, by denying the claim that the theses are not self-satisfying. To make this case, it must be argued that the attributes mentioned in the theses are 'physicalistically acceptable',[19] that the RTs and physical/non-physical relations called for by the theses are determined by the physical bases, and that the truth of the theses themselves is likewise determined by the physical bases. In what follows, I shall address these obligations by showing that arguments to the contrary are not compelling and by indicating the directions that a thoroughgoing defence of physicalism must take.

The first argument for the claim that physicalist theses are not self-satisfying alleges that the attributes alluded to in the theses (viz., *realization, truth, explanation, determination*) do not satisfy the requirement that they be associated with an RT which reveals how they are determinately realized in the physical world. As a result, the physicalist, while purporting to make claims regarding conditions that every objectively real attribute must satisfy, is alluding to attributes that do not satisfy those conditions and hence are not objectively real.

Focusing on *truth*, the physicalist requires, while the objector denies, that *truth* can be fitted into the physicalist framework. If *truth* is a genuine attribute of (for example) sentences, then this means that a realization theory exists which would show how configurations of physical attributes and objects suffice to realize the tokening of a true sentence in nature. Such an RT would also

---

[19] That is, they must be associated with RTs which reveal how they are determinately realizable in the physical world and how they can play their various explanatory roles.

play the required role in explanations involving *truth*. There are many physicalists (for example, Devitt 1984; Field 1972) who embrace a line such as this, and there are many critics (for example, McDowall 1978; Putnam 1983: chapter 12) who disdain their efforts. In my opinion, there are to date no clear winners on this point. On the other hand, an alternative approach to *truth* is to abandon the idea that *truth* is a genuine attribute of anything, and hence to shed the need for an RT. On such a 'deflationary' approach (cf. Field 1986), *truth* has no trouble fitting into a physicalist system, and the critic's objection loses its force on this count anyway.

At this point, I see no reason why the physicalist needs to take sides on the question of whether *truth* is a genuine attribute or not. If it is not, then all is well. If it is, then the physicalist does take on the burden of establishing that an RT for *truth* exists and that *truth* is both realizable and explainable in accord with physicalist demands. Whether such a burden can be carried is the question of whether *truth* is an insurmountable obstacle to the successful working out of the programme, a topic to be discussed below when I examine the prospects for success of the programme. According to this line of response, the reply to the self-applicability objection is dependent upon the prospects for success of the programme. That is, assuming that a deflationary account of *truth* is not viable, then it is an empirical presupposition of the physicalist programme that a physicalistically acceptable account of *truth* can be developed. Although it is unsatisfactory to leave the issue open like this, it is the best the physicalist can do pending the successful working out of one or other of the types of approach to *truth* compatible with physicalism. At the least, the objection is inconclusive at this time.

And so it is with *explanation* as well. Putnam (1983: chapter 12; 1978) has argued that the 'interest-relativity' and 'context-sensitivity' of *explanation* mean that it cannot be fitted into a physicalist framework because the concepts of *causality*, *rationality*, and *interests* cannot be so incorporated. The arguments he offers in support of this claim are, in my opinion, question-begging and inconclusive; hence the physicalist can and should stand pat with regard to the commitment to fit *explanation* into the system in accord with the theses. The charge of inability to self-satisfy can be held at bay pending either more conclusive negative arguments or a successful physicalist account. But physicalists need to acknowledge that, in so far as they are concerned with *explanation*, they take on an empirical commitment to develop a physicalistically acceptable account of explanation which appropriately deals with the role of interests, context, and so on. Again, the acceptability of the programme depends to a certain extent upon its prospects for success.

With respect to *realization* and *determination*, the situation is similar. That is, the physicalist has a commitment to develop cogent and viable accounts of these relations. However, at the present time there is no need to give up on the programme because the theses are not self-satisfying *vis-à-vis* such relations. With respect to *realization*, there is, as I have acknowledged above, a pressing need to clarify this relation in a way that allows us to see how non-physical attributes are realized by physically-based attributes (i.e. we need an RT for the *realization* relation itself). Like the relations of *identity* and *mereological fusion, realization* is a part of the general metaphysical framework within which physicalist system-building takes place. And all relations within that framework must be philosophically defensible and readily incorporable within the physicalist programme. My earlier discussion of *realization* (in Chapters 1 and 4) were aimed at meeting these requirements. It is a presupposition of the acceptability of the programme that such an account can be properly completed. But this relation is no worse off than *identity, nomological sufficiency*, and *mereological fusion*, which have been featured in other formulations of physicalism.

In the case of determination relations, there are no additional points to be made beyond the acknowledgement that physicalist philosophers must meet the requirement of showing these relations to be defensible and usable within their programme. With respect to *truth determination*, a proper fit with a physicalistically acceptable account of *truth* must be attained. However, there are no further difficulties that the relation poses for the physicalist. Much recent work (for instance, Kim 1984; 1987) regarding *determination* (or, *supervenience*) has shown it to be both a rich and a defensible relation, usable within many different sorts of philosophical programme. I would add that, in so far as the special demands made upon *determination* by physicalism are concerned, there are no difficulties that I can see for considering it as physicalistically acceptable beyond the aforementioned concern about *truth*. The determination of one state of affairs by another is, in my opinion, both philosophically clear and physicalistically acceptable.

I conclude that, although the physicalist does have a number of accrued commitments for defending the use of such notions as *realization, determination, truth,* and *explanation,* there is no reason to suppose either that these commitments cannot be kept or that, as a consequence of employing such notions, the physicalist programme is not self-satisfying. This is not to say that the commitments are easily kept. It is to say that, like any significant philosophical programme, physicalism must defend itself against difficulties, including various forms of self-referential inconsistency, which result from too uncritical deployment of semantic, epistemological, and metaphysical concepts.

A second line of argument, alleging that physicalism is not self-satisfying, focuses on the 'vertical' relations that are supposed to hold between the lower-level and higher-level elements of the system. In what follows, I shall consider two versions of this objection. The first concerns the status of the realization theories that are supposed to coordinate the explanatory relations between the physical and the non-physical domains, while the second is concerned with the relation of *determination*. In both cases, the objection is that, relative to a given set of physical bases, the facts concerning RTs and determination relations are not themselves determined and hence do not satisfy physicalists' demands on objective matters of fact.

The first such indeterminacy objection is to the effect that RTs, and hence the theses (i.e. T4 and T5) and vertical explanations that invoke them, are defective because they are not objectively true *by physicalist standards*. The argument for this charge is as follows. The theses presuppose that, for each attribute, there exists an RT which plays a central role in explaining the specific instantiations of the attribute. However, there are no objective facts of the matter, *from a physicalist point of view*, concerning RTs (i.e. the physical truths do not determine the truths regarding RTs). Thus the alleged vertical explanatory relations between lower-level and higher-level attributes are not objectively real, by the physicalist's own lights.

In approaching this objection, I assume that there exist both a well-established physics and well-established higher-level theories in other disciplines. These assumptions help to focus attention on the issue of the existence of RTs, which are supposed to provide explanatory accounts of the realization of higher-level attributes. Specifically, the critical question is this: Is it the case that, for each attribute, there exists a unique RT which provides a true account of what it is in virtue of which the attribute is realized? The idea behind the objection is that, if there are alternative and equally acceptable, but incompatible, RTs for a given attribute and if there are no objective, physical grounds for choosing among them, then the answer to the critical question is, 'No!'. Thus there are no objective facts concerning RTs or the vertical explanations they underwrite.

According to the objection, it is entirely possible to introduce RTs into a physicalist system of knowledge, but which RTs are introduced is not an objective matter. Hence what physically-based attributes constitute the realization of a given non-physical attribute is not an objective matter. Thus there are no facts of the matter regarding correct vertical explanations. The grounds for such claims are that there are no empirical, methodological, or theoretical considerations which would decide between otherwise equally acceptable alternative RTs and there are no physical considerations which determine which of the competing RTs is true and which is false. Hence, by the physicalist's own standard of objectivity, the choice between competing RTs does not involve an objective matter of fact.

Putnam suggests something along these lines in the following passages:

suppose that whenever I have a blue sense datum a particular event E takes place in the visual cortex. Was the event of my having a blue sense datum just now *identical* with the event E in my visual cortex just now? Or was it rather identical with the *larger* event of E plus signals to the speech center? If you say the *latter*, then very likely you will deny that patients with a split corpus collosum have blue sense data when blue is presented to the right lobe only; if you say the former, then very likely you will say they *do* . . . In this way, even the decision about 'token identity' in a particular case is inextricably bound up with what one wants to say about general issues, as fact is always bound up with theory.

The *special* trouble about the mind/body case is that it will never make an empirical difference whether we say the right lobe is conscious in the split-brain case or not. That the right lobe is only, so to speak, simulating consciousness and that it is 'really' conscious are observationally indistinguishable theories as far as observers with unsplit brains are concerned. My own view is that there is an element of legislation or posit that enters here; the idea of a firm fact of the matter, not at all made by us, in the area of mind/brain relations is illusory. (Putnam 1979: 609.)

In considering the significance of these passages it is crucial to set aside the talk about *identity*: the points made apply equally to claims about *realization*. And it is troubling that Putnam illustrates his point with what he presents as a problem that is 'special' to the mind/brain case. It is troubling because he goes on to claim a general moral:

If I am right, then this is another theme to which Goodman constantly returns: that even where reduction is possible it is typically non-unique. Ontological identification is just another form of reduction and shares the non-uniqueness and the dependence upon legislation and posit which are characteristic of *all* reduction. (Putnam 1979: 609, emphasis added.)

Realization is also another form of reduction. And the Goodmanian point is the quite general one that all reduction is non-unique and, to some extent, a matter of legislation and posit, not a matter of physically-based objective fact. Thus, in my terms, the idea is that there are equally acceptable ways of conceiving of how non-physical attributes are realized by physical attributes, and there are no further grounds, physical or otherwise, for preferring one way of conceiving of how a given attribute is realized to a different way. Thus there is indeterminacy of the realization relation and of realization theories. In a nutshell, the objector poses the following question: What are the empirical and factual grounds that determine which among competing RTs is true? The objector's answer is that there are none and that the introduction of RTs into a physicalist system of knowledge is not an objective matter. Thus the key notion upon which the theses concerning explanation rest (i.e. the idea of an RT) collapses, as do the theses themselves.

In reply to this line of objection, my strategy is to begin by claiming that RTs are in no worse shape than any other kind of explanatory theory with regard to the considerations suggested. There is no reason to believe, for example, that RTs are not subject to empirical test in the same way and to the same extent as other theories. And the same sorts of evidential considerations (for example, compatibility with existing theory, simplicity) which apply to other theories also apply to RTs in the same ways and to the same extent. Thus RTs are *a posteriori*. And then, as Putnam and Goodman point out, it is indeed possible that there are evidentially indistinguishable RTs for any given attribute. So for each attribute there may be an equivalence class of theories that could play the role of an RT for the attribute and are equally correct by all evidential criteria. But this, by itself, is not sufficient to establish the claim that there is no fact of the matter regarding which theory is the correct RT for some attribute.

I maintain that how we view the situation depends upon the broader philosophical environment we inhabit. If we are metaphysical realists, then there is a fact of the matter regarding which of competing, incompatible RTs is correct. Whether the correct theory can be identified depends upon further epistemological considerations. Such matters do not undermine physicalism: it is the existence of a uniquely correct RT that is at issue, not the question of our actually having knowledge of it. On the other hand, if we are a species of non-realist, then the existence of a plurality of alternative, incompatible RTs will not be troubling to us, nor will it compromise the physicalist programme. Indeed, it is here that 'legislation and posit' appropriately enter the fray. Relative to the selection of one or another of the competing RTs, a physicalist system can be developed, and within each system there will be uniquely correct accounts of the realization of attributes.[20] A plurality of equally viable, but incompatible, physicalist systems is not a problem for non-realists: each system is 'true' or, alternatively perhaps, each provides a representation of a different world.

An alternative line of reply essentially denies the possibility of there being alternative RTs for some given attribute which have equal claims to the truth. If we reflect upon Putnam's example, it is clear that what he is pointing to is the possibility of developing theories in terms of different psychological attributes that might be given the same name. In my terms, the realization of the event of having a blue sense datum, B, could be viewed in two different ways: viz., activity in the visual cortex (V) or activity in the visual

---

[20] Though they are *a posteriori* in character, RTs are also necessary. Since any necessary truth is trivially determined by the physical truths, such truths satisfy physicalist demands regarding determination by the physical basis. Within a given world, then, the truth values of RTs are trivially determined by the physical truths.

cortex plus signals to the speech centre (VS). Rather than showing the indeterminacy of 'reduction' relative to the physical basis, this shows that different psychological theories are equally possible and, in fact, not incompatible: both could be true, albeit different. There is not one psychological property or event, B, with two possible realization theories between which there is no basis for choosing. Rather, there are two different properties going by the same name (viz., 'having a blue sense datum'). The correct description of the situation is that there are two mental events/attributes each with its own RT and each a candidate for a role in some psychological theory. In so far as the RTs, V and VS, describe the nature of the respective attributes, they are both necessary and are both determined by the physical base. What all this means is that there are different ways of carving up the world for the purpose of theory construction. There are, within the physicalist view of knowledge, different systems of knowledge about the natural order that can be erected upon a physical basis. This should not be cause for alarm but rather it should constitute a further insight into physicalism in particular and systems of knowledge in general.

With respect to the objection that RTs and related explanations are indeterminate relative to the physical bases, I conclude that the case has not been made. I incline towards the second, less radical, line of response (i.e. that in cases of the sort described by Putnam there are two properties going by the same name). However, in the event of more penetrating examples by critics, I want to keep in reserve the first line of reply: i.e. a genuine case of evidentially equivalent, but incompatible, RTs can be dealt with regardless of the philosophical environment one inhabits.

The second line of argument aimed at establishing the physicalist unacceptability of vertical relations within the physicalist system concerns the relation of *determination* that is invoked in theses T2 and T3. It might be argued that the relations of truth and fact *determination* are indeterminate relative to the physical bases since the physical truths and facts do not fix the truths and facts regarding the alleged determination of all truths and facts by those physical bases. Hence, by the physicalist's own standard of objectivity, there are no facts of the matter regarding what the objective facts and truths are. In light of those criteria, the theses T2 and T3 themselves lack objective truth values, and the relations to which they allude are not objectively real. Again, it would seem, the physicalist is hoist with his own petard.

In replying to this objection, it is important to keep in mind that it assumes the physical bases are fixed and determinate and thus that the problem here is not due to any alleged indeterminacy of the determiners. Rather, within a system having a fixed basis, the problem is that *determination* relations are indeterminate relative to those determiners. Given this stipulation, I note that there has never been an argument offered by critics that attacks the determinacy of the *determination* relations relative to a fixed physical basis. This is, perhaps, surprising since *truth determination* is a species of semantic relation and would be a natural target for those who are suspicious of the physicalist acceptability of such relations.

On this matter, Post makes the following claim:

the world determines truth (even if it does not determine reference). A fortiori, the world determines not only physical truth but also whether it is true that physical truth determines all truth, hence whether TT [T3] is true. (Post 1987: 185.)

Post would add that, given the equivalence of T2 and T3, the same point holds regarding the truth of T2 and the associated determination relations: viz., that the world makes it the case that the physical facts determine all the facts, and hence that T2 is true.

However, although I agree with Post that the physicalist has no immediate cause for concern in this area, I doubt whether a proponent of the current objection would be satisfied with the bald assertion that the world resolves all indeterminacy of the determination relations. The importance of the objection is that it also points to some of the commitments of the physicalist programme. For there to be physicalist determinacy of *truth*, *determination* relations, and of the theses that invoke them, a specific understanding of the nature of *truth*, *determination*, and *realization* is required. Thus there must be acceptable RTs for these attributes that show what makes it the case that some claim is true and that show how it is that the physical facts and truths fix the non-physical facts and truths. Therefore a successful reply to the current objection, like a successful reply to the earlier objection concerning the attributes invoked by the theses, requires that the physicalist should be able to deliver with respect to the empirical and theoretical commitments of the programme.[21] A full development of the physicalist programme involves an acceptable theory of *truth, realization,* and *determination* sufficient to allow for the determinacy of determination relations relative to a physical base.

The third line of objection to physicalism, which also alleges a failure of self-satisfaction, claims that the truth of the theses of physicalism is not determined by the physical basis for truth. Thus they are not objective truths about the natural order. The response here also requires that the phenomena that the theses concern (i.e. *realization, truth, determination, explanation*) be adequately accounted for within the framework of the system. Once those phenomena are made physicalistically acceptable, and if the theses are true, then it will be a trivial matter that they are determined by the physical bases. This is because the theses, if true, are metaphysically necessary. Hence they will be true in every metaphysically possible state of nature and they will satisfy the demands of the determination relation (i.e. once the physical truths are fixed, so are the truth values of the theses). Assuming that the commitment to properly embed *truth,* etc. within a physicalist system can be met, there will be no problem of the indeterminacy of the theses.

To summarize the issues raised by the objection that physicalism is unacceptable because it does not satisfy the very demands that it makes upon any objective truth or fact about nature: if viable theories of the key attributes invoked by physicalist theses can be developed in a way that is both philosophically tenable and physicalistically acceptable, then all the objections alleging failure of self-satisfaction can be met. As I have already acknowledged, this is no small order; it requires that physicalists recognize that the programme takes on a number of empirical and theoretical commitments that must be discharged if the programme is to succeed.

Physicalist Objectivity and Indeterminacy of the Bases

The next objection to physicalism, briefly discussed in Chapter 1, attacks more deeply the physicalist's conception of objectivity than did the previous objection. Recall that one of the physicalist's

---

[21] Note that in his book, Post (1987) develops accounts of *truth, determination,* and vertical relations that take him substantially along the path of meeting this requirement.

goals is to formulate standards of objective fact and truth, and that theses T2 and T3 are offered as formulations of such standards. The key to objectivity, in this approach, is bearing the right relations to the physical bases (viz., that of being determined by those bases). The present objection is that this approach is seriously defective for one of two reasons: either the whole idea of objectivity that informs the approach is problematic, or the bases relative to which objectivity is measured are themselves not sufficiently objective in the relevant sense. Either way, the physicalist's effort to achieve the stated goal fails. It is either profoundly misguided to begin with or, although a laudable goal, it is not attainable within the physicalist's framework.

The attack upon the objectivity of the physical bases alleges that there is no objective fact of the matter regarding what is in the physical bases for either ideology, ontology, or doctrine as developed in Chapter 3. The objection is not one based solely on the *epistemic* indeterminacy also discussed in that chapter. That is, it is not predicated exclusively upon the evidential equivalence of incompatible physical theories. Rather, the current objection strikes directly at the physicalist's standard of objectivity and leads to the conclusion that the physicalist is in no position to claim objective reality or objective truth for the elements in the various physical bases which ground the physicalist system and which define the limits of objective matters of fact. Thus, the objector concludes that physicalism is bankrupt as a programme since it is built on defective foundations.

One version of this objection is suggested by Putnam in his review of Goodman's *Ways of Worldmaking*. In discussing an idea of Hartry Field's regarding the possibility of 'reducing' psychology to physics in more than one way, he writes:

The trouble with this move is that the version chosen as basic—the physicalist version—has *incompatible reductions to itself*... In Newtonian physics fields can be reduced to particles acting at a distance, and particles exerting forces at a distance can be replaced by particles interacting lo cally with fields. Even general relativity, long thought to be inseparable from curved space-time, has an equivalent version...which dispenses with warped space. Nor will it help to hope for an ideal limit in which uniqueness will finally appear... (Putnam 1979: 612–13.)

As with discussions of evidentially equivalent theories, the argument here needs to be supplemented in certain ways to yield the desired problematic results. Specifically, it must be assumed that the inter-definable theories are indeed different theories, that they are logically incompatible, and, that they are 'irreconcilable' with each other, in the sense discussed in Chapter 3.

Given such assumptions, Putnam's argument here is based on the claim that it is inevitable that any fundamental physical theory will have a counterpart that is mathematically equivalent, theoretically distinct, logically incompatible, and irreconcilable with it in any way. Assuming the truth of this claim, it follows that for any set of bases that are generated from the original theory, there is a set of bases generable from the counterpart and that these two sets of bases are incompatible with each other. However, given the mathematical equivalence of the two original theories, it also follows that the truths and phenomena in the one set are determined by the truths and phenomena in the other set. Hence, whichever theoretical framework is accepted, an incompatible framework will be judged as equally objective and acceptable relative to it. As a consequence, there would seem to be no fact of the matter, from the physicalist's point of view, regarding which set of bases is the correct one. Thus the physicalist's standard of objectivity depends upon matters that are not themselves objective in the relevant sense.

One reply to this argument, suggested by the remarks of Post cited earlier, is that 'the world' determines which of the competing theories is in fact the true theory and that the correct physicalist bases are those that are generable in terms of that theory. The other theory, though mathematically equivalent, is theoretically and logically incompatible with the true theory, and hence it is both false and unacceptable for providing the bases for objective truth. As is evident, this reply is fuelled by some version of metaphysical realism as well as a robust conception of theoretical equivalence. Although the latter may well be defensible, the former assumption is bogged down in metaphysical debates of the sort I have discussed elsewhere and in which the physicalist need not be an interested party.

A second line of reply is to accept the premises of the objection and to allow that physicalism can be embedded in either a realist or a non-realist framework in such a way as to avoid the conclusion that the physical bases are not sufficiently objective to provide a standard of objective truth and reality. In a realist framework, the first line of reply just rehearsed can be endorsed, while in a non-realist framework the physicalist can allow that, within a world constituted by the theories and concepts of one of the 'competing' theories, the bases generable from those theories provide the ground of objectivity, while, within a world constituted by the other theories and concepts, an alternative set of bases provides the relevant ground. In no world are both theories and associated sets of bases admissible. The one precludes the other because, although the theories are mathematically equivalent, they are theoretically distinct, logically incompatible, and irreconcilable. Physicalism provides a standard of objectivity in each world, albeit in terms of a different set of bases in the two worlds.

As I have emphasized above, I endorse this strategic approach (i.e. that of allowing that physicalism can survive in either philosophical framework) in order to disengage physicalism from disputes that are tangential to its main purposes. Of course, if realism is the only viable framework, then physicalism will ultimately align itself with realism and respond to the current objection as outlined above; *mutatis mutandis* for a viable species of non-realism. If both sorts of framework are viable, then the physicalist can pick and choose with impunity. On either view, a substantial notion of physically-based objectivity can be developed, a notion that is world relative. In the case of the realist, there is only one actual world, whereas in the case of the non-realist there are many. But in all cases, what is an objective fact of the matter is internally pinned down by the physical basis operative in a given actual world.

I see no good reason why a physicalist must be committed to there being only one *actual* world. *Qua* physicalist, what is important is that, relative to the actual world one lives in, the physical basis suffices as a ground for all that is real and true and possible in the ways called for by the theses. If there are other actual worlds that are structured relative to a different physical basis, this no more undermines physicalist standards of objectivity than does the existence of other *possible* worlds which have a radically different (or no) physical basis. What is an objective matter of fact is a world relative matter, whether the range of worlds one considers covers alternative possible worlds or alternative actual worlds. Hence, for the physicalist, the debate regarding metaphysical realism is neutral *vis-à-vis* the issue of the significance for physicalism of the problem of inter-definable and co-determining, but irreconcilable, physical theories.

Finally, as I observed in Chapter 3, there are two other lines of argument, based on important Quinean doctrines, which lead to the conclusion that there is no matter of fact regarding the contents of the physicalist bases. The first is predicated upon the indeterminacy of translation and alleges that there is no fact of the matter regarding the referential apparatus of any given physical theory from which the bases are to be generated. The second argument is predicated upon the inscrutability of reference and alleges that, even if the referential apparatus of the language of physical theory is pinned down, there is no fact of the matter regarding what the terms of the language refer to within a given domain of objects and attributes.[22] Both these arguments strike at one or another of the bases alleged to be generable in accordance with the procedures outlined in Chapter 3.

However, as my discussion of the objections in this section and the previous one suggest, neither of these arguments poses any clearly insurmountable problems for the physicalist. What they probably show is that to advance a significant physicalist doctrine one necessarily accrues certain debts that must be paid off. In this case, the debt involves the development of a physicalistically acceptable account of the syntactic and semantic features of language. It is ironic that Quine, an arch physicalist, should also be the author of doctrines that potentially impact negatively upon his physicalism. The arguments based upon those doctrines perhaps signal some tension in Quine's views: can he legitimately use physicalism as a premiss in arguing for there being no fact of the matter regarding translation and reference if physicalism presupposes that there is a fact of the matter regarding translation and reference?

The next objection alleges that the physicalist's conception of objectivity is a defective one: even if the physical bases can be isolated in a determinate way, there is no realm of pure fact and truth that can be separated out from the contributions of the knowing subject. Hence there are no facts or truths that can stand

---

[22] There may well be a third 'Quinean' argument that proceeds from the doctrine of ontological relativity and also concludes that, given a theory and a parsing of referential apparatus, there is no fact of the matter regarding what the ontology of the theory is.

independent of subjective factors and be determined by physical facts and truths exclusively. If anything, allowing that physicalist objectivity is in some sense 'system-relative' implicitly acknowledges this point and entirely undermines the physicalist approach.

As noted earlier, I agree that all knowledge involves contributions from knowing subjects. Conceptual or representational resources, for example, are required for human knowledge, and the particular choice of such resources involves a certain amount of arbitrariness[23] of selection. Since how the world is conceived or represented depends upon such selections, there is unavoidable relativity of knowledge based upon them. Further, there are many sorts of cognitive interest and methodological canons that may guide the development of systems of knowledge. And certainly the relevance of various objects, attributes, processes, etc. to an investigation depends upon which cognitive interests and methods of inquiry are in play. I take all these points to be common ground shared by both physicalists and their opponents.

However, this common ground fails to underwrite an objection to the physicalist theses concerned with objectivity. Those theses are aimed at capturing a particular way of viewing nature and knowledge. *Within* that view, nature and knowledge exhibit the sharp distinction between objective and subjective factors described above, despite the fact that such a view rests upon a selection of conceptual or representational resources and despite the fact that the programme for developing such a view is guided by certain cognitive interests and methodological canons. It is part of the working out of the physicalist programme to distinguish the subjective from the objective in the ways expressed by the theses. Success in this task would vindicate the physicalist's conception of objectivity 'from the inside'.

The objector will be likely to make the rejoinder at this point that I have not met the real thrust of the objection. To relativize the distinction between the objective and the subjective to a particular view or system does not come to terms with the deeper problem of whether the system itself is grounded in objective, mind independent facts. Simply to make distinctions within the physicalist system does nothing to establish that those distinctions

---

[23] Though it does not involve total arbitrariness: some resources are not useful for some purposes. The point is that it is unlikely that there is always just 'one right tool for the job'.

are themselves objective (i.e. to show that they are grounded in a mind or system independent world of fact). Such an internalist reply as I have given, the objector urges, does not establish either that the physicalist's account of objectivity is itself objective or that there are not other accounts of objectivity, diverging radically from the physicalist's, that are equally tenable. Hence a purely internal account of objectivity fails to establish what the physicalist really wants, namely, an 'absolutist' account of objectivity.

I think that this line of argument does not succeed in showing that the physicalist's conception of objectivity is incoherent or defective. A perfectly coherent conception can be had without compromising the insights that all knowledge involves a contribution from the knowing subject and that distinctions between objectivity and subjectivity can be made *only* from within a system of concepts and objects. *Any conception* of this distinction, or of any other for that matter, is made in this way. The most that has been shown is that an internal defence of the physicalist's conception (i.e. a defence made in terms of the successful development of a physicalist framework that draws the relevant distinctions) does not establish that physicalist objectivity is absolutist in some metaphysical realist sense. In particular, such a defence does not establish that a physicalist conception of objectivity is *unique* or that it concerns a world of mind, language, or theory independent objects and attributes. Such trappings require further argumentation on the physicalist's part if they are to be defensible as aspects of the position. Again, my own view is that such trappings are gratuitous and that a viable and valuable physicalism need have nothing to do with them, *one way or the other*.

The Problem of Downward Incorporation

The next objection, preliminarily discussed in Chapter 3, concerns the problem of 'downward incorporation' of 'patently non-physical entities' into the physical basis. Given the methodological role, R2, of the theses and given the conception of physics and the physical bases advanced earlier, it is conceivable that, in a situation of apparent conflict between the demands of physicalism and some higher-level phenomenon, the decision might be made to revise physics by downwardly incorporating the putatively recalcitrant higher-level phenomenon. As discussed in Chapters 3 and 5, the ground rules on such a move require that it should not be strictly an *ad hoc* one designed to save physicalism. Further, it should be acknowledged by both friends and foes of physicalism alike that such a scenario is not always unreasonable (for example, the electromagnetic theory case). However, the current objection alleges that there is nothing in the general situation as described to preclude the downward incorporation of mental, semantic, or even normative phenomena, and that surely such eventualities would be an extreme embarrassment to the physicalist, if not downright defeat of the programme by trivialization of the demands it must satisfy.

There are several considerations that bear upon an adequate response to this objection. First, a distinction must be drawn between the ways in which 'the physical' has been identified in the past and the more theoretically-based manner in which it is identified according to contemporary versions of physicalism. The former relied essentially upon *a priori* criteria, for example location, extension, and impenetrability, whereas the latter leaves it to physics to identify the physical domain subject to various general and specific constraints upon physical theory construction, as described in Chapter 3. Past identifications took it as a fixed *a priori* fact that the mental was not part of the physical domain because of differences in essential properties possessed by the two sorts of phenomena. A theoretically-based conception of the physical and of the mental does not, in general, take there to be any such fixed *a priori* distinctions. Therefore, although past physicalists often conceived of their view within a framework which featured the mental/physical dichotomy as fixed and irrevocable, an alternative framework exists in which this and other distinctions do not have that character.

Second, although physical theory construction is subject to general methodological principles (for example, principles concerning the testing and evaluation of theories) and to some more specific constraints (for example, that theories be responsive to a certain set of questions, that they exhibit a certain sort of comprehensiveness), it is widely acknowledged that specific, *a priori* substantive constraints upon physical theories are ill-advised (for example, the independence of space and time, four-dimensional space-time, action-at-a-distance, local contact action, separability of systems). This is not to say that one or another of the supposed constraints might not be theoretically valuable; it is to say that the value of such a constraint is not an *a priori* matter. In general, physicists proceed as though there are *no* substantive constraints upon the theoretical resources available to them in their efforts to do the work of physics.[24] Historically significant assumptions and theoretical frameworks have sometimes been confused with irrevocable fixtures in the framework of theoretical physics, but time and again physicists have recognized that it is counterproductive to hang on to the fashions of the past. For physicalists, the moral is that, to be in tune with physics, they should not introduce constraints upon physics that physicists do not acknowledge. Since it is special recognition of physics that motivates the physicalists, it is probably a good idea not to distort the object of that recognition.

Finally, I cannot emphasize strongly enough the significance of the restriction on *ad hoc* revisions of physics for the problem in hand. To say, for example, that mental phenomena might be downwardly incorporated into the physical basis is not to say that, whenever physicalists run into recalcitrant mental phenomena, they need only re-label the phenomena as 'physical' to escape a potentially serious predicament. Downward incorporation means just that: *incorporation* of the phenomenon into the existing physical framework. This requires significant theoretical motivation from the point of view of physical theory construction, as well as preservation, rather than destruction, of the essential theoretical unity of the existing body of theory. Downward incorporation is no mean feat, and should not be viewed as such by critics of physicalism.

With these three points in mind, my reply to the objection is to bite the bullet and allow that indeed it is conceivable that physics might be revised to incorporate mental, and other phenomena previously identified as non-mental, into the physical basis. Thus it is conceivable that, in the face of a recalcitrant higher-level 'non-physical' phenomenon, it could be appropriate to relieve the tension by an act of downward incorporation. But such incorporations are subject to the ground rules I have discussed above. Thus they

---

[24] The most compelling example of this, from the point of view of the current objection, is the hypothesis of acts of 'pure consciousness' (Wigner 1961), introduced for the purpose of understanding the collapse of the wave function (i.e. the so-called 'measurement problem' in quantum mechanics). It is not because it involves an introduction of mental phenomena for the purposes of understanding measurement that the hypothesis is not now taken too seriously.

are not *ad hoc* attempts to save physicalism, although they are surprising from the point of view of prior conceptions of what was in the physical basis. To be surprising, however, is not to be illegitimate. It was surprising to many when Einstein announced that space and time are not independent and when experiments concerning the Bell inequalities established that physical systems are not separable in the way previously believed. However, neither of these developments were illegitimate. If the choice is between allowing that 'the mental' could be incorporated into the physical and imposing *a priori* substantive constraints upon what physicists may permissibly appeal to for theoretical purposes, the physicalist must lean towards the former. That those who have highly evolved sensitivities to previous distinctions and theories are offended by this is not much of an objection to the physicalist programme.

To soften this blow, I might add that I think our ancestors did indeed latch on to important distinctions and problems for the physicalist programme (for example, how can mind, meaning, and values be realized in a physical world?). And I certainly tend to agree with those who think such problems are not going to be solved via downward incorporation. Indeed, if anything is likely to be an objection to the programme, it is that these phenomena, although real, do not fit in. But this acknowledgement is not to say that the distinction between the mental (etc.) and the physical and the problems this distinction poses are *a priori* fixed features of the physicalist programme. The objection that downward incorporation of the mental (etc.) undermines the programme can only originate from the belief that such distinctions are such features, and this belief is, I maintain, antagonistic to the spirit of the programme, even if it is consistent with its contingent history.

The Theses are not a posteriori

In light of the discussion of the first five objections, I want to return now to the issue of whether physicalist theses lack the *a posteriori* epistemological status attributed to them by M2. It might occur to many that physicalism is not genuinely testable by empirical evidence if, for example, downward incorporation of recalcitrant phenomena is a possibility, or if the theses are 'self-applicable', or if the problems outlined in Chapter 5 concerning confirmatory and disconfirmatory evidence are intractable. The deeper key to understanding why the theses appear to some critics as not subject to empirical test is to see that there is significant tension between the descriptive and normative functions of the theses. As I shall argue, such tension is not sufficient to undermine the *a posteriori* status of the theses, but it is something that must be duly noted and taken into account by circumspect physicalists.

One important objection against the empirical status of the theses proceeds by focusing on their apparently 'self-supporting' character. Since, according to M3', they play a normative role in the conduct of inquiry, they can protect themselves from counter-example in either of two ways. First, given some apparently recalcitrant phenomenon, they can force revisions somewhere in the physicalist system (for example, by 'downward incorporation' into physics, by revision in our conception of the connections between higher- and lower-level phenomena), thereby relieving the tension. Or, second, they can screen out the putative counter-example to the theses by underwriting its rejection as 'unreal' and hence not a problem for the programme. Either way, the theses are spared. Combining this especially pressing difficulty with the objection that there are intractable problems, surveyed in Chapter 5, regarding confirmatory and disconfirmatory evidence bearing on the theses leads to the conclusion that their *a posteriori* status is seriously in doubt. They seem to function more like 'regulative ideals' than like empirical hypotheses.

As mentioned earlier, at one time I thought that the best course for the physicalist here is to drop M2 and to view the theses as regulative ideals with empirical presuppositions. However, I now think that M2 can plausibly be maintained despite the heat generated by the objection. First, with regard to the problems with confirmatory evidence, they are no more difficult in the case of physicalist theses than they are with regard to any other hypothesis for which inductive arguments are adduced. In general, the problem of generalizing from a sample to a larger domain requires some theoretical underpinning regarding representativeness of the sample and features of that larger domain. Physicalists, when offering inductive support for their theses, must, like everyone else, *justify* the contention that success in one area generalizes to others. Again like everyone else, they must do this without assuming the very hypothesis for which evidence is being collected. Second, the problem posed by difficulties experienced in distinguishing apparently recalcitrant phenomena which in fact fit into the system (for instance, a mystery or a difficult problem) from a genuine counter-example is also a problem affecting all domains of empirical inquiry, not just physicalism. And in no domain of inquiry is there a general algorithm for making the desired distinctions. Rather, such difficulties must be dealt with on a case-by-case basis. Although both of the above are problems to be faced by physicalists, neither is a compelling objection to the metathesis M2.

Finally, the idea that the theses are self-supporting in the way described above (viz., to force revisions elsewhere in the system or to screen out recalcitrant phenomena) is based on an incorrect understanding of how the theses function in their regulative capacity. Like any principle taken to be true in a system, the theses constrain all other truth claims. And like all other claims to truth, when there is conflict, a specific assessment must be made of the balance of power. Which of the conflicting claims is appropriately retained depends upon its place and status in the system. In my discussion of M3' in Chapter 5, some of the details of such considerations were outlined. It was pointed out that there are no simple rules of priority, only ways of attempting to achieve the best balance. Sometimes downward incorporation is indicated, sometimes rejection of the recalcitrant phenomena, and sometimes physicalism itself must be held to account. The problems in this area for physicalist theses are no more severe or undermining of empirical status than they are for other empirical claims. The normative role of the theses derives from their place of relative importance in the overall system.

In addition to this reply, I want to add that, as with the problem of downward incorporation of the mental, this problem should not be dealt with in any way that introduces inappropriate *a priori* constraints upon inquiry. To remove the *a posteriori* status from the theses would be, in effect, to impose substantial *a priori* constraints upon the evolution of scientific theories among others, and this is quite contrary to both the spirit of science as well as the spirit of physicalism. That theories are currently developed in a physicalist environment means that physicalist principles express a current vision of how things hang together; but acceptance of that vision only signals the current level of commitment to the truth of the theses, not their *a priori* status.

It is Unreasonable to be a Physicalist

There are those who object to physicalism on the grounds that it is unreasonable to be a physicalist. The grounds for this charge range across a wide range of considerations and have often been tainted by polemical flourishes contributing little to the debate. It has been suggested that physicalism is a matter of 'blind faith' or prejudice, that it is 'utopian', that it is incompatible with more central epistemological doctrines (for example, empiricism), and that it flies in the face of the facts. As I shall argue below, when these charges are suitably clarified and subjected to scrutiny they are not compelling and they do not lead to the conclusion that physicalism is unreasonable.

As a preliminary, I want to observe that there are at least three different ways in which a programme can be reasonable or unreasonable to pursue. Since these ways are more or less independent of each other, it is wise to distinguish them and to keep track of them while advancing or responding to objections. First, as I have emphasized in Chapter 1, a programme can be reasonable to pursue in virtue of the values and purposes which it would promote if it were to be successful: a potentially quite valuable programme is one it is reasonable to pursue, at least up to the point at which the cost of implementation or the spectre of diminishing prospects of success reach certain levels. Thus promise of a rich return provides a strong *prima facie* case for the reasonableness of the programme. Second, a programme can be reasonable to pursue if there is evidence that it has been and will continue to be useful for certain valued purposes. And, third, a programme can be reasonable to pursue if its central doctrinal elements are likely to be true, in the light of relevant evidential considerations. Such considerations include directly confirmatory or disconfirmatory evidence as well as the status of various critical objections. Since useful programmes need not involve true doctrinal elements and since truth of a doctrine does not guarantee utility for many purposes, it is important to keep these different senses of reasonableness in mind.

The first objection to the reasonableness of being a physicalist is suggested by the following passage from the work of Nelson Goodman:

But the evidence for such physicalist reducibility is negligible, and even the claim is nebulous since physics itself is fragmentary and unstable and the kind and consequences of reduction envisaged are vague.[25] (Goodman 1978: 5.)

Goodman here appears to take it as obvious that there is no good reason to suppose that physicalist claims are likely to be true: the evidence is alleged to be 'negligible'. However, the objection lacks force since very little systematic study has been made of the evidential status of physicalism, especially the status of more recent formulations of physicalism that are not tied to classical reductionism. As I have noted previously, it is a shortcoming of both proponents and opponents of the physicalist programme that neither has pursued the task of systematically assessing the current evidential status of the theses. In the absence of such an assessment, it is difficult to evaluate the force of the charge of unreasonableness. Indeed, in the absence of such an evidential assessment, proponents of physicalism will tend to focus on the full or partial successes of the programme (for instance, on examples in chemistry and biology), whereas critics will tend to focus on areas of difficulty (for instance, mental phenomena, values, cultural phenomena). But at this stage of inquiry, Goodman's bald claim comes to little. There is an impasse that cannot be resolved by anything other than actually developing the relevant evidence and working on the areas of difficulty. And I do not believe that, at this time, either proponents or opponents have made compelling cases with respect to the key issues. In addition to there being no survey of the evidence, proponents have not provided definite demonstrations of how to incorporate the mental and other recalcitrant phenomena into the physicalist framework, while opponents have provided no definitive demonstration that such incorporation is impossible. Thus, given such a lack of conclusive evidence, I do not see how one could conclude that it was unreasonable to be a physicalist.

A second line of argument for the unreasonableness of physicalism alleges that physicalism is incompatible with the radical empiricism that is characteristic of scientific practice. It is charged that physicalism not only fails to offer any advantages not enjoyed by such empiricism, but also constitutes a serious potential liability

---

[25] It is the evidential issue that concerns me here. I have attempted to respond to the other two of Goodman's complaints elsewhere in this project.

when it plays an active role in the conduct of inquiry. In Chapter 3, I discussed Cartwright's view that a basic empiricist outlook does not, in effect, have a serious place for physicalist constraints upon theory construction; whereas physicalism requires unity and consistency among our theoretical models, real science deals in a plurality of incompatible models each suited for the specific context dependent purposes for which it was developed. To the extent that it limits the development of such a plurality of models, physicalist unification would interfere with, rather than promote, the basic goals of scientific theorizing and model-building.

It might be suggested in further support of this view that there is no significant scientific purpose that could not be served if physicalism were abandoned and therefore physicalism is unnecessary in science (Van Fraasen 1980; Wagner unpublished). With respect to the empirical adequacy of the full constellation of our scientific theories, there is no essential advantage enjoyed by a physicalistically structured system. And with respect to the degree of unity enjoyed by such a system of theories, a non-physicalist is not precluded from seeking and developing as much unity as is required to guarantee empirical adequacy: any more unity would be gratuitous from a scientific point of view. On the other hand, it is not at all evident, according to the objection, that a commitment to physicalism would not lead scientists down numerous garden paths and into making numerous faulty judgements about the acceptability of various scientific models. For example, legitimate models of high-level phenomena might be rejected on the grounds that they do not fit in with a currently accepted, although ultimately defective, physics.

In reply, I first want to underscore what I think is right about Cartwright's criticism: viz., that physicalism does indeed go beyond a radical empiricism which asserts that theories and models are justifiably introduced, accepted, and believed exclusively on the basis of how well they fit some specific domain of empirical data in some particular context of scientific problem-solving. But, then, why should one believe either that such a form of empiricism is philosophically tenable or that it is an appropriate characterization of the methodological practices of scientists? Incompatibility with this view of justification and acceptance is not a compelling problem for physicalism, since such a view is clearly question-begging.

I would also remind critics of this bent that before such criticisms are to be taken seriously, the question of whether there are intelligible alternatives to physicalism which can be justifiably attributed to actual scientific and other inquiry must be considered in more detail. I seriously question a description of current scientific practice that views science as broken down into fully autonomous, isolated empirical inquiries concerned with specific local problems the solutions to which are unconstrained by, for example, consistency with models and theories in other domains. And I wonder whether, for instance, an autonomous psychology (relative to neuroscience) or an autonomous biology (relative to chemistry) or an autonomous sociology (relative to psychology) are either plausible or likely to be empirically adequate. So the critics who argue that it is unreasonable to be a physicalist have, it seems to me, the burden of presenting an intelligible and appropriate alternative account of how various research programmes are (or are not) related. The idea that the application of high-level theories does not have low-level factual presuppositions is questionable at best. And if the account of those factual presuppositions is not physicalistic, what is it?

With respect to the claim that science conducted in a non-physicalist environment is at no disadvantage relative to physicalism because it will yield a system of theories that is empirically adequate to the extent that this is possible and that is unified to the extent that is appropriate for empirical adequacy, I suggest the following. Physicalism, unlike the empiricist alternative, requires that we seek unity. Therefore it is more likely that such unity, if it is there to be discovered, will be discovered within a physicalist framework. Scientific practice, unguided by physicalist principles, could hit upon various sorts of underlying unity, but it is unclear that it must do so in order to achieve its limited empiricist objectives. As a consequence, if physicalism is true, then it seems that a radical empiricist science will be less likely than a physicalist science to hit upon that truth. Thus, in so far as physicalist unity is desirable and a likely possibility, it is reasonable to be a physicalist, at least until the odds lengthen on physicalism.

With respect to the suggestion that physicalism might lead to mistakes resulting from inappropriate reliance on a defective extant physics, the discussion in Chapter 5 concerning the conditions under which physicalist principles play their methodological roles should dispel much of this concern. Although it is always possible that errors can be made by physicalists and non-physicalists alike, there is no reason to suppose that physicalism makes researchers more prone to error. A proper understanding of how physicalism exerts its regulative influence should make it clear that physicalism does not always give priority to physics when a crisis arises, that physicalism only comes into play when the relevant sciences are reasonably mature, and so on. The critic here needs to provide a more compelling case for the likelihood of errors on the part of physicalists relative to the likelihood of errors on the part of non-physicalist empiricists. In the absence of such a case, it is hard to conclude that physicalism is a liability within the scientist's arsenal.

In conclusion, I want it to be clear that my purpose here has not been to argue that physicalism is reasonable in either of the senses of reasonableness concerning actual utility or truth. It has only been to argue that the reasons offered by critics for the unreasonableness of physicalism are not compelling. As I have already indicated, in addition to the need for a review of the current state of the evidence bearing on physicalism, the primary focus should be on the various areas of difficulty facing a successful working out of the programme. Whether the programme is ultimately one that it is reasonable to pursue depends on getting a clearer handle upon both the state of the existing evidence and the prospects for successfully incorporating various problematic phenomena within the physicalist framework. Having said this, I do of course believe that physicalism is reasonable to pursue from the point of view of the pay-off it holds out for us, and I am optimistic about the prospects for paying off the various debts that physicalism has incurred and for overcoming the various obstacles it must negotiate.

A Physicalist System must Inevitably be Incomplete

The final objection to physicalism that I shall mention now is that, given the constraints imposed by the theses, a physicalist system structured in accordance with those constraints will inevitably be incomplete. Such a system, it is claimed, will inevitably leave some aspect of existence out of account. Indeed, the list of those objects, attributes, and types of 'truth' that allegedly cannot be fitted into a physicalist system is quite long, as we shall see below. It is, however, inappropriate for me to take on here all the various objections and challenges to physicalism, based on specific types of phenomena or truth, that allegedly cannot be fitted into the system. That would be a book in itself. I suggest, however, that none of the *a priori* arguments designed to show that physicalist theses are false are compelling. At most they point to potentially serious obstacles. In the next section I shall identify a number of obstacles to the successful working out of the programme and discuss some considerations bearing upon the likelihood of their being effectively negotiated.

6.3. PROSPECTS FOR SUCCESS OF THE PHYSICALIST PROGRAMME

In the previous sections we considered the first two dimensions of assessment of physicalism. In this section, we will turn to the third: *success* in the working out of the programme. Recall that by 'the working out of the programme' I mean the real work of the programme, which is construction of a system of knowledge that represents the world as being structured in accord with physicalist principles and that actually delivers the truths and explanations called for by those principles. Success in this enterprise is clearly a very stringent standard of assessment, and is clearly dependent upon a wide range of factors, many of which seem overwhelmingly likely to be unfavourable. None the less, it is important for the physicalist to keep in mind that the goal of the programme is not merely the enunciation of physicalist doctrine and the development of supportive evidence: little pay-off results from such activities. Rather, the goal is to build a system which yields the pay-off described in Chapter 1.

There are several ways one might go about attempting to assess the prospects for success of the programme. One might begin by assessing the adequacy and acceptability of the theses via an examination of a wide range of objections and relevant evidence. A partial effort along such lines was made earlier in this chapter.

Further, one might study various *general* considerations bearing on success. For example, the availability of certain resources and patterns of communication are quite relevant to the success of such a comprehensive research programme. Finally, one might look at specific domains likely to offer up resistance to efforts at incorporation within a physicalist framework. I shall label these 'obstacles' to the successful working out of the programme. In what follows I shall make a preliminary assessment of the prospects of success of the programme by discussing both general concerns and specific areas of difficulty. It is not, however, within the scope of the present project to review systematically all potential obstacles in order to determine how seriously they threaten the success of the programme. My aim is to identify the problem areas and to make some general, strategic comments concerning how the programme for dealing with them might be pursued.

The first general consideration bearing upon the prospects for success is that it is quite likely that some (possibly a great deal) of what is true of the world may be forever beyond our ken. Such factors as size, complexity, and accessibility, for example, may preclude our ever gaining a clear and precise grasp of some phenomena (for instance, phenomena occurring inside the human brain or inside a galaxy sixty trillion light years away or at one billionth of a second after the Big Bang). The complexity of a phenomenon may mean that no systematic effort at description or explanation will ever be successful. Some phenomena may require resources beyond those available to us (for example, sufficient energy for observation of the 'top quark'). And yet other phenomena may resist any efforts at linguistic description (certain subjective states, for instance). Thus private phenomena, although real, may not be readily integrated into a physicalist system if only because the representational resources of the system are inadequate for expressing such phenomena. As a consequence, if there are instances of any of these sorts of difficulty, the programme may not be fully successful, even if the theses are true. But clearly any programme must work within the limitations imposed by its subject matter and the available resources and capacities. Failure due to such limitations is contingent failure at most, and not indicative of deep flaws in the programme.

A second sort of reason for potential failure is based upon the scope of the programme. The programme is an enormous under-taking that subsumes all particular forms of inquiry and human activity, and calls for the development of a scaffolding within which all such pursuits are integrated. That is, the programme calls for integrative work bearing upon every domain; it is clearly a multi-disciplinary effort calling for specialization of all sorts. The required integrative work, as seen in our discussion of M3, depends upon satisfaction of various conditions by the domains involved in the integration. Specifically, the successful completion of the programme turns upon the mature development, if not successful completion, of all such other enterprises. This in itself makes the success of the programme doubtful. In addition, inter-domain coordination is never very easy and often something that practitioners in a given area are not always particularly motivated to pursue. Thus the demands of the programme, though quite wide-ranging, are subject to various highly local factors and con-straints. For such reasons, then, it is very unlikely that the pro-gramme will be evenly and effectively pursued in all areas.

Following this point further, it seems a general fact of current research institutions that success of a research programme often means that many people are aggressively pursuing the same prob-lem, often in competition with each other, and usually with con-siderable economic and other support. Within the larger sociological framework in which research is pursued, there are considerable incentives and pay-offs associated with successful research activ-ity. However, it is not at all evident that every aspect of physicalist system-building is favourably situated with respect to personnel, resources, incentives, and so on. Many problems (for example, the physical bases of poetry) may seem odd or arcane to most people, and surely not so relevant to deeply important human concerns as to justify the mobilization of considerable economic and social resources. Without support, many hard problems may go un-attended, not because they don't have some importance within the larger architectonics of the system, but because circumstances are simply unfavourable for their being productively pursued.

Putting aside problems concerning limitations of time, space, and resources and abstracting away from the sociological context of research, it must be fully recognized that some problems are just plain difficult, even when circumstances are favourable. Insights may be missing, standard strategies and modes of understanding may be inappropriate, and researchers may be simply unlucky. Thus physicalism, like any other programme, depends upon factors which are often lacking, but which are essential for solution to certain problems. Thomas Nagel's lamentations (1979; 1986) regarding subjectivity and qualitative states may perhaps be understood as pointing to such a shortfall.

Finally, pursuit of the physicalist programme is impeded by a failure of participants and critics alike to understand exactly what the programme requires of them. Equating physicalism with classical reductionism is, to this day, a widespread and lamentable confusion that only muddies the waters. On the other hand, as we saw in Chapter 2, too deep a retreat from reductionism obviates the necessity of focusing on how attributes are realized in nature. Purely non-reductive physicalism undercuts successful pursuit of the programme by understating its demands. One important theme of this book is that it is important for philosophers to think more intensively about what physicalism does involve. And my contention is that there should be agreement that it involves something more than supervenience and something other than classical reduction.

Turning to more specific areas of concern, critics and friends of physicalism alike have identified numerous entities that are candidates for being serious, if not insurmountable, obstacles to the successful working out of the programme. As I indicated in the last section, there are, to my knowledge, no arguments which establish that any sort of entity necessarily falls outside a physicalist framework. But this by no means precludes the possibility of some type of phenomenon actually constituting a counter-example to the theses, and hence an insurmountable obstacle. Physicalist theses are not provable by *a priori* means. And, as I have already suggested, the success of the programme is far from guaranteed, even if the theses are true.

It is well beyond the scope of the present work to review the many potential obstacles and to discuss the work that has been done to date concerning them. This would, of course, be desirable. Instead, I shall offer a catalogue of the various potential obstacles and make some general remarks concerning the prospects of the programme. The following is an outline of what I take to be the main obstacles to the successful construction of a physicalist system.

OBSTACLES TO THE PROGRAMME

Mind

- Qualitative content.
- Semantic/intentional/informational content.
- Mental causation.
- Subjectivity, points of view, self-reflexive thought.
- Rationality.
- Action, habits, behavioural style.
- Freedom and autonomy.
- Personhood, personality, character.

Meaning

- Truth and reference.
- Linguistic meaning.
- Metaphor.
- Vagueness.
- Symbolism, myth.

Values

- Moral: rightness, goodness, responsibility.
- Aesthetic: beauty, ugliness.
- Epistemic: knowledge, justification, explanation.
- Pathology concepts: illness, disease, disorder.

Society and Culture

- Social systems, processes, and institutions.
- Cultural artifacts.
- Literature, paintings, music.
- Customs.

Each of the entries in the outline subsumes more specific cat egories of potential obstacle (for example, 'social institutions' subsume legal, political, and economic entities of various sorts). Thus the effort required for both working out the programme and appraising its prospects for success is enormous. It is important, however, to be clear on exactly what is and what is not required for achieving these objectives.

Physicalism, although requiring that certain constraints be satisfied by all elements of the system, does not legislate any particular account of mind, meaning, values, or culture. The belief that physicalism, for example, implies conventionalism in ethics is a widespread misapprehension.[26] What is called for is the development of appropriate realization theories which show how physically-based objects and attributes realize the myriad types of entity that populate the world. Recall that a particular realization theory may appeal to objects and attributes that are not themselves in the physical basis but which are grounded in the physical basis via their own realization theory. Thus, an account of the realization of, say, moral properties might appeal to various non-physical properties that are themselves physically realized (for example, interests, pain, pleasure). There is no one type of realization theory that applies to all higher-level phenomena, and realization theories need not provide direct links between physical and non-physical phenomena.

In assessing the prospects of the programme, as well as in working it out directly, judgements are called for, at appropriate points, regarding whether a putative entity that resists incorporation via a realization theory must be either downwardly incorporated (a rare occurrence), rejected as unreal and hence not within the scope of the programme (also a rare occurrence), treated as a counter-example to physicalism, or viewed as simply posing a recalcitrant problem for which solutions do exist. In the usual case, the situation is unclear. Thus sceptics will typically offer up 'impossibility' arguments of one sort or another, while proponents will either work directly upon the development of realization theories or, short of that, offer up 'possibility' arguments designed to defuse scepticism and suggest directions of productive inquiry. Again in the usual case, neither side is particularly impressed by the activities of the other.[27] A good example of just this sort of dynamic is to be found in discussions concerning mental representation where neither sceptics nor advocates of physicalism are in short supply and where the situation is evidently deadlocked.

---

[26] See Post (1987) for an excellent discussion of how physicalism is compatible with moral realism.

[27] I might add that, in the usual case, there is unclarity and disagreement at least partly because the requirements upon the appropriate normative employment of the theses are not satisfied.

I cannot pursue discussion of the outlook of the programme further without delving into the specific issues bearing upon each of the many potential obstacles to success, and this I have sworn not to do in this work. I must, therefore, leave this aspect of the assessment of the programme in a somewhat dissatisfying state. However, my own view of matters here is quite optimistic, although not without recognition of the enormity and difficulty of the task facing physicalists. Since physicalism is not the monopolistic doctrine imagined by some, and since building a physicalist system yields considerable cognitive and non-cognitive pay-off, pursuit of the programme is, in my opinion, a way of trying to have one's cake and eat it too. The full richness of life and the world is not ignored or devalued, the full range of inquiry and activity into what is real, true, and possible is enhanced rather than impeded, and a unified vision of things may yet be developed which yields many of the gains outlined in Chapter 1.

7

Significance of the Physicalist Programme

> the materialist's desire, which is ... to show that people are
> not all that special, not such remarkably unique elements of
> the universe.
>
> A. Sussman (1981: 106)

> one of my claims will be that often the pursuit of a highly
> unified conception of life and the world leads to philosophi-
> cal mistakes—to false reductions or to the refusal to recog-
> nize part of what is real.
>
> T. Nagel (1986: 3)

The aim of this chapter is to say something about why the
physicalist programme is significant. It is not a topic to which
physicalists in recent years have given much attention;[1] and it is
not a topic that lends itself readily to systematic treatment.
Nevertheless, it is a topic that requires more discussion, even if
only in a preliminary way, in order to clarify what the programme
is and is not and to deepen understanding of why it matters. Many
critics of physicalism, likewise, have not attended to the potential
or actual virtues of the programme, and hence they have never
effectively joined issue with physicalists at deeper levels.

To organize discussion, I shall distinguish between the significance of physicalism with respect to philosophy, science, culture,
and individuals. My point of view in each of these areas is essentially that, although physicalism has more significance than most
critics acknowledge, it is by no means a total philosophy. It does
not constitute a set of solutions to a wide range of technical philosophical problems, although it does provide a framework within
which to work. Nor does it determine every aspect of the framework of scientific activity, although, again, its presence is felt more
or less clearly and it is a relatively lasting feature of the current
framework within which science is pursued in Western societies.

---

[1] Hellman and Thompson (1977) and Post (1987) are exceptions here.

Finally, the physicalist programme does not determine solutions to all cultural or individual life problems. How lives are lived in a community and how individuals confront the problems of their individual existence are largely open questions relative to a physicalist framework. As I shall argue shortly, there is no monopoly on how to confront problems in philosophy, science, culture, and personal experience that follows from physicalist theses, although they impose certain constraints.

7.1. PHYSICALISM AND PHILOSOPHY

Physicalism provides a vision, a way of seeing and understanding reality, and a way of pursuing knowledge. As I have suggested above, it need not be conceived of as a monopolistic vision, although it can be and has been so conceived. Again, as I have suggested throughout this work, it is a programme that is neutral on a number of important philosophical disputes: metaphysical realism and non-realism, nominalism and Platonism, actualism and modal realism. Further, I would suggest that, although physicalism is not compatible with the radical empiricism discussed in the last chapter, it is compatible with a wide range of empiricist, pragmatist, and even some rationalist approaches to the pursuit of knowledge.[2] On the other hand, physicalism is a form of naturalism, and a rather stringent one at that. And, as I shall discuss below, physicalism is not compatible with scientific anti-realism in any of its various forms (for example, verificationism, instrumentalism).

Now, the main point I want to make here is that physicalism should be understood as a component of a larger philosophical system, and that there are many possible systems within which physicalism can be embedded. Physicalism is a philosophical 'module'. In piecing together a philosophical system, one may fit physicalism together with other components to form an integrated whole that does a certain sort of philosophical work. Alternatively put, to develop a philosophical system one cannot simply embrace physicalism since that would leave too many aspects of the system undefined. Thus it is a false understanding of physicalism to think

[2] Although I hasten to point out that, historically, physicalists have generally been moderate empiricists of one sort or another.

that the programme determines all of one's philosophical commitments. Rather, a specific version of physicalism takes its shape from those other commitments that are joined with it to form a comprehensive philosophical system.[3]

Historical formulations of physicalism have, therefore, always been presented in such larger philosophical contexts, and one of the reasons for the differences among such formulations is this variation in other commitments (for instance, compare the physicalism of Quine with that of David Lewis). My explanation for why some versions of physicalism have had more monopolistic pretensions than others concerns the background philosophical frameworks within which the different versions were pursued. Someone in the grip of a crude 'scientism' will pursue physicalism in a rabid, scientistic way, whereas one who recognizes that physicalism is not committed to scientism and who is more permissive regarding the forms of knowledge and the range of legitimate methods for its acquisition may well pursue the programme in a more open-minded way.

But the point goes beyond the historical differences among physicalists regarding larger commitments. Rather, it extends to a wide range of possible differences as well. Although most physicalists have been metaphysical realists and moderate empiricists, there is nothing in the nature of physicalism itself to preclude developing the programme in a non-realist and pragmatist framework, for example. There are, as it turns out, many ways to assemble views to form a philosophical system, and there is no reason to saddle physicalism with a narrow handful just because most physicalists in the past have tended to make certain other commitments in addition to their physicalism. Such commitments may well represent philosophical fashions, or the biases of a particular philosophical movement which are only gratuitously associated with the core ideas and values of the physicalist programme.

The point is important because unification relative to a physical basis can have value in many different philosophical environments, not just the realist, empiricist, and nominalist environments that are often associated with physicalism. The physicalism component concerns a certain brand of unification; it does not concern the other commitments, although the larger interpretation of physicalist

---

[3] Thanks to Richard Boyd for helpful discussion of this point.

unity will vary from context to context as a result of such commitments.[4] If one imagines that physicalism in a non-realist, or Platonist, or pragmatist setting is not 'real' physicalism, it is possibly because either one is in the grip of some other 'ism', or one wishes to paint physicalism as being committed to one of those others for polemical purposes, or one is too easily impressed by historical associations of physicalism with certain other views. Physicalists and critics alike should free themselves from the idea that there is only one way to be a physicalist and should recognize that the programme can be pursued in a wide variety of ways, the merits or defects of which do not bear upon the value of physicalism *per se.*

There are, of course, many other ways in which physicalism is significant for philosophy (for example, with respect to the constraints it imposes on approaches to understanding mind, meaning, and values). Some of these have been mentioned in earlier chapters, and there is a substantial literature bearing upon each of them that it is not appropriate to review here. In a spirit similar to that concerning philosophical systems generally, I want only to observe that, although physicalism constrains particular approaches to specific problems, it does not of itself entail specific solutions to many of those problems. For example, being a physicalist involves being open to what a correct analysis of moral properties involves. Given such an analysis, it is then a contingent matter whether the physicalist programme can be successfully pursued *vis-à-vis* morality. Thus physicalism itself entails neither conventionalism in moral matters, nor relativism, nor utilitarianism, nor egoism. What physicalism does entail is that whatever analysis of moral facts is correct, it must be in conformity with physicalist theses, a very different sort of entailment. The situation is similar for other sorts of values, subjective experience, human rationality, human freedom, personhood, and meaningfulness. And, in light of the discussion in Chapter 5 concerning the methodological employment of physicalist principles, if there is a conflict (for example, if the account of the nature of moral facts does not fit well with physicalist principles), then resolution of the conflict proceeds in the ways outlined: viz., one of four options might be pursued depending upon various

---

[4] This point is, I hope, demonstrated by my discussion of the neutrality of physicalism with regard to metaphysical realism and its alternatives. See Chs. 3 and 6.

considerations, and there is no guarantee that physicalism will be preserved at the end.

At the head of this chapter, Nagel suggests that among the philosophical mistakes committed by those who pursue unification programmes of the sort being discussed here are 'false reductions' and the 'refusal to recognize part of what is real'. Both are indeed mistakes, but there is no reason to suppose that they are inevitable, especially given the character of the methodological deliberations in which a sophisticated system-builder should engage. However, Nagel seems to suggest otherwise, especially for those who are enamoured by the importance of physics in the scheme of our understanding of the world:

For many philosophers the exemplary case of reality is the world described by physics, the science in which we have achieved our greatest detachment from a specifically human perspective on the world. But for precisely that reason physics is bound to leave undescribed the irreducibly subjective character of conscious mental processes, whatever may be their intimate relation to the physical operation of the brain. The subjectivity of consciousness is an *irreducible feature of reality*—without which we couldn't do physics or anything else—and it must occupy as fundamental a place in any credible world view as matter, energy, space, time, and numbers. (Nagel 1986: 7.)

Now, there is much in this passage with which a physicalist can agree: the success of physics, the importance of the world as described by physics, our inability to describe subjective experience using the language of physics, the dependence of our ability to pursue physics upon our conscious mental capacity, the irreducibility of consciousness, and its fundamental importance in our view of the world. But the ambiguity of the terms 'irreducible' and 'fundamental' masks the objectionable features of Nagel's claims and the misguided character of his dispute with physicalist unification programmes.

Although it is doubtful that there is agreement with the suggestion that physics has achieved maximal detachment from human experience, I shall bypass this point.[5] Rather, I want to emphasize that physicalism does not require that all that exists can be

---

[5] Consider the apparently essential role of human observation and the rather rigid retention of classical physical concepts within quantum mechanics. Both suggest that physics is still tied to human experience at the most fundamental level, although not perhaps in the sense that Nagel intends.

described in the language of physics. That fact alone does not provide much leverage against physicalist unification, as I am sure Nagel would allow. Thus Nagel's case against unification programmes rests solely on his repeated talk of the 'irreducibility' of subjectivity, consciousness, and the first person point of view. The contention is that either such programmes falsely claim such reductions or they illicitly ban these phenomena from reality. But what is the relevant sense of 'reducibility' if the criticism is to have a sufficiently wide scope to cover all significant physicalist programmes? It should not be either the claim of describability within the language of physics or any version of classical reductionism (for example, CR, CR', or their generalizations as discussed in Chapter 2). I think Nagel would agree with this also. What this means is that Nagel's attack is broadly enough directed to include among its targets the sort of explanatory reduction advocated in this book.[6]

However, being on target does not guarantee that one can deliver a telling blow, and I think there are several reasons why Nagel's discussion of the shortcomings of physicalist unification programmes is ineffective. Assuming that what is required for a successful reduction of consciousness (etc.) to the physical base is a realization theory, then the issue is whether or not such a theory exists and suffices to reveal how such phenomena are realized within the physical universe. In effect, for the past twenty years or so Nagel (1974; 1979; 1986) has been complaining that we know of no such theory, but his complaints do not translate into effective arguments. Yes, the burden for the physicalist is to produce an acceptable account of how mental phenomena are realized, and no, no such theory has emerged. But this does not mean that either false reductions or illicit denials of reality are inevitable. It means that there is more work to be done.

Nagel's claim that subjective experience enjoys a fundamental place in the scheme of things is plagued by vagueness and ambiguity. To assert that something is fundamental always requires one to specify the respect in which it is fundamental. For all that Nagel has stated in the passage cited, we can only conclude that such phenomena enjoy a fundamental place with respect to both

---

[6] How Nagel responds to the various forms of non-reductive physicalism surveyed in Ch. 2 is less clear to me, although I suspect he would concur with my criticisms of those views.

descriptive purposes (i.e. they cannot be described using a purely physical language) and epistemological purposes (i.e. we cannot engage in physical inquiry unless we are conscious). But neither of these claims of priority are incompatible with such phenomena being metaphysically and explanatorily non-fundamental. Here the issue turns on the question just discussed of whether a realization theory for mental phenomena exists and whether physics can be successful without downward incorporation of mental phenomena into the physical basis. The considerations he cites do not bear upon either of these issues.

Thus Nagel's attacks upon physicalist unification programmes do not conclusively establish that physicalist unification is inevitably doomed to commit philosophical errors. Therefore the importance of physicalism with respect to philosophical understanding of how mind, meaning, and values are located in the Universe is not necessarily highly limited, nor is it necessarily negative in character. The physicalist programme is open to further developments in important areas of analytical work, and it allows that its fate depends upon how well it can accommodate mature findings in those areas. But it does not at the present time need to adopt a darkly sceptical view of what those findings will be or how they will affect the prospects of the programme.

The moral of these deliberations is that physicalists and critics of physicalism alike must be circumspect in how they characterize and critically assess a version of the physicalist programme. Teasing apart what is peculiar to physicalism and what is an additional commitment that is only contingently related to physicalist theses is crucial for accurate assessment. And it must be allowed by both sides that physicalism must be compatible with what is real and true in other domains beyond the purely physical domain. Whether physicalism is acceptable and whether it will be successful depends upon how well it fits with knowledge we may not yet possess. These points are also crucial for more general understanding of the process of building comprehensive philosophical systems. Such systems are constructed from more or less independent component views and from findings in widely disparate areas of inquiry. If one does not understand the relations between the parts of the total system, one cannot build or modify or effectively use it. It is my hope that this study of physicalism has brought home some of these points about the nature of philosophical work.

7.2. PHYSICALISM AND SCIENCE

In a brief discussion of the significance of physicalism it is not possible to cover all the bases, or even some of the bases, in detail. With regard to science, I have already discussed the independence of physicalism from crude forms of scientism (Chapter 6) and I have discussed the methodological roles of physicalism (Chapter 5), roles that are especially evident in scientific inquiry. Further, I have indicated that physicalism is not compatible with a radical form of empiricism which claims that compatibility with empirical data in a given context is the only constraint upon the acceptability of hypotheses in science. Physicalism imposes additional evidential considerations which are operative in the sciences, although their deployment at any given time is subject to conditions of appropriateness that are not always satisfied (Chapter 5).

The importance of the roles R1 and R2 may be seen best in contexts in which, for example, types and possibilities of influence are at issue. Such contexts include the study of claims regarding the paranormal (for example, the use of screening techniques designed to preclude known sorts of influence), the study of the therapeutic effects of various medical and psychiatric treatments (for example, the effect of mental states upon the course of a disease), and the integration of research that crosses standard divisions in areas of knowledge (for example, the integration of disciplines within cognitive science). That there is any question that physicalism plays such roles in these and many other contexts is I think quite surprising.

Further, the range of scientific studies in which physicalist assumptions and principles clearly operate is quite broad, even when it is not necessarily part of such studies to pursue inquiry into the physical realization of the phenomena of central concern. In the social sciences, for instance, there may be little impetus to develop a full-blown account of the physical realization of, say, an economic system. But there is good reason to believe that physicalist assumptions are operative when the time comes to apply models of such a system to the real world: namely, it is assumed that the underlying mechanisms that realize economic processes are physically based, and that they constrain the class of economic models that can fit the real world phenomena of interest.

It should be clearly acknowledged that physicalism makes no all-encompassing prescriptions regarding the types of model-building or patterns of explanation that can be developed in the various sciences. The relevance and impact of physicalist principles is more subtle than that of turning all scientists into researchers concerned with the physical domain, or of implying that physicalist principles are always consciously appealed to by every scientist at every stage of inquiry, or, especially, of implying that there is only one kind of model or explanation that is appropriate (i.e. explanation as it appears in physics). Thus I differ with Field (unpublished) and Campbell (1988) who both suggest that physicalism requires that the explanatory structures in the special sciences must be preserved under 'reduction' to physics. Although physicalism requires that all explanation is physically based and that all attributes appealed to in such explanations are physically based, the structure of higher-level explanations can be lost as one follows the chain of vertical explanations that ground the non-physical in the physical domain.

Finally, physicalism does imply a form of scientific realism. An instrumentalist or fictionalist account of scientific theories in general and of physics in particular does not square with the physicalist's concerns about ontological matters (i.e. dependence, supervenience, and realization), since such concerns require the existence of elements in both the physical domain and in the ontologically dependent domains studied by higher-level scientific activity. The basis for all objective fact and truth and of all entities and influences must be real, not just a convenient fiction and not something about whose existence we need have no beliefs. Similarly, if the physical domain is to play any role at all in the explanation of the realization of all other phenomena, the entities in that domain must exist.

I caution the reader to distinguish between scientific realism (SR) and the so-called 'metaphysical realism' (MR) discussed above. As I see it, SR is an internal doctrine which, like physicalism, can be embedded in a number of larger philosophical environments without loss of significance. SR concerns the interpretation of science, and especially of how scientists understand their own activity. It also concerns how scientific activity is to be understood as being embedded in and related to its own object of inquiry: for example, scientists causally interact with the entities that they study (cf. Boyd 1984 and 1985). But all such interpretation can be trumped by a determined opponent of metaphysical realism. For example, the existence of physical entities can be one or another sort of 'dependent' existence, be it dependence upon mind or theory or language.

Physicalism, then, plays a significant role in the conduct of scientific inquiry, and physicalism implies scientific realism without being committed thereby to metaphysical realism. Such claims are not examples of normative philosophy of science, and should not be construed as such. The empirical status of the theses means that they had better stand up to a rigorous empirical study of the institution of science. Part of my aim in this book has been to clarify what the physicalist hypothesis regarding science is and what it is not, and to elicit some of the complications involved in testing that hypothesis.

7.3. PHYSICALISM AND CULTURE

As with philosophy and science, so it is with culture conceived of in very broad terms. Although physicalism does impose certain kinds of constraints, the constraints leave considerable room for wide variation in the types of cultural forms, institutions, and practices that can evolve. In particular, it should be readily recognized that physicalism is quite compatible with the existence and effective functioning of cultures which explicitly reject physicalist principles. Seeing why this is so is important for a proper understanding of the significance of the programme.

The nature and range of cultures that can possibly be realized in this world is constrained by physicalist principles only in the sense that any actually realized culture must be physically grounded in the ways called for by those principles. But the range of possible human (or other) experience, behaviour, interaction, meaningfulness, and cultural forms is none the less multifaceted and wide. Exploration and understanding of the range of such possibilities is an important, although currently immature, area of inquiry. We can perhaps gain a sense of how physicalist principles might significantly limit such possibilities by considering, not the wide diversity of actual cultural forms which fall well within physicalist constraints, but the productions of those among us who test the limits of such possibility: for example, the work of science fiction-writers and other authors who imagine cultures radically different from any with which we are familiar. In their most extreme form, such imaginings may go beyond what is physically possible (for instance, a culture of beings made of amorphous jelly that consume all that comes into their path as they move around the environment, or beings that punish their criminal elements by infusing them into mirrors that hurtle through intergalactic space).

But although physicalism may place some limits upon possible cultural forms and practices, it in no way dictates the way or ways in which every culture must of necessity be organized or how everyone must act and live within such cultures. From the 'basic building blocks of the universe' many wonderful—and some not so wonderful—cultural forms can be realized. Of course, if various norms (moral and otherwise) have objective application in the physical universe, then the physical facts will determine which of the possible cultures are, say, morally valuable and which are not. However, this is not a matter of physicalist prescription, but rather a matter of what the objective facts are.

In addition to its bearing on the range of *possible* cultures, physicalism does constrain what is objectively real and true within cultures. That is, since the principles of physical fact and truth determination (i.e. T2 and T3) set limits upon what is objectively true and real in the world, they do apply to the claims and postulations made within any culture that is actually realized. Physicalism does not, except possibly in certain non-literal or trivial senses, endorse the idea of truth as 'culture relative'. Thus with regard to many religious doctrines, myths, forms of experience, and customs, although each is a possibly realizable cultural form, each may well step beyond the limits of what is objectively and literally true. Physicalism implies that there are no supernatural gods or other beings and that the claims of many religious texts are literally false. But physicalists can make such judgements while not necessarily condemning the religious beliefs and practices involved. Literal truth is by no means always what is important in a given context, and physicalists, like everyone else, must honour that fact of life. Similar points can readily be made regarding medical, educational, and legal practices as well (i.e. literally false claims can promote health and reduce suffering, they can promote the acquisition of knowledge and skills, and they can lead to effective regulation of social interaction).

In general, there may be many possible solutions to cultural problems and many possible responses to cultural needs. Although each must be grounded in the physical domain, they need not be explicitly guided by objective truths and facts as judged by physicalist lights. They may, rather, be based upon literal falsehoods and the postulation of nonexistent entities and processes. And, of course, the physicalist must allow for, and provide an account of, the possibility that the beliefs of a culture, which guide its practices, although literally false, are non-literally 'true' in other ways.[7]

Finally, physicalism does suggest, without implying it across the board, that some cultural practices can be improved (from the point of view of those who live in the culture)[8] by taking into account physically-based facts and truths. Thus medical practices within a culture may be only partially effective in achieving the goals of medical practice in that culture (for example, healing the sick, easing suffering, providing a meaningful context for pain and death, and so on). *All things considered*, a given practice might well be improved if it were more explicitly based on what is literally true by physicalist standards. For example, in some 'primitive' cultures, allowing that there are bacterial agents that cause an illness and that can be eradicated by use of an antibiotic *might* promote the medical purposes of that culture. And, for example, in our own culture, the creation of medical environments that emphasize hope and meaningfulness and in which persons rather than 'cases' are valued might better promote a number of medical purposes that we acknowledge.

But I want to emphasize that all things must be considered in making such a judgement. Such importations into a culture as the ones mentioned may be highly destructive of deeper and more important cultural values and goals than those cited. The use of antibiotics, for example, could fail to solve many of the cultural problems that need to be solved. Physicalists, like everyone else, must remember (or learn) that life's problems are not always easily identified, let alone solved, and that technological expertise can sometimes solve narrowly conceived problems at the expense of what matters most. On the other hand, such expertise can be quite valuable and should not be pointlessly ignored. The potential

[7] See Post (1987: chs. 5 and 8) for discussion of these issues.
[8] That is, with respect to the purposes those cultural practices are supposed to serve according to the inhabitants of the culture.

importance of literal fact and truth by physicalist lights is enormous, but the exploitation of such facts and truths is invariably a more delicate and complicated matter than many physicalists and their critics may suppose.

An example of such complexity involves the use of psychotropic medication in contemporary psychiatric practice in our own culture. It is an important discovery that the physically-based facts regarding our brain functioning reveal ways of chemically manipulating such functioning to our potential benefit (for instance, to gain relief from distressing psychotic symptoms). But if one only focuses on the narrow goal of symptom reduction, one could be led to a rather simplistic understanding of the role of such medications. Their broader impact upon the lives of individuals and upon the culture at large must inevitably be taken into account if the practice of using them is to be properly comprehended. The impact upon us all of too eager a readiness to medicate and to intervene directly in our brain chemistry is ill-understood, to say the least. Likewise, our understanding of the full clinical impact of our medication practices upon individuals is still quite immature. It is an important aspect of the evolution of such practices within our culture that we learn about such greater impacts and that we shape those practices accordingly. The physically-based facts regarding how we can alter our brains and minds are only a part of what is involved in deciding what will work best.

In summary, physicalism has at least a fourfold impact upon culture: it constrains the range of possible cultures; it underlies the assessment of what is objectively true and real within a cultural framework; it underlies the relative effectiveness of specific cultural forms and practices; and it underlies a range of possible ways in which cultural practices and forms can be altered, for better or worse. But such impacts do not involve legislation of either physicalist doctrine or particular cultural practices. Physical facts and truths circumscribe the limits of cultural possibilities, they realize all actual cultural forms, they ground all assessment of such forms regarding moral and other worth, literal truth, and objective reality, and they delineate the range of possible changes of which a culture may avail itself. But none of this carries categorical normative force. More is required for such normative assessment, and what more is required goes well beyond the implications of physicalism.

It is, of course, also conceivable that physicalism could be specifically embraced within a culture and that, as a consequence, cultural practices and institutions are explicitly constrained by physicalist doctrine. It follows from what I have said above that such specific endorsement of physicalism would not, of itself, imply specific cultural practices. The role that physicalist principles might play could vary widely depending on the interests and values embraced by members of the culture. The lack of unconditional priority of physical truth coupled with the many different ways that cultural problems can be solved would mean that any pair of 'physicalist cultures' could differ enormously, though they would be in agreement on the physical bases and the physicalist theses. How facts and values are integrated to solve cultural and life problems can vary widely even when there is agreement on what the non-valuational facts are and on how physicalism constrains the facts. Again, the complexity involved in such integration should not be overlooked.[9]

7.4. PHYSICALISM AND THE INDIVIDUAL

I shall close my discussion of physicalism with a brief excursion into what I take the implications of the programme to be for individual persons. Such implications arise at two levels: how individuals are in fact embedded in the physical fabric of the universe and how recognition of such embedding influences a person's cognitive, affective, and behavioural responses to life. Lucretius, of course, thought that if one saw deeply enough into the nature of things one would see many things that would shape the course of one's existence. More importantly, one *wouldn't* see many things (for instance, gods, Hades, etc.) which lead one away from forms of existence that are good and meaningful. Whether one quarrels with Lucretius's conception of what is good and meaningful and whether one disputes that there is such a thing as 'the' nature of things is not as important as the general point that physicalism (or

---

[9] I note that this variability among physicalist cultures is compatible with the satisfaction of physicalist theses in all cases. Indeed, for each such culture the physical facts determine all the facts regarding its practices, their efficacy, their worth, etc. And of course the physical facts determine whether or not physicalist doctrine is accepted or rejected in any given culture.

atomism) does seem to have some significant consequences for an individual's life. What those consequences are and what they are not is surely deserving of more attention than most physicalist philosophers in this century have given them. Indeed, who we are and how we live our lives depends to some extent upon how we think of ourselves in relation to the rest of the world.

On the assumption that physicalist theses are true, then, as in the case of culture, serious constraints are imposed upon the possible forms that an individual life can take. The kinds of thoughts, experiences, patterns of action, creative activity, desires, and so on that an individual can have are limited by what goes on in the physical domain. But, again as with culture, within such limits there is room for wide-ranging variation, and there is absolutely no implication that a person must embrace physicalist doctrine if their life is to be possible or if it is to be worthy. Physicalism is quite compatible with there being many different responses to the demands of life and many different patterns of growth and adaptation that are all equally legitimate and of value. The only restrictions imposed by the doctrine are those concerning possibility as circumscribed by the physical bases and the theses. Of course, if there are objective norms that are applicable to what goes on in a person's life, then the physical facts will determine which lives are more or less worthy in some respect (for example, moral worth, degree of rationality, level of effectiveness in pursuing identified objectives). But this is not the same as saying that being a physicalist commits one to a certain pattern of existence as being the most worthy. As with culture, physicalism underlies both the relative effectiveness of one's style of life and the possible revisions that would lead to more (or less) effective pursuit of certain sorts of goals in one's life.

Note that this description of the implications of physicalism freely employs various personal, mental, and normative terms (for instance, 'thoughts, experiences', 'moral worth'). The best way of describing ourselves and our place in the physical universe will, of course, depend upon acceptable philosophical and theoretical analysis of the relevant language and concepts (for example, the development of appropriate RTs). Although the physicalist is not committed to particular developments in these areas (for example, physicalism is compatible with both eliminative materialism with respect to the mental as well as a full-blown mentalistic psychology), I incline towards what appear to be the obvious facts of life: viz., that there are persons with mental lives who engage in actions with various degrees of moral and instrumental worth. I think the deepest and most important burden of the physicalist is to reveal how these phenomena are possible within a physically-based universe. Thus the implications of physicalism for individual persons are to be understood in terms qualified by this caveat on what the appropriate conceptual framework is and hence what the non-physical facts are. The caveat aside, physicalism does have implications regarding possibility, assessment, effectiveness, and interventions, however such things are described.

Beyond these general implications, the more interesting issue is: If one explicitly embraces physicalist doctrine as a part of one's system of belief (or, if you will, part of one's personal philosophy of life), what follows? What impact will it have upon our experience, our behaviour, our outlook, and our relations with others? Will we become immediately demeaned, unnerved, and depressed? Or will we become appropriately humbled, realistic, and empowered? What sort of person will we be? What kind of life will we lead? It is this sort of implication that was central to Lucretius's and Epicurus's understanding of atomism and of why they believed one ought to be an atomist. Ought we to adopt the same attitude? Ought we to be physicalists in part, at least, because *if we embrace it, it will have salutary effects on who we are and how we live*?

The answers to these questions are not necessarily specific. Physicalism is likely to be compatible with a variety of human responses to the problems involved in living a life, although it does prescribe constraints which can make a difference if taken into account. And how physicalist doctrine impacts upon a person will depend, to some extent, upon the actual working out of the programme with respect to the *obstacles* identified in Chapter 6. A full conceptualization of one's place in the scheme of things depends upon the choice of a conceptual framework, and the impact of the physicalist system on a person depends upon what concepts are employed (i.e. what concepts, in addition to physical concepts, are employed within the system). Further, the significance of the programme for individuals who embrace it also depends upon the course of research in physics itself. Thus, in the absence of the details of physical theory, physicalism only constrains, without

pinning down, specific consequences about, say, the evolution of
the universe or the range of possible influences that bear upon
a person (i.e. in the absence of theoretical details, a physicalist is
largely in the dark concerning these matters). None the less, there
are a number of things that can be said about what physicalism
means to an individual in light of the directions in which physics
appears to be moving and in light of an optimistic appraisal of
how the obstacles will be dealt with (i.e. mind, meaning, and
values will be incorporated within the system). I shall break down
the likely significance for the individual into three categories: the
view of self, the view of others, and the view of the world.

To embrace physicalism is to see oneself as a tiny and relatively
powerless speck embedded in a massive cosmic structure; one is a
tiny ripple in the physical fabric of existence. It is to see oneself
as a highly contingent and vulnerable creature with clearcut spa-
tial and temporal boundaries. And it is to see oneself as subject to
disease and inevitable extinction. Death for the physicalist means
the end of one's physical *and* mental existence. To see oneself as
an expression of the physical fabric of the universe also means
that one understands that there are limitations upon what one can
do and who one can be. Both the influences upon us and our
power to influence the world are due to, and limited by, physical
processes. Indeed, all that we are capable of is constrained by the
character of physical structures and processes.

Is this a demeaning and depressing picture to embrace? The
answer is 'yes' for those who have a tendency always to see half-
empty glasses and, perhaps, for those who have radically inflated
egos. But if one looks further into the physicalist vision, one sees
the positive side. Within the limitations imposed by the physical
domain, there is an enormous range of possibilities for development. There is much that one can do and be as a creature in the
physical universe. Understanding that one is embedded in the
physical fabric of things is to understand *both* that one is limited
by physical realities and that one has enormous potential for
meaningful, creative, and worthy activity of all sorts. Life for the
physicalist is a challenge, given one's place in the cosmic physical
structure and given one's consequent limitations. For many endeavours, understanding that one is an expression of the physical
can lead one to exploit one's possibilities more effectively than if
one had a different understanding of one's place in the scheme of

things (for example, specific understanding of the physical under-pinnings of mental functioning could lead to the development of more effective techniques for improving mental performance in personal, clinical, and educational settings).

Physicalism, remember, does not imply that there are no minds, no values, and no meaning in the world; it implies only that if there are such things then they are realized subject to physicalist constraints. My own optimistic approach to physicalism is that, as the programme is further worked out, it will become increasingly evident just how those things which lead us to believe that our lives matter are realized in the physical world. As I see it, the physicalist view of oneself is that of a being with, *inter alia*, interests, richness of experience, creative inspirations, capacity for self-control, and moral worth. And it is to see oneself as a being who, if motivated, can attempt to be more of the kind of person he or she wants to be. This picture of life is in no way demeaning, although one may well be humbled by an appreciation of the limitations and vulnerabilities that qualify it. But such humility also opens the doors to a deepened understanding that can improve the prospects for achieving one's goals. Such understanding can also underwrite a realistic and morally significant approach to the problems of living and dying. For example, acknowledging and appreciating that death is indeed the end of one's mental life, and incorporating such an understanding into an approach to living, can lead to the enrichment and deepening of personal experiences and one's outlook regarding oneself, others, and the world.

Although physicalism implies a certain picture of how we are embedded in the physical universe and of what some of our limitations are, it most certainly does not imply that our lives are worthless and without meaning, that we are incapable of making an impact upon the world or of having a say in what happens to us, that we are incapable of improving our understanding, our character, and our actions, that moral, aesthetic, and instrumental excellence are impossible, or that we must be depressed and humiliated once the physicalist truth is known. Although we may be 'insignificant' specks in the physical universe, our lives are still capable of worth and meaning. Indeed, we are 'meaning makers' *par excellence*, and this is determined by the physical realm, not obviated by it.

A critic might interject that this rosy picture is undermined by two problems. First, if our existence does indeed end with the death of the body and if the universe will in all likelihood evolve to a state in which all human existence will cease and all our creations be destroyed, then how can life be meaningful at all for the physicalist? How can one view oneself as *anything but* a tiny and insignificant speck in the physical universe?

The argument seems to suppose that a life can be meaningful only if it goes on for ever, or only if one's creations do, or only if the chain of human existence does. Why should one believe any of these things? The meaningfulness of one's life, of what one creates, or of what one does, is not simply a matter of endurance or of relationship to some other being who can marvel at one's accomplishments. Significance in the universe is realized by qualities and relations of persons, or acts, or products that may be quite time-limited and beyond the recognition of anyone else (for example, the act of saving a life even if no one ever knows about it). Since this reply depends upon my optimistic view of the future course of physicalism and upon a certain view of meaningfulness,[10] views I shall not be arguing for further in this book, I will limit myself to challenging the critic to mount a plausible case for why a physicalist view of oneself must be that of a merely meaningless speck, given the aforementioned limitations on our existence.

A second possible problem for the physicalist view I have sketched is as follows: if we are truly embedded in the physical fabric of things as the physicalist claims, then one's self-concept must inevitably be that of a being who is a 'mere cog in a giant cosmic wheel' and thus of a being with no say in what happens to it. How could such a self-concept be other than demeaning and depressing? Perhaps it is such thinking that leads Sussman (in the passage quoted at the beginning of this chapter) to attribute to the materialist a desire to show that people are not all that special. And in his deliberations upon the implications of determinism, Earman reasons as follows:

[10] Of course, the question of what makes a life (or anything else) meaningful and of value is a question that physicalists must address in the full working out of the physicalist programme. Being both a physicalist and someone who believes that there is meaning and value in the universe and in our lives commits me to the view that it is possible to understand how this is so within a physicalist framework.

The more precisely science locates man in nature the more difficult it becomes to sustain a sense of autonomy for human actions. As autonomy shrinks so does our sense of uniqueness and worth as well as the basis for a moral perspective on human action. (Earman 1986: 249–50.)

It is not my aim here to take on the free-will problem at the very end of a book on the foundations of physicalism. Human freedom and autonomy are both, in my view, obstacles to the physicalist programme which must be adequately handled if the programme is to succeed. Thus this sort of objection raises an important set of issues that deserve a full hearing and treatment. However, I think the objection misses the mark in suggesting that somehow a physicalist *must* see him- or herself as a mere cog in a cosmic wheel. Since physicalism by itself does not imply a particular understanding of mind and human action, there is, at least potentially, room for alternative views. That is, given a physicalist perspective, how one reacts will depend upon what other things one believes and how one weighs beliefs in terms of relative importance. For example, I think it evident that humans do make their way around in the world possessing both an ever-increasing understanding of how the world works and a capacity to exploit such understanding to suit their actions to their purposes and to prevailing contexts. The degree to which we are effective in making decisions and pursuing courses of action that are effective in realizing our goals is a measure of the control we have over ourselves and our environment. If we are cogs in a wheel in some sense, we are pretty influential cogs. Sufficiently so that it is hard to see why we must become depressed when we realize that all our capacities are physically grounded.

But the critic will not be impressed, since these considerations do not undo the point that all aspects of our mind and behaviour are realized by the physical. And doesn't this mean that we are compelled to do and think whatever the physical makes us do and think? I do not think so, but I shall not defend myself further here. Since my main point concerns the physicalist self-concept, I shall restrict myself to the claim that more argumentation is required before we must conclude that physicalist doctrine inevitably crowds out a view of oneself as autonomous and free, and hence leads to a demeaning and depressing self-concept. Rather, to see deeply into how we are embedded in nature is to see how it is that we have the enormous capacities that we have. Such knowledge empowers us further inasmuch as it increases our understanding of ourselves and enables us to exploit physically-based processes in order to pursue our various purposes further. Why, then, get depressed?

With regard to the physicalist's view of other beings and the nature of our relationships to them, there is, again, room for variation. Physicalism does not imply a single party line on how to conceive of one's neighbour or one's family, nor does it imply how to relate to them. There is, in particular, no prescription that we view others in purely physical terms or as 'mere cogs in a wheel' any more than we should view ourselves in those ways.[11] And there is surely no reason for relationships to be undermined or narrowly conceived because one views them as grounded in physical interactions. Rather, within the limits of possibility defined by the physical basis, a physicalist can both view others, and relate to them, in widely diverse, human terms.

Although it is possible to demean someone by viewing them as a purely physical system, this is not a requirement of physicalism. Rather, it suggests the crude monopolistic pretension, rejected above, that the most important description is always a physical description. Not only can the physicalist view others in many ways, but also the physicalist acknowledges and can pursue a wide range of possible forms of relationship exhibiting various degrees of richness and quality. Physicalist beliefs about other people always occur in a context of other beliefs, values, and purposes, and it is the total mix that leads us to act in certain ways and to engage in certain kinds of relationship. It is, therefore, much too simplistic to think that acceptance of physicalist doctrine alone determines how one acts or how one treats one's neighbour.

---

[11] Not only does physicalism allow that there are beings who can be conceived in the same terms in which we conceive of ourselves, but also it is likely that we are not alone in the universe in the sense that there are probably alien beings capable of thought and intelligent behaviour. The limits upon such possibilities are imposed by the physical facts, but within such limits those possibilities are, to many thinkers, real. As a consequence, physicalists have no cause to imagine that 'we' are unique or special within creation. We are one among many forms of being that can evolve in nature, and there is no guarantee that other forms do not share our most prized features. However, this is not to say that we are not important, or that our lives have no meaning, or that we are somehow demeaned.

Whether or not one is a physicalist is probably not going to be the deciding factor in what kind of person you will be, how you will view others, or how you will relate to them. The physical facts make such openendedness possible! Further, physicalism does not imply how one *ought* to view and treat others or what kinds of relationship one *ought* to enter into with them. There may be moral or instrumental facts concerning such matters, but those do not follow from physicalism alone.

Finally, the contemporary physicalist's view of the world is of an enormous totality defined in terms of space-time structure and very likely bounded (for example, at singularities in the space-time manifold). If we focus on the time dimension, the universe is probably evolving from a state of high energy towards a state that will not support life or organization of the sort with which we are familiar (for instance, a 'big crunch' or a cold, dead universe). Thus neither we nor our successors nor our products will survive as the universe unfolds. And, from the physicalist point of view, the world does not appear to be inhabited by deities or other supernatural beings, nor are there any places answering to the description of either heaven or hell (i.e. places where one's spirit goes in an afterlife). Classical atomism is therefore echoed in twentieth-century physicalism.

But although this world will not ultimately sustain us, it is, in the physicalist view, rich in possibilities; it is a goldmine of resources for us to exploit as we make our way around in it, if only we are clever enough and determined enough to take advantage of them. For better or worse, this is the only world we have, and the physicalist vision portrays it as one which, although definitely hostile and threatening in many ways, is also the arena of our survival and our flourishing, however temporary. Whether one emphasizes the hostility of the world to our interests or its supportive aspects is a matter of experience and temperament. Neither attitude is dictated by the physicalist programme.

Given this all too brief characterization of the physicalist view of our self, our neighbours, and the world we inhabit, it should be evident that with regard to motivation and direction in our lives, none is implied. One can adopt a positive attitude and use physicalist understanding in pursuit of one's goals in life or one can adopt a negative attitude, feel demeaned, and see life as pointless and not worth the trouble of living. And then there is a range of different stances in between. Physicalism does imply some things about ourselves and the world, but it does not impose a single system of beliefs about what is important and what one ought to do. Physicalism is compatible with many different ways of being.

References

BLOCK, N. (1980), 'Troubles With Functionalism', in N. Block (ed.), *Readings in the Philosophy of Psychology*, vol. 1 (Cambridge, Mass., Harvard University Press).

BOYD, R. (1980), 'Materialism Without Reductionism: What Physicalism Does Not Entail', in N. Block (ed.), *Readings in the Philosophy of Psychology*, vol. 1 (Cambridge, Mass., Harvard University Press).

—— (unpublished), 'Materialism Without Reductionism: Non-Humean Causation and the Evidence for Physicalism'.

—— (1984), 'The Current Status of Scientific Realism', in J. Leplin (ed.), *Scientific Realism* (Berkeley, Calif., University of California Press).

—— (1985), '*Lex Orandi est Lex Credendi*', in P. Churchland and C. Hooker (eds.), *Images of Science* (Chicago, University of Chicago Press).

BROMBERGER, S. (1966), 'Why Questions', in R. Colodny (ed.), *Mind and Cosmos* (Pittsburgh, University of Pittsburgh Press).

CAMPBELL, K. (1988), 'Review of *The Faces of Existence: An Essay in Nonreductive Metaphysics*', *Philosophy and Phenomenological Research* 49: 358–62.

CAMPBELL, N. (1953), *What Is Science?* (New York, Dover).

CARNAP, R. (1967), *The Logical Syntax of Language* (London, Routledge & Kegan Paul).

—— (1969), 'Logical Foundations of the Unity of Science', in O. Neurath, R. Carnap, and C. Morris (eds.), *Foundations of the Unity of Science*, vol. 1 (Chicago, University of Chicago Press).

CARTWRIGHT, N. (1983), *How the Laws of Physics Lie* (Oxford, Oxford University Press).

CAUSEY, R. (1977), *Unity of Science* (Dordrecht, Reidel).

—— (1981), 'Reduction and Ontological Unification: Reply to McCauley', *Philosophy of Science* 48: 228–31.

CHOMSKY, N. (1968), *Language and Mind* (New York, Harcourt, Brace, & World).

—— (1975), *Reflections on Language* (New York, Pantheon).

—— (1980), *Rules and Representations* (New York, Columbia University Press).

—— (1988), *Language and the Problems of Knowledge* (Cambridge, Mass., MIT Press).

CHURCHLAND, P. M. (1984), *Matter and Consciousness* (Cambridge, Mass., MIT Press).

—— (1985), 'Reduction, Qualia, and the Direct Introspection of Mental States', *Journal of Philosophy* 82: 8–28.

CHURCHLAND, P. S. (1986), *Neurophilosophy* (Cambridge, Mass., MIT Press).

CORNMAN, J. (1971), *Materialism and Sensations* (New Haven, Conn., Yale University Press).

CUMMINS, R. (1989), *Meaning and Mental Representation* (Cambridge, Mass., MIT Press).

CURRIE, G. (1984), 'Individualism and Supervenience', *British Journal of the Philosophy of Science* 35: 345–58.

DARDEN, L., and MAULL, N. (1977), 'Interfield Theories', *Philosophy of Science* 44: 43–64.

DAVIDSON, D. (1970), 'Mental Events', in L. Foster and J. Swanson (eds.), *Experience and Theory* (Amherst, Mass., University of Massachusetts Press).

—— (1973), 'Radical Interpretation', *Dialectica* 27: 313–28.

DAVIES, P. C. W., and BROWN, J. (eds.) (1988), *Superstrings: A Theory of Everything* (Cambridge, Cambridge University Press).

DEPAUL, M. (1987), 'Supervenience and Moral Dependence', *Philosophical Studies* 51: 425–39.

DEVITT, M. (1984), *Realism and Truth* (Princeton, NJ, Princeton University Press).

DUMMETT, M. (1978), *Truth and Other Enigmas* (Cambridge, Mass., Harvard University Press).

DUPRE, J. (1983), 'The Disunity of Science', *Mind* 92: 321–46.

EARMAN, J. (1975), 'What Is Physicalism?', *Journal of Philosophy* 72: 565–7.

—— (1986), *A Primer on Determinism* (Dordrecht, Reidel).

ENC, B. (1976), 'Identity Statements and Microreductions', *Journal of Philosophy* 73: 285–306.

FEIGL, H. (1963), 'Physicalism, Unity of Science, and the Foundations of Psychology', in P. Schilpp (ed.), *The Philosophy of Rudolf Carnap* (La Salle, Ill., Open Court).

—— (1969), 'The Origin and Spirit of Logical Positivism', in P. Achinstein and S. Barker (eds.), *The Legacy of Logical Positivism* (Baltimore, Md., The Johns Hopkins Press).

FIELD, H. (1972), 'Tarski's Theory of Truth', *Journal of Philosophy* 69: 347–75.

—— (1975), 'Conventionalism and Instrumentalism in Semantics', *Nous* 9: 375–405.

—— (1980), *Science Without Numbers* (Princeton, NJ, Princeton University Press).

FIELD, H. (1982), 'Realism and Relativism', *Journal of Philosophy* 79: 553–67.
—— (1986), 'The Deflationary Conception of Truth', in G. Macdonald and C. Wright (eds.), *Fact, Science, and Morality* (Oxford, Basil Blackwell).
—— (unpublished), 'Physicalism'.
FINE, A. (1986), *The Shaky Game: Einstein, Realism, and the Quantum Theory* (Chicago, University of Chicago Press).
FODOR, J. (1975), *The Language of Thought* (New York, Thomas Y. Crowall).
—— (1978), 'Computation and Reduction', in C. J. Savage (ed.), *Minnesota Studies in the Philosophy of Science*, vol. 9 (Minneapolis, University of Minnesota Press).
—— (1987), *Psychosemantics* (Cambridge, Mass., MIT Press).
—— (1991), 'You Can Fool Some of the People All of the Time, Everything Else Being Equal: Hedged Laws and Psychological Explanations', *Mind* 100: 19–34.
FRIEDMAN, M. (1974), 'Explanation and Scientific Understanding', *Journal of Philosophy* 71: 5–19.
—— (1975), 'Physicalism and the Indeterminacy of Translation', *Nous* 9: 353–73.
—— (1981), 'Theoretical Explanation', in R. Healey (ed.), *Reduction, Time and Reality* (Cambridge, Cambridge University Press).
—— (1983), *Foundations of Space-Time Theories* (Princeton, NJ, Princeton University Press).
GIERE, R. (1988), *Explaining Science: A Cognitive Approach* (Chicago, University of Chicago Press).
GILLESPIE, N. (1984), 'Subvenient Identities and Supervenient Differences', *Southern Journal of Philosophy* 22, Supplement: 111–16.
GLYMOUR, C. (1971), 'Theoretical Realism and Theoretical Equivalence', in R. C. Buck and R. S. Cohen (eds.), *Boston Studies in the Philosophy of Science*, vol. 8 (Dordrecht, Reidel).
GOODMAN, N. (1972), *Problems and Projects* (Indianapolis, Bobbs-Merrill).
—— (1978), *Ways of Worldmaking* (Indianapolis, Hackett).
—— (1979), 'Matter Over Mind', *New York Review of Books*, 17 May: 42.
—— (1984), *Of Mind and Other Matters* (Cambridge, Mass., Harvard University Press).
—— and ELGIN, C. (1988), *Reconceptions in Philosophy and Other Arts and Sciences* (Indianapolis, Hackett).
GRIMES, T. (1988), 'The Myth of Supervenience', *Pacific Philosophical Quarterly* 69: 152–60.
GRUNBAUM, A. (1976), '*Ad Hoc* Auxiliary Hypotheses and Falsificationism', *British Journal of the Philosophy of Science* 27: 329–62.
HACKING, I. (1983), *Representing and Intervening* (Cambridge, Cambridge University Press).

HAUGELAND, J. (1982), 'Weak Supervenience', *American Philosophical Quarterly* 19: 93–103.

—— (1984*a*), 'Ontological Supervenience', *Southern Journal of Philosophy* 22, Supplement: 1–12.

—— (1984*b*), 'Phenomenal Causes', *Southern Journal of Philosophy* 22, Supplement: 63–70.

HAWKING, S. (1988), *A Brief History of Time* (New York, Bantam Books).

HEALEY, R. (1978), 'Physicalist Imperialism', *Proceedings of the Aristotelian Society* 79: 191–211.

HELLMAN, G. (1983), 'Realist Principles', *Philosophy of Science* 50: 227–49.

—— (1985), 'Determination and Logical Truth', *Journal of Philosophy* 82: 607–16.

—— and THOMPSON, F. W. (1975), 'Physicalism: Ontology, Determination, and Reduction', *Journal of Philosophy* 72: 551–64.

—— (1977), 'Physicalist Materialism', *Nous* 11: 309–45.

HEMPEL, C. (1949), 'The Logical Analysis of Psychology', in H. Feigl and W. Sellars (eds.), *Readings in Philosophical Analysis* (New York, Appleton-Century-Crofts).

—— (1969), 'Reduction: Ontological and Linguistic Facets', in S. Morgenbesser, P. Suppes, and M. White (eds.), *Philosophy, Science, and Method: Essays in Honor of Ernest Nagel* (New York, St Martin's Press).

—— (1980), 'Comments on Goodman's *Ways of Worldmaking*', *Synthese* 45: 193–9.

HOOKER, C. A. (1981*a*), 'Towards a General Theory of Reduction, Part I, Historical Framework', *Dialogue* 20: 38–59.

—— (1981*b*), 'Towards a General Theory of Reduction, Part II, Identity and Reduction', *Dialogue* 20: 201–306.

—— (1981*c*), 'Towards a General Theory of Reduction, Part III, Cross-Categorial Reduction', *Dialogue* 20: 496–529.

—— (1987), *A Realist Theory of Science* (Albany, NY, State University of New York Press).

HORGAN, T. (1982), 'Supervenience and Microphysics', *Pacific Philosophical Quarterly* 63: 29–43.

—— (1984), 'Supervenience and Cosmic Hermeneutics', *Southern Journal of Philosophy* 22, Supplement: 19–38.

—— (1987), 'Supervenient Qualia', *Philosophical Review* 96: 491–520.

HORWICH, P. (1986), 'A Defence of Conventionalism', in G. MacDonald and C. Wright (eds.), *Fact, Science and Morality* (Oxford, Basil Blackwell).

HULL, D. (1974), *Philosophy of Biological Science* (Englewood Cliffs, NJ, Prentice Hall).

JACKSON, F. (1986), 'What Mary Didn't Know', *Journal of Philosophy* 83: 291–5.

KEMENY, J. (1959), *A Philosopher Looks at Science* (Princeton, NJ, D. Van Nordstrom).

KIM, J. (1984), 'Concepts of Supervenience', *Philosophy and Phenomenological Research* 45: 153–76.

—— (1987), '"Strong" and "Global" Supervenience Revisited', *Philosophy and Phenomenological Research* 48: 315–26.

—— (1988), 'Supervenience for Multiple Domains', *Philosophical Topics* 16: 129–50.

KINCAID, H. (1986), 'Reduction, Explanation and Individualism', *Philosophy of Science* 53: 492–514.

—— (1987), 'Supervenience Doesn't Entail Reducibility', *Southern Journal of Philosophy* 25: 343–56.

—— (1988), 'Supervenience and Explanation', *Synthese* 77: 251–81.

—— (1990), 'Molecular Biology and the Unity of Science', *Philosophy of Science* 57: 575–93.

KITCHER, P. (1981), 'Explanatory Unification', *Philosophy of Science* 48: 507–31.

—— (1982), 'Genes', *British Journal of the Philosophy of Science* 33: 337–59.

—— (1984), '1953, A Tale of Two Sciences and All That', *Philosophical Review* 43: 335–73.

—— and SALMON, W. (1987), 'Van Fraasen on Explanation', *Journal of Philosophy* 84: 315–30.

KRIPKE, S. (1972), *Naming and Necessity* (Cambridge, Mass., Harvard University Press).

LAUDEN, L., and LEPLIN, J. (1991), 'Empirical Equivalence and Underdetermination', *Journal of Philosophy* 88: 449–72.

LEVINSON, J. (1984), 'Aesthetic Supervenience', *Southern Journal of Philosophy* 22, Supplement: 93–110.

LEWIS, D. K. (1983), 'New Work for a Theory of Universals', *Australasian Journal of Philosophy* 61: 343–77.

—— (1986), *On the Plurality of Worlds* (New York, Basil Blackwell).

LITTLE, D. (1991), *Varieties of Social Explanation: An Introduction to the Philosophy of Social Science* (Boulder, Colo., Westview Press).

McDOWALL, J. (1978), 'Physicalism and Primitive Denotation: Field on Tarski', *Erkenntnis* 13: 131–52.

MALINAS, G. (1973), 'Physical Properties', *Philosophia* 3: 17–31.

MEEHL, P., and SELLARS, W. (1956), 'The Concept of Emergence', in H. Feigl and M. Scriven (eds.), *Minnesota Studies in the Philosophy of Science*, vol. 1 (Minneapolis, University of Minnesota Press).

NAGEL, T. (1974), 'What is it like to be a Bat?', *Philosophical Review* 83: 435–50.

—— (1979), *Mortal Questions* (Cambridge, Cambridge University Press).

—— (1986), *The View From Nowhere* (Oxford, Oxford University Press).

OPPENHEIM, P., and PUTNAM, H. (1958), 'Unity of Science as a Working Hypothesis', in H. Feigl, M. Scriven, and G. Maxwell (eds.), *Minnesota Studies in the Philosophy of Science*, vol. 2 (Minneapolis, University of Minnesota Press).

PETRIE, B. (1987), 'Global Supervenience and Reduction', *Philosophy and Phenomenological Research* 48: 119–30.

POLAND, J. (1983), 'Problems and Prospects for the Physicalist Program in Science', Doctoral Dissertation, MIT.

POST, J. (1984), 'On the Determinacy of Truth and Translation', *Southern Journal of Philosophy* 22, Supplement: 117–35.

—— (1987), *The Faces of Existence* (Ithaca, NY, Cornell University Press).

—— (1991), *Metaphysics: A Contemporary Introduction* (New York, Paragon House).

—— (unpublished), 'Versus Asymmetric Determination/Supervenience'.

PUTNAM, H. (1970), 'On Properties', in N. Rescher (ed.), *Essays in Honor of Carl Hempel* (Dordrecht, Reidel).

—— (1975), *Mind, Language, and Reality: Philosophical Papers, Volume 2* (Cambridge, Cambridge University Press).

—— (1978), *Meaning and the Moral Sciences* (London, Routledge & Kegan Paul).

—— (1979), 'Reflections on Goodman's *Ways of Worldmaking*', *Journal of Philosophy* 76: 603–18.

—— (1981), *Reason, Truth and History* (Cambridge, Cambridge University Press).

—— (1983), *Realism and Reason: Philosophical Papers, Volume 3* (Cambridge, Cambridge University Press).

—— (1987), *The Many Faces of Realism* (La Salle, Ill., Open Court).

QUINE, W. V. (1951), 'Ontology and Ideology', *Philosophical Studies* 2: 11–15.

—— (1960), *Word and Object* (Cambridge, Mass., MIT Press).

—— (1966), *The Ways of Paradox* (New York, Random House).

—— (1969a), 'Ontological Relativity', in *Ontological Relativity and Other Essays* (New York, Columbia University Press).

—— (1969b), 'Reply to Chomsky', in D. Davidson and J. Hintikka (eds.), *Words and Objections: Essays on the Work of W. V. Quine* (Dordrecht, Reidel).

—— (1970), 'On the Reasons for Indeterminacy of Translation', *Journal of Philosophy* 67: 178–83.

—— (1975), 'On Empirically Equivalent Systems of the World', *Erkenntnis* 9: 313–28.

—— (1978), 'Otherworldly', *New York Review of Books*, 23 November: 25.

—— (1979), 'Facts of the Matter', *Southwestern Journal of Philosophy* 9: 155–69.

QUINE, W. V. (1981*a*), *Theories and Things* (Cambridge, Mass., Harvard University Press).

—— (1981*b*), 'What Price Bivalence?', *Journal of Philosophy* 78: 90–5.

RAILTON, P. (1986*a*), 'Moral Realism', *Philosophical Review* 95: 163–207.

—— (1986*b*), 'Facts and Values', *Philosophical Topics* 14: 5–31.

RORTY, R. (1979), *Philosophy and the Mirror of Nature* (Princeton, NJ, Princeton University Press).

ROSENBERG, A. (1985), *The Structure of Biological Science* (Cambridge, Cambridge University Press).

SALMON, W. (1984), *Scientific Explanation and the Causal Structure of the World* (Princeton, NJ, Princeton University Press).

SCHAFFNER, K. (1977), 'Reduction, Reductionism, Values, and Progress in the Biomedical Sciences', in R. J. Colodney (ed.), *Logic, Laws, and Life* (Pittsburgh, University of Pittsburgh Press).

SCHEFFLER, I. (1967), *Science and Subjectivity* (Indianapolis, Bobbs-Merrill).

SCHIFFER, S. (1991), 'Ceteris Paribus Laws', *Mind* 100: 1–17.

SEAGER, W. (1988), 'Weak Supervenience and Materialism', *Philosophy and Phenomenological Research* 48: 697–709.

SKLAR, L. (1985), 'Methodological Conservatism', in *Philosophy and Spacetime Physics* (Berkeley, Calif., University of California Press).

SPECTOR, M. (1978), *Concepts of Reduction in Physical Science* (Philadelphia, Temple University Press).

STITCH, S. (1983), *From Folk Psychology to Cognitive Science: The Case Against Belief* (Cambridge, Mass., MIT Press).

STROUD, B. (1987), 'The Physical World', *Proceedings of the Aristotelian Society*, Supplement: 263–77.

SUSSMAN, A. (1981), 'Reflections on the Chances for a Scientific Dualism', *Journal of Philosophy* 78: 95–118.

TELLER, P. (1984), 'A Poor Man's Guide to Supervenience and Determination', *Southern Journal of Philosophy* 22, Supplement: 137–62.

—— (1985), 'Is Supervenience Just Disguised Reduction?', *Southern Journal of Philosophy* 23: 93–9.

—— (1986), 'Relational Holism and Quantum Mechanics', *British Journal of the Philosophy of Science* 37: 71–81.

VAN FRAASEN, B. (1980), *The Scientific Image* (Oxford, Clarendon Press).

VAN GULICK, R. (1985), 'Physicalism and the Subjectivity of the Mental', *Philosophical Topics* 13: 51–70.

VON ECKARDT, B. (1978), 'Inferring Functional Localization from Neurological Evidence', in E. Walker (ed.), *Explorations in the Biology of Language* (Montgomery, Vt., Bradford Books).

—— (1993), *What Is Cognitive Science?* (Cambridge, Mass., MIT Press).

WAGNER, S. (unpublished), 'Truth, Physicalism, and Ultimate Theory'.

WHITTAKER, E. (1960), *A History of the Theories of Aether and Electricity*, vol. 1 (New York, Harper & Brothers).

WIGNER, E. P. (1961), 'Remarks on the Mind–Body Question', in I. J. Good (ed.), *The Scientist Speculates* (London, Heinemann).

WILSON, M. (1980), 'The Observational Uniqueness of Some Theories', *Journal of Philosophy* 77: 208–33.

WIMSATT, W. (1976), 'Reductionism, Levels of Organization, and the Mind–Body Problem', in G. Globus, G. Maxwell, and I. Savodnik (eds.), *Consciousness and Brain: A Scientific and Philosophical Inquiry* (New York, Plenum).

—— (1978), 'Reduction and Reductionism', in P. Asquith and H. Kyberg (eds.), *Current Issues in the Philosophy of Science* (East Lansing, Mich., Philosophy of Science Association).

INDEX